Problems and Solutions for
Groups, Lie Groups,
Lie Algebras
with Applications

Problems and Solutions for Groups, Lie Groups, Lie Algebras with Applications

$$\exp\left(tx\frac{d}{dx}\right)\exp\left(t\frac{d}{dx}\right)x = xe^t + t$$

$$\exp\left(t\frac{d}{dx}\right)\exp\left(tx\frac{d}{dx}\right)x = e^t(x + t)$$

Willi-Hans Steeb
Igor Tanski
Yorick Hardy
University of Johannesburg, South Africa

World Scientific

NEW JERSEY · LONDON · SINGAPORE · BEIJING · SHANGHAI · HONG KONG · TAIPEI · CHENNAI

Published by

World Scientific Publishing Co. Pte. Ltd.

5 Toh Tuck Link, Singapore 596224

USA office: 27 Warren Street, Suite 401-402, Hackensack, NJ 07601

UK office: 57 Shelton Street, Covent Garden, London WC2H 9HE

British Library Cataloguing-in-Publication Data
A catalogue record for this book is available from the British Library.

ISBN-13 978-981-4383-90-5
ISBN-10 981-4383-90-2

Printed in Singapore by World Scientific Printers.

Preface

The purpose of this book is to supply a collection of problems in group theory, Lie group theory and Lie algebras. Furthermore, chapter 4 contains applications of these topics. Each chapter contains 100 completely solved problems. Chapters 1, 2 and 3 give a short but comprehensive introduction to the topics providing all the relevant definitions and concepts. Chapters 1, 2 and 3 also contain two solved programming problems and eight supplementary problems. Chapter 4 contains 10 solved programming problems and 10 supplementary problems. Chapter 4 covers mainly applications in mathematical and theoretical physics as well as quantum mechanics, differential geometry and relativity. Problems cover beginner, advanced and research topics. The problems are self-contained.

Accompanying problem books for this book are:

Problems and Solutions in Introductory and Advanced Matrix Calculus

by Willi-Hans Steeb
World Scientific Publishing, Singapore 2006
ISBN 981 256 916 2
http://www.worldscibooks.com/mathematics/6202.html

Problems and Solutions in Quantum Computing and Quantum Information, third edition

by Willi-Hans Steeb and Yorick Hardy
World Scientific, Singapore, 2006
ISBN 981-256-916-2
http://www.worldscibooks.com/physics/6077.html

The International School for Scientific Computing (ISSC) provides certificate courses for this subject. Please contact the author if you want to do this course or other courses of the ISSC.

e-mail addresses of the authors:

steebwilli@gmail.com
yorickhardy@gmail.com

Home page of the authors:

http://issc.uj.ac.za

Contents

Notation

:=	is defined as
$\in$	belongs to (a set)
$\notin$	does not belong to (a set)
$T \subset S$	subset T of set S
$S \cap T$	the intersection of the sets S and T
$S \cup T$	the union of the sets S and T
$\emptyset$	empty set
$\mathbb{N}$	set of natural numbers
$\mathbb{Z}$	set of integers
$\mathbb{Q}$	set of rational numbers
$\mathbb{R}$	set of real numbers
$\mathbb{R}^+$	set of nonnegative real numbers
$\mathbb{C}$	set of complex numbers
$\mathbb{R}^n$	n-dimensional Euclidean space
	space of column vectors with n real components
$\mathbb{C}^n$	n-dimensional complex linear space
	space of column vectors with n complex components
$\mathcal{H}$	Hilbert space
i	$\sqrt{-1}$
$\Re z$	real part of the complex number z
$\Im z$	imaginary part of the complex number z
$|z|$	modulus of the complex number z
	$|x + iy| = (x^2 + y^2)^{1/2}, \ x, y \in \mathbb{R}$
$f(S)$	image of the set S under the mapping f
$f \circ g$	composition of two mappings $(f \circ g)(x) = f(g(x))$
G	group
$Z(G)$	center of the group G
$\mathbb{Z}_n$	cyclic group $\{0, 1, \ldots, n-1\}$
	under addition modulo n
G/N	factor group
D_n	nth dihedral group
S_n	symmetric group on n letters, permutation group
A_n	alternating group on n letters, alternating group
L	Lie algebra
$\mathbf{x}$	column vector in the vector space $\mathbb{C}^n$
$\mathbf{x}^T$	transpose of $\mathbf{x}$ (row vector)
$\mathbf{0}$	zero (column) vector
$\|\cdot\|$	norm

$\mathbf{x} \cdot \mathbf{y} \equiv \mathbf{x}^*\mathbf{y}$	scalar product (inner product) in $\mathbb{C}^n$
$\mathbf{x} \times \mathbf{y}$	vector product in $\mathbb{R}^3$
S^2	two sphere
A, B, C	$m \times n$ matrices
$\det(A)$	determinant of a square matrix A
$\mathrm{tr}(A)$	trace of a square matrix A
$\mathrm{rank}(A)$	rank of a matrix A
A^T	transpose of the matrix A
$\overline{A}$	conjugate of the matrix A
A^*	conjugate transpose of matrix A
$A^\dagger$	conjugate transpose of matrix A (notation used in physics)
A^{-1}	inverse of the square matrix A (if it exists)
I_n	$n \times n$ unit matrix
I	unit operator
0_n	$n \times n$ zero matrix
AB	matrix product of an $m \times n$ matrix A and an $n \times p$ matrix B
$[A, B] := AB - BA$	commutator of square matrices A and B
$[A, B]_+ := AB + BA$	anticommutator of square matrices A and B
$A \otimes B$	Kronecker product of matrices A and B
$A \oplus B$	Direct sum of matrices A and B
δ_{jk}	Kronecker delta with $\delta_{jk} = 1$ for $j = k$ and $\delta_{jk} = 0$ for $j \neq k$
λ	eigenvalue
ϵ	real parameter
t	time variable
$\hat{H}$	Hamilton operator
$\hat{N}$	Number operator
g	metric tensor field
ϵ	real parameter
$\wedge$	exterior product
d	exterior derivative

The *Pauli spin matrices* are used extensively in the book. They are given by

$$\sigma_x := \begin{pmatrix} 0 & 1 \\ 1 & 0 \end{pmatrix}, \quad \sigma_y := \begin{pmatrix} 0 & -i \\ i & 0 \end{pmatrix}, \quad \sigma_z := \begin{pmatrix} 1 & 0 \\ 0 & -1 \end{pmatrix}.$$

In some cases we will also use σ_1, σ_2 and σ_3 to denote σ_x, σ_y and σ_z .

Chapter 1

Groups

A *group* G is a set of objects $\{\, a, b, c, \dots \,\}$ (not necessarily countable) together with a binary operation which associates with any ordered pair of elements a, b in G a third element ab in G (closure). The binary operation (called group multiplication) is subject to the following requirements:

1) There exists an element e in G called the *identity element* (also called *neutral element*) such that $eg = ge = g$ for all $g \in G$.

2) For every $g \in G$ there exists an *inverse element* g^{-1} in G such that $gg^{-1} = g^{-1}g = e$.

3) *Associative law.* The identity $(ab)c = a(bc)$ is satisfied for all $a, b, c \in G$.

If $ab = ba$ for all $a, b \in G$ we call the group *commutative.*

If G has a finite number of elements it has *finite order* $n(G)$, where $n(G)$ is the number of elements. Otherwise, G has infinite order. *Lagrange theorem* tells us that the order of a subgroup of a finite group is a divisor of the order of the group.

If H is a subset of the group G closed under the group operation of G, and if H is itself a group under the induced operation, then H is a *subgroup* of G.

Let G be a group and S a subgroup. If for all $g \in G$, the *right coset*

$$Sg := \{\, sg \,:\, s \in S \,\}$$

1

is equal to the *left coset*

$$gS := \{\, gs \,:\, s \in S \,\}$$

then we say that the subgroup S is a normal or invariant subgroup of G. A subgroup H of a group G is called a *normal subgroup* if $gH = Hg$ for all $g \in G$. This is denoted by $H \lhd G$. We also define

$$gHg^{-1} := \{\, ghg^{-1} \,:\, h \in H \,\}.$$

The *center* $Z(G)$ of a group G is defined as the set of elements $z \in G$ which commute with all elements of the group, i.e.

$$Z(G) := \{\, z \in G \,:\, zg = gz \text{ for all } g \in G \,\}.$$

Let G be a group. For any subset X of G, we define its *centralizer* $C(X)$ to be

$$C(X) := \{\, y \in G \,:\, xy = yx \text{ for all } x \in X \,\}.$$

If $X \subset Y$, then $C(Y) \subset C(X)$.

A *cyclic group* G is a group containing an element g with the property that every other element of G can be written as a power of g, i.e. such that for all $h \in G$, for some $n \in \mathbb{Z}$, $h = g^n$. We then say that G is the cyclic group generated by g.

Let $(G_1, *)$ and $(G_2, \circ)$ be groups. A function $f : G_1 \to G_2$ with

$$f(a * b) = f(a) \circ f(b), \qquad \text{for all } a, b \in G_1$$

is called a *homomorphism*. If f is invertible, then f is an *isomorphism*.

Groups have matrix representations with invertible $n \times n$ matrices and matrix multiplication as group multiplication. The identity element is the identity matrix. The inverse element is the inverse matrix. An important subgroup is the set of unitary matrices U, where $U^* = U^{-1}$.

Let $GL(n, \mathbb{F})$ be the group of invertible $n \times n$ matrices with entries in the field $\mathbb{F}$, where $\mathbb{F}$ is $\mathbb{R}$ or $\mathbb{C}$. Let G be a group. A *matrix representation* of G over the field $\mathbb{F}$ is a homomorphism ρ from G to $GL(n, \mathbb{F})$. The degree of ρ is the integer n. Let $\rho : G \to GL(n, \mathbb{F})$. Then ρ is a representation if and only if

$$\rho(g \circ h) = \rho(g)\rho(h)$$

for all $g, h \in G$.

Problem 1. Let $x, y \in \mathbb{R}$. Does the composition

$$x \bullet y := \sqrt[3]{x^3 + y^3}$$

define a group?

Solution 1. Yes. We have $\sqrt[3]{x^3 + y^3} \in \mathbb{R}$. The neutral element is 0. The inverse element of x is $-x$. The associative law

$$(x \bullet y) \bullet z = (\sqrt[3]{x^3 + y^3}) \bullet z = \sqrt[3]{(\sqrt[3]{x^3 + y^3})^3 + z^3}$$
$$= \sqrt[3]{x^3 + y^3 + z^3} = \sqrt[3]{x^3 + (\sqrt[3]{y^3 + z^3})^3}$$
$$= x \bullet (y \bullet z)$$

also holds. The group is commutative since $x \bullet y = y \bullet x$ for all $x, y \in \mathbb{R}$.

Problem 2. Let $x, y \in \mathbb{R} \setminus \{0\}$ and $\cdot$ denotes multiplication in $\mathbb{R}$. Does the composition

$$x \bullet y := \frac{x \cdot y}{2}$$

define a group?

Solution 2. Yes. We have $(x \cdot y)/2 \in \mathbb{R} \setminus \{0\}$. The neutral element is 2 and $4/x$ is inverse to x. The associative law also holds. The group is commutative.

Problem 3. Let $x, y \in \mathbb{R}$. Is the composition $x \bullet y := |x + y|$ associative? Here $|.|$ denotes the absolute value.

Solution 3. The answer is no. We have

$$0 = |(|1 + (-1)|) + 0| \neq |1 + (|(-1) + 0|)| = 2.$$

Problem 4. Consider the set

$$G = \{ (a, b) \in \mathbb{R}^2 \ : \ a \neq 0 \}.$$

We define the composition

$$(a, b) \bullet (c, d) := (ac, ad + b).$$

Show that this composition defines a group. Is the group commutative?

Solution 4.   The composition is associative. The neutral element is $(1,0)$. The inverse element of (a,b) is $(1/a, -b/a)$. The group is not commutative since

$$(a,b) \bullet (c,d) = (ac, ad+b) \quad \text{and} \quad (c,d) \bullet (a,b) = (ca, cb+d).$$

Problem 5.   Show that the set $\{+1, -1, +i, -i\}$ forms a group under multiplication. Find all subgroups.

Solution 5.   The multiplication table is

$\cdot$	1	-1	i	$-i$
1	1	-1	i	$-i$
-1	-1	1	$-i$	i
i	i	$-i$	-1	1
$-i$	$-i$	i	1	-1

From the table we find that the group is commutative. We can also deduce this property from the commutativity of complex multiplication. From the table we find that the inverse elements are

$$1^{-1} = 1, \quad (-1)^{-1} = -1, \quad (i)^{-1} = -i, \quad (-i)^{-1} = i.$$

The associative law follows from the associative law for multiplication of complex numbers. Thus the set $\{+1, -1, +i, -i\}$ forms a group under multiplication. We classify subgroups by their orders. There is only one subgroup of order 1, the trivial group - $\{1\}$. Subgroups of order 2 contain two elements - the identity element $e = 1$ and one additional element a. There are two possibilities: $a^2 = e$ or $a^2 = a$. The first possibility provides one subgroup: $\{1, -1\}$ which is commutative. The element (-1) is inverse to itself, it is in involution. The second possibility reduce to the trivial group $\{1\}$. There are no subgroups of order 3, because the order of a subgroup must divide the order of the group (*Lagrange's theorem*). There is only one subgroup of order 4 - this is the group itself. Therefore the group has in total $1 + 1 + 1 = 3$ subgroups with one proper subgroup.

Problem 6.   Let $i = \sqrt{-1}$. Let S be the set of complex numbers of the form $q + pi\sqrt{5}$, where $p, q \in \mathbb{Q}$ and are not both simultaneously 0. Show that this set forms a group under multiplication of complex numbers.

Solution 6.   Consider the product of two numbers $q_1 + ip_1\sqrt{5}$ and $q_2 + ip_2\sqrt{5}$.

$$q_3 + ip_3\sqrt{5} = (q_1 + ip_1\sqrt{5})(q_2 + ip_2\sqrt{5}) = q_1q_2 - 5p_1p_2 + i\sqrt{5}(q_1p_2 + p_1q_2).$$

Thus

$$q_3 = q_1 q_2 - 5 p_1 p_2 \in \mathbb{Q}, \qquad p_3 = q_1 p_2 + p_1 q_2 \in \mathbb{Q}$$

and

$$q_3^2 + 5 p_3^2 = (q_1^2 + 5 p_1^2)(q_2^2 + 5 p_2^2) > 0.$$

Therefore the product of two numbers of the form $q + i p \sqrt{5}$ belongs to the same set. The identity element is 1, i.e. $q = 1$ and $p = 0$. The inverse element is

$$\frac{q - i p \sqrt{5}}{q^2 + 5 p^2} = \frac{q}{q^2 + 5 p^2} + \left(\frac{-p}{q^2 + 5 p^2} \right) i \sqrt{5}.$$

The existence of the inverse element follows from the fact that $p, q \in \mathbb{Q}$ and are not both simultaneously 0. Thus the inverse element belongs to the same set. The associative law follows from the same law of multiplication of complex numbers. Thus the set of numbers of the form $q + i p \sqrt{5}$, where $p, q \in \mathbb{Q}$ forms an abelian group under multiplication.

Problem 7. Let p be a prime number with $p \geq 3$. Let r and s be rational numbers ($r, s \in \mathbb{Q}$) with $r^2 + s^2 > 0$. Show that the set given by the numbers $r + s \sqrt{p}$ form a commutative group.

Solution 7. Associativity and commutativity follow from the multiplication of real numbers. We have

$$(r_1 + s_1 \sqrt{p})(r_2 + s_2 \sqrt{p}) = (r_1 r_2 + p s_1 s_2) + (r_1 s_2 + r_2 s_1) \sqrt{p}.$$

Since $r_1 r_2 + p s_1 s_2 \in \mathbb{Q}$ and $r_1 s_2 + r_2 s_1 \in \mathbb{Q}$ the operation is closed. The neutral element is 1, i.e. $r = 1$ and $s = 0$. The inverse element is

$$\frac{r - s \sqrt{p}}{(r + s \sqrt{p})(r - s \sqrt{p})} = \frac{r}{r^2 - s^2 p} + \left(\frac{-p}{r^2 - s^2 p} \right) \sqrt{p}.$$

Problem 8. Show that the set

$$\{ e^{i \alpha} : \alpha \in \mathbb{R} \}$$

forms a group under multiplication. Note that $|e^{i \alpha}| = 1$.

Solution 8. We have

$$e^{i \alpha} e^{i \beta} = e^{i(\alpha + \beta)}.$$

The neutral element is given by $\alpha = 0$, i.e $e^0 = 1$. The inverse element of $e^{i \alpha}$ is $e^{-i \alpha}$. The associative and commutative laws follow from the same laws of multiplication for complex numbers. Thus the set $\{e^{i\alpha} : \alpha \in \mathbb{R}\}$ forms a 1-dimensional group under multiplication of complex numbers.

Problem 9.   Consider the additive group $(\mathbb{Z}, +)$. Give a proper subgroup.

Solution 9.   An example of a proper subgroup would be all even numbers, since the sum of two even numbers is again an even number.

Problem 10.   Let $S := \mathbb{R} \setminus \{-1\}$. We define the binary operation on S

$$a \bullet b := a + b + ab.$$

Show that $\langle S, \bullet \rangle$ forms a group. Is the group commutative?

Solution 10.   When a, b are elements of S, then $a \bullet b$ is an element of S. Suppose instead that $a + b + ab = -1$. Then $a(1 + b) = -(1 + b)$. Thus $b = -1$ or $a = -1$. Since $a \neq -1$ and $b \neq -1$ we have $a + b + ab \neq -1$. The neutral element is $e = 0$ since

$$a \bullet e = a + 0 + a0 = a, \qquad e \bullet b = 0 + b + 0b = b.$$

To find the inverse of an arbitrary element g we consider right multiplication by its inverse g^{-1}. We obtain

$$g \bullet g^{-1} = g + g^{-1} + gg^{-1} = 0 \Rightarrow g^{-1} = -g/(1 + g).$$

All elements are invertible, because $g \neq -1$. The associative law holds

$$(a \bullet b) \bullet c = (a + b + ab) \bullet c = (a + b + ab) + c + (a + b + ab)c$$
$$= a + b + c + (ab + bc + ca) + abc = a \bullet (b \bullet c).$$

The group is commutative since $a \bullet b = b \bullet a$.

Problem 11.   Let G be a group and $x, y \in G$. Show that $(xy)^{-1} = y^{-1}x^{-1}$.

Solution 11.   Let e be the neutral element of G. We have

$$xyy^{-1}x^{-1} = xex^{-1} = xx^{-1} = e$$

and

$$y^{-1}x^{-1}xy = y^{-1}ey = y^{-1}y = e.$$

Problem 12.   Let S be the set of all rational numbers $\mathbb{Q}$ in the interval $0 \le q < 1$. Define the operation $(q, p \in S)$

$$q \bullet p := \begin{cases} q + p & \text{if } 0 \le q + p < 1 \\ q + p - 1 & \text{if } \quad q + p \ge 1 \end{cases}.$$

Show that S with this operation is an abelian group.

Solution 12.   The group neutral element is 0, i.e. $p + 0 = 0 + p = p$. The inverse element is $p^{-1} = 1 - p$ since

$$p + p^{-1} = 1 \ge 1 \Rightarrow p \bullet p^{-1} = 1 - 1 = 0.$$

The inverse element exists for all p, because $p < 1$. Associativity follows from the associative law for addition of rational numbers. Commutativity follows from the commutativity of addition of rational numbers. We can write each nonzero element of the group as the proper fraction

$$p = \frac{a}{b}, \quad a, b \in \mathbb{N}, \quad a < b$$

where the fraction is irreducible (a and b have no common divisors).

Problem 13.   Show that the finite set

$$\mathbb{Z}_n := \{\, 0, 1, \dots, n - 1 \,\}$$

for $n \ge 1$ forms an abelian group under addition modulo n. The group is referred to as the group of integers modulo n.

Solution 13.   The neutral element is 0. The inverse of 0 is 0. For any $j > 0$ in $\mathbb{Z}_n$ the inverse of j is $n - j$. Obviously the associative law holds. The group law is

$$a + b < n \Rightarrow a \bullet b = a + b$$

$$a + b \ge n \Rightarrow a \bullet b = a + b - n, \quad a, b \in \mathbb{Z}_n.$$

The expressions are symmetric in the two arguments a, b. Thus the group is abelian.

Problem 14.   Let $n \in \mathbb{N}$. Let $U(n)$ be the set of all positive integers less than n and relatively prime to n. Then $U(n)$ is a commutative group under multiplication modulo n. Find the group table for $U(8)$.

Solution 14.   Obviously 1 and $n - 1$ are elements of $U(n)$, where 1 is the neutral element. For $n = 8$ we have $U(8) = \{\, 1, 3, 5, 7 \,\}$. The group table is

mod 8	1	3	5	7
1	1	3	5	7
3	3	1	7	5
5	5	7	1	3
7	7	5	3	1

Problem 15.   Consider the subset of odd integers

$$\{1, 3, 7, 9, 11, 13, 17, 19\}.$$

Show that this set forms an abelian group under multiplication modulo 20.

Solution 15.   The multiplication table modulo 20 reads

·	1	3	7	9	11	13	17	19
1	1	3	7	9	11	13	17	19
3	3	9	1	7	13	19	11	17
7	7	1	9	3	17	11	19	13
9	9	7	3	1	19	17	13	11
11	11	13	17	19	1	3	7	9
13	13	19	11	17	3	9	1	7
17	17	11	19	13	7	1	9	3
19	19	17	13	11	9	7	3	1

The table shows that products of each two elements of the set is again an element of the set. The identity element is 1 - implied by the same property of modulo 20 multiplication for all integer numbers (or from the multiplication table). The inverse elements are

$$1^{-1} = 1, \quad 3^{-1} = 7, \quad 7^{-1} = 3, \quad 9^{-1} = 9,$$

$$11^{-1} = 11, \quad 13^{-1} = 17, \quad 17^{-1} = 13, \quad 19^{-1} = 19.$$

The associative law for modulo 20 multiplication for all integer numbers implies the same law for the set in question.

Problem 16.   Give the group table for the cyclic group $\mathbb{Z}_6$ of 6 elements.

Solution 16.   The neutral element is 0. The group is commutative. The group table is

+	0	1	2	3	4	5
0	0	1	2	3	4	5
1	1	2	3	4	5	0
2	2	3	4	5	0	1
3	3	4	5	0	1	2
4	4	5	0	1	2	3
5	5	0	1	2	3	4

Problem 17. Consider the finite group $\mathbb{Z}_2 \times \mathbb{Z}_3$ which has $2 \cdot 3 = 6$ elements given by $(0,0)$, $(0,1)$, $(0,2)$, $(1,0)$, $(1,1)$, $(1,2)$. The neutral element is $(0,0)$. Show that $\mathbb{Z}_2 \times \mathbb{Z}_3$ is cyclic.

Solution 17. It is only necessary to find a generator. We start with $(1,1)$. Then

$$
\begin{aligned}
(1,1) &= (1,1) \\
2(1,1) &= (1,1) + (1,1) = (0,2) \\
3(1,1) &= 2(1,1) + (1,1) = (1,0) \\
4(1,1) &= 3(1,1) + (1,1) = (1,0) + (1,1) = (0,1) \\
5(1,1) &= 4(1,1) + (1,1) = (0,1) + (1,1) = (1,2) \\
6(1,1) &= 5(1,1) + (1,1) = (1,2) + (1,1) = (0,0).
\end{aligned}
$$

Therefore the element $(1,1)$ generates all elements of the commutative group $\mathbb{Z}_2 \times \mathbb{Z}_3$.

Problem 18. Consider the functions defined on $\mathbb{R} \setminus \{0,1\}$

$$
f_1(x) = x, \quad f_2(x) = \frac{1}{x}, \quad f_3(x) = 1 - x,
$$

$$
f_4(x) = \frac{x}{x-1}, \quad f_5(x) = \frac{1}{1-x}, \quad f_6(x) = 1 - \frac{1}{x}.
$$

Show that these functions form a group with the *function composition* $f_j \circ f_k$, where

$$
(f_j \circ f_k)(x) := f_j(f_k(x)).
$$

Solution 18. The neutral element is f_1. The group table is

$\circ$	f_1	f_2	f_3	f_4	f_5	f_6
f_1	f_1	f_2	f_3	f_4	f_5	f_6
f_2	f_2	f_1	f_5	f_6	f_3	f_4
f_3	f_3	f_6	f_1	f_5	f_4	f_2
f_4	f_4	f_5	f_6	f_1	f_2	f_3
f_5	f_5	f_4	f_2	f_3	f_6	f_1
f_6	f_6	f_3	f_4	f_2	f_1	f_5

For example

$$(f_3 \circ f_4)(x) = f_3(f_4(x)) = 1 - \frac{x}{x-1} = \frac{1}{1-x} = f_5(x)$$

$$(f_4 \circ f_3)(x) = f_4(f_3(x)) = \frac{1-x}{1-x-1} = 1 - \frac{1}{x} = f_6(x).$$

Each element has an inverse. The associativity law holds for function composition. The group is not commutative.

Problem 19. An *isomorphism* of a group G with itself is an *automorphism*. Show that for each $g \in G$ the mapping $i_g : G \to G$ defined by

$$x i_g := g^{-1} x g$$

is an automorphism of G, the inner automorphism of G under conjugation by the group element g. We have to show that i_g is an isomorphism of G with itself. Thus we have to show it is one to one, onto, and that

$$(xy)i_g = (x i_g)(y i_g)$$

for all $g \in G$.

Solution 19. If $x i_g = y i_g$, then $g^{-1} x g = g^{-1} y g$. Thus $x = y$. For onto, if $x \in G$, then applying the associative law yields

$$(g x g^{-1}) i_g = g^{-1}(g x g^{-1}) g = x.$$

Now $(xy)i_g = g^{-1} x y g$ and

$$(x i_g)(y i_g) = (g^{-1} x g)(g^{-1} y g) = g^{-1} x y g$$

since $g g^{-1} = e$. Thus $(xy)i_g = (x i_g)(y i_g)$.

Problem 20. Let C_n be the *cyclic group*. Show that $C_6 \simeq C_3 \times C_2$.

Solution 20. The cyclic group C_3 contains three elements and the cyclic group C_2 contains two elements. Let $b = c^2$, generating the cyclic group $C_3 = \{e, b, b^2\} = B$ where e is the neutral element. Let $a = c^3$ generating the cyclic group $C_2 = \{e, a\} = A$. Now b and a commute and every element of the cyclic group C_6 can be written as $a^r b^s$. Thus $c = ab^2$, $c^2 = b$, $c^3 = a$, $c^4 = b^2$ and $c^5 = ab$. Therefore $C_6 = A \times B = C_3 \times C_2$.

Problem 21. (i) Let $Z(G)$ be the *center* of the group G. Show that $Z(G)$ is a commutative subgroup of G.
(ii) Let $a \in Z(G)$ and $g \in G$. Then $ag = ga$. Show that $a^{-1}g = ga^{-1}$.

Solution 21. (i) The neutral element e is in $Z(G)$. If $z_1, z_2 \in Z(G)$ we have

$$z_1 z_2 g = z_1 g z_2 = g z_1 z_2.$$

Thus $Z(G)$ is closed under the group operation. It is therefore a subgroup. Furthermore it is commutative since we can take $g = z_2$ in the defining relation $z_1 g = g z_1$. We multiply both sides of identity $gz = zg$ by z^{-1} from left and right. This gives

$$z^{-1}g = gz^{-1} \Rightarrow z^{-1} \in Z(G).$$

The associative law follows from the same law for the group G. Thus the set $Z(G)$ forms a subgroup of group G.
(ii) Let e be the neutral element. From $ag = ga$ we obtain using the associative law

$$a^{-1}(ag)a^{-1} = a^{-1}(ga)a^{-1}$$
$$(a^{-1}a)(ga^{-1}) = (a^{-1}g)(aa^{-1})$$
$$ega^{-1} = a^{-1}ge$$
$$ga^{-1} = a^{-1}g.$$

Problem 22. Show that the center $Z(G)$ of a group G is a *normal subgroup* of the group G.

Solution 22. Let z_1 and z_2 be two elements of the center $Z(G)$. For each $g \in G$ we have

$$g z_1 z_2 = z_1 g z_2 = z_1 z_2 g \Rightarrow z_1 z_2 \in Z(G).$$

Multiplying both sides of $gz = zg$ by z^{-1} from left and right yields

$$z^{-1}g = gz^{-1}.$$

Thus $z^{-1} \in Z(G)$. The associative law follows from the same law for group G. Hence the set $Z(G)$ forms a subgroup of group G. The commutative law follows from

$$zg = gz \quad \text{for all } g \in G \Rightarrow zg = gz \quad \text{for all } g \in Z(G).$$

Thus $Z(G)$ is a commutative subgroup of the group G. Let $H = Z(G)$. We must show that $H = gHg^{-1}$ for all $g \in G$. Let z be in H. Then we know that $gz = zg$. Thus $gzg^{-1} = zgg^{-1} = z$. Hence z is in gHg^{-1}. On the other hand, suppose z is in gHg^{-1}. Then $z = zgg^{-1} = gzg^{-1}$ so that $zg = gz$ and hence z is in H. Hence the center $Z(G)$ is a normal subgroup of G.

Problem 23.   Let H be a subgroup of a group G. The *normalizer* $N(H)$ of H in G is defined by

$$N(H) := \{ g \in G : gHg^{-1} = H \}.$$

(i) Show that $N(H)$ is a subgroup of G.
(ii) Show that H is a normal subgroup of $N(H)$.

Solution 23.   (i) First we show that the normalizer is closed under multiplication. Let g, k be two elements of $N(H)$. Then for all $h \in H$ we have

$$(ghg^{-1} \in H) \wedge (khk^{-1} \in H).$$

Note that for any $h \in H$

$$(gk)h(gk)^{-1} = (gk)h(k^{-1}g^{-1}) = g(khk^{-1})g^{-1}.$$

Since $k \in N(H)$ we have $khk^{-1} = h'$ for some $h' \in H$. Thus

$$(gk)h(gk)^{-1} = gh'g^{-1}.$$

Since $g \in N(H)$, $gh'g^{-1}$ is also in H. Therefore, we conclude that $(gk)h(gk)^{-1}$ is in H for all $h \in H$. This implies that the element gk is also in $N(H)$. Let g be in $N(H)$. Now for any $h \in H$

$$ghg^{-1} = h' \Leftrightarrow g^{-1}h'g = h.$$

Therefore when h runs over all elements of H, h' also runs over all elements of H. Thus for all $h' \in H$

$$g^{-1}h'g = h \in H.$$

However

$$g^{-1}h'g = (g^{-1})h'(g^{-1})^{-1}.$$

Therefore, g^{-1} is also in $N(H)$. Thus $N(H)$ is a subgroup of G.
(ii) Let h be arbitrary element of H and n be arbitrary element of $N(H)$. Conjugation of h by n is $c = nhn^{-1} \in H$ by definition of the normalizer. Thus H is a normal subgroup of $N(H)$.

Problem 24.   Consider a group G. We define the *group commutator* $(x, y \in G)$

$$[x, y] := xyx^{-1}y^{-1}$$

Consider the two 2×2 matrices with determinant 1

$$x = \begin{pmatrix} t & 1 \\ -1 & 0 \end{pmatrix}, \qquad y = \begin{pmatrix} 1 & b \\ c & 1 + bc \end{pmatrix}$$

where $t, b, c \in \mathbb{R}$. This means x and y are elements of $SL(2, \mathbb{R})$. Find the group commutator.

Solution 24.   We have

$$x^{-1} = \begin{pmatrix} 0 & -1 \\ 1 & t \end{pmatrix}, \qquad y^{-1} = \begin{pmatrix} 1 + bc & -b \\ -c & 1 \end{pmatrix}.$$

Obviously these matrices are also elements of $SL(2, \mathbb{R})$. Thus

$$xyx^{-1}y^{-1} =$$

$$\begin{pmatrix} b^2 c(c+t) - b(c^2 t + c(t^2 - 2) - t) + c^2 + 1 & -b^2(c+t) + b(ct + t^2 - 1) - c \\ -b^2 c + b(ct - 1) & b^2 - bt + 1 \end{pmatrix}.$$

This matrix also has determinant equal to 1.

Problem 25.   Let G be a group. Given two elements $g_1, g_2 \in G$. One defines the *group commutator* of g_1 and g_2 to be the element

$$g_1 g_2 g_1^{-1} g_2^{-1}.$$

Consider the compact Lie group $SO(2, \mathbb{R})$ with

$$g_1(\alpha) = \begin{pmatrix} \cos \alpha & \sin \alpha \\ -\sin \alpha & \cos \alpha \end{pmatrix}, \qquad g_1(\beta) = \begin{pmatrix} \cos \beta & \sin \beta \\ -\sin \beta & \cos \beta \end{pmatrix}.$$

Find the group commutator.

Solution 25.   Since the group is commutative we find

$$g_1(\alpha)g_2(\beta)g_1^{-1}(\alpha)g_2^{-1}(\beta) = \begin{pmatrix} 1 & 0 \\ 0 & 1 \end{pmatrix}.$$

Problem 26.   (i) Show that the set of all group commutators $aba^{-1}b^{-1}$ of a group G generates a normal subgroup G' (the so-called *commutator subgroup*) of G.
(ii) Show that G/G' is abelian.

Solution 26.   (i) Obviously the commutators generate a subgroup G' of G. The inverse of a commutator is again a commutator since

$$(aba^{-1}b^{-1})^{-1} = bab^{-1}a^{-1}.$$

Furthermore the identity element e is a commutator since $e = eee^{-1}e^{-1}$. Then G' consists of all of all finite products of commutators. Next we show that for $x \in G'$ we have $g^{-1}xg \in G'$ for all $g \in G$. We have to show that if x is a product of commutators, so is $g^{-1}xg$ for all $g \in G$. By inserting the neutral element $e = gg^{-1}$ between each product of commutators occurring in x, we find that it is sufficient to show that for each commutator $cdc^{-1}d^{-1}$ that $g^{-1}(cdc^{-1}d^{-1})g$ is in the subgroup G'. We have

$$\begin{aligned}
g^{-1}(cdc^{-1}d^{-1})g &= (g^{-1}cdc^{-1})(e)(d^{-1}g) \\
&= (g^{-1}cdc^{-1})(gd^{-1}dg^{-1})(d^{-1}g) \\
&= ((g^{-1}c)d(g^{-1}c)^{-1}d^{-1})(dg^{-1}d^{-1}g)
\end{aligned}$$

which is in G'. Thus G' is normal in G.
(ii) From

$$\begin{aligned}
(aG)(bG') = abG' &= ab(b^{-1}a^{-1}ba)G' \\
&= (abb^{-1}a^{-1})baG' = baG' \\
&= (bG')(aG')
\end{aligned}$$

we find that G/G' is abelian.

Problem 27.   Let H and K be two subgroups of a group G. Show that the relation

$$g \sim g' \text{ if } g' = hgk \text{ for some } h \in H, \; k \in K$$

is an *equivalence relation*, partitioning G into double cosets Hg_jK for $g_j \in G$.

Solution 27.   The relation is reflexive. Choose $h = k = e$ in the definition

$$g' \simeq g \quad \text{if} \quad g' = hgk.$$

This is possible, because the subgroup H is a group. Thus $e \in H$ and the same for K. It follows that $g \simeq g$. The relation is symmetric. Let

$h' = h^{-1}$, $k' = k^{-1}$. This is possible, because the subgroup H is a group, so $h \in H \Rightarrow h' \in H$ and the same for K. It follows that

$$g' \simeq g \Leftrightarrow g' = hgk \Rightarrow g = h^{-1}g'k^{-1} = h'g'k' \Rightarrow g \simeq g'.$$

The relation is transitive. Suppose that (for appropriate $h_1, h_2 \in H$ and $k_1, k_2 \in K$)

$$a \simeq b \Leftrightarrow a = h_1 b k_1, \qquad b \simeq c \Leftrightarrow b = h_2 c k_2.$$

Then using $h' = h_1 h_2$ and $k' = k_1 k_2$ we find

$$a = h_1 b k_1 = h_1(h_2 c k_2)k_1 = (h_1 h_2)c(k_2 k_1) = h'ck' \Rightarrow a \simeq c.$$

Here we used the associative law and the group property

$$h_1, h_2 \in H \Rightarrow h' = h_1 h_2 \in H$$

and the same for K.

Problem 28. Let G be a group with the composition $\bullet$. Let H be a subgroup of G. We define the relation

$$a \sim b \Leftrightarrow a \bullet b^{-1} \in H \quad \text{for} \quad a, b \in G.$$

Is this relation an *equivalence relation*?

Solution 28. Yes. We have

$$a \sim a \Leftrightarrow a \bullet a^{-1} = e \in H$$

where e is the identity element of the group. If $a \sim b$ we have $a \bullet b^{-1} \in H$. Thus $(b \bullet a^{-1})^{-1} \in H$. Therefore $b \bullet a^{-1} \in H \Rightarrow b \sim a$. Let $a \sim b$ and $b \sim c$. Then $a \bullet b^{-1} \in H$ and $b \bullet c^{-1} \in H$. It follows that

$$(a \bullet b^{-1}) \bullet (b \bullet c^{-1}) = a \bullet c^{-1} \in H.$$

Hence $a \sim c$.

Problem 29. Let G be a group. Let G_1 and G_2 be two subgroups of G. Show that the intersection of G_1 and G_2 is itself a subgroup.

Solution 29. Obviously the identity e lies in both G_1 and G_2. Thus $G_1 \cap G_2$ is non-empty. If x, y are elements of the intersection $G_1 \cap G_2$, they are both elements of G_1 and both elements of G_2. Since G_1 and G_2 are subgroups the product xy^{-1} lies in G_1 and G_2. Therefore $xy^{-1} \in G_1 \cap G_2$.

Problem 30.   The set $\{1, -1\}$ forms a group G_1 under multiplication. The set $\{0, 1\}$ forms a group G_2 under the *XOR-operation* $\oplus$, i.e.

$$0 \oplus 0 = 0, \quad 0 \oplus 1 = 1, \quad 1 \oplus 0 = 1, \quad 1 \oplus 1 = 0.$$

Show that the two groups are *isomorphic*.

Solution 30.   Since the neutral element of the group with the XOR-operation is 0, we have the isomorphism $\phi : G_1 \to G_2$ with

$$\phi(1) = 0, \qquad \phi(-1) = 1.$$

Note that all groups with 2 elements are isomorphic.

Problem 31.   Show that any infinite cyclic group G is isomorphic to the group $\mathbb{Z}$ of integers under addition.

Solution 31.   Let g be the generator of G. Then

$$G = \{ g^n : n \in \mathbb{Z} \}.$$

We define the map $\phi : G \to \mathbb{Z}$ by $\phi(g^n) = n$ for all $g^n \in G$. If $\phi(g^n) = \phi(g^m)$, then $n = m$ and $g^n = g^m$. Thus the map ϕ is one to one if G is not finite. For any $n \in \mathbb{Z}$, the element $g^n \in G$ is mapped onto n by the map ϕ. Thus ϕ is onto $\mathbb{Z}$. Now

$$\phi(g^n g^m) = \phi(g^{n+m}) = n + m$$

and

$$\phi(g^n) + \phi(g^m) = n + m.$$

Thus

$$\phi(g^n g^m) = \phi(g^{n+m}) = n + m = \phi(g^n) + \phi(g^m).$$

Problem 32.   Let $\phi_1 : G_1 \to G_2$ and $\phi_2 : G_2 \to G_1$ be homomorphisms such that $\phi_1 \phi_2 = \phi_2 \phi_1 = i$, where i is the identity map. This means $\phi_1 \phi_2 : G_1 \to G_1$ and $\phi_2 \phi_1 : G_2 \to G_2$ are both the identity map. The *First Isomorphism Theorem* states that the image of a group homomorphism, $\mathrm{im}(\phi_1)$ is isomorphic to the quotient group $G_1 / \ker(\phi_1)$.
(i) Show that both ϕ_1 and ϕ_2 are isomorphisms of G_1 with G_2.
(ii) Show that $\phi_1 = (\phi_2)^{-1}$.

Solution 32.   (i) One can deduce that ϕ_1 maps the identity element e_1 of G_1 to the identity element e_2 of G_2, and it also maps inverses to inverses as

$$\phi_1(u^{-1}) = \phi_1(u)^{-1}.$$

Therefore ϕ_1 is compatible with the group structure. We define the kernel of ϕ_1 to be the set of elements in G_1 which are mapped to the identity in G_2

$$\ker(\phi_1) = \{ u \in G : \phi_1(u) = e_2 \}$$

and the image of ϕ_1 to be

$$\mathrm{im}(\phi_1) = \{ \phi_1(u) : u \in G_1 \}.$$

Consider the condition $\phi_1 \phi_2 = \phi_2 \phi_1 = i$. It follows that $\ker(\phi_1) = e_1$, $\mathrm{im}(\phi_1) = G_2$. This means that G_2 is isomorphic to G_1.
(ii) It follows that both ϕ_1 and ϕ_2 are isomorphisms. Then $\phi_1 = (\phi_2)^{-1}$.

Problem 33. Show that the finite groups $G_1 := \{ +1, -1 : \cdot \}$ and

$$G_2 = \left\{ \begin{pmatrix} 1 & 0 \\ 0 & 1 \end{pmatrix}, \begin{pmatrix} 0 & 1 \\ 1 & 0 \end{pmatrix} : * \right\}$$

are isomorphic, where $*$ denotes matrix multiplication.

Solution 33. The identity of G_1 must map to the identity in G_2. We have

$$\phi(+1) = \begin{pmatrix} 1 & 0 \\ 0 & 1 \end{pmatrix}, \qquad \phi(-1) = \begin{pmatrix} 0 & 1 \\ 1 & 0 \end{pmatrix}.$$

Problem 34. Consider the matrix groups with matrix multiplication

$$G = \left\{ \begin{pmatrix} a & b \\ 0 & a^{-1} \end{pmatrix} : a, b \in \mathbb{R}, \ a > 0 \right\}, \quad N = \left\{ \begin{pmatrix} 1 & b \\ 0 & 1 \end{pmatrix} : b \in \mathbb{R} \right\}.$$

(i) Show that N is a *normal subgroup* of G.
(ii) Show that the factor group G/N is isomorphic to the additive group $(\mathbb{R}, +)$.

Solution 34. (i) Matrix multiplication yields

$$\begin{pmatrix} a_1 & b_1 \\ 0 & a_1^{-1} \end{pmatrix} \begin{pmatrix} a_2 & b_2 \\ 0 & a_2^{-1} \end{pmatrix} = \begin{pmatrix} a_1 a_2 & a_1 b_2 + b_1 a_2^{-1} \\ 0 & a_1^{-1} a_2^{-1} \end{pmatrix}.$$

The inverse of an element of G is given by

$$\begin{pmatrix} a & b \\ 0 & a^{-1} \end{pmatrix}^{-1} = \begin{pmatrix} a^{-1} & -b \\ 0 & a \end{pmatrix}.$$

Setting $a = 1$ we see that N is a subgroup. Now let

$$\begin{pmatrix} 1 & \beta \\ 0 & 1 \end{pmatrix} \in N.$$

It follows that

$$\begin{pmatrix} a & b \\ 0 & a^{-1} \end{pmatrix}^{-1} \begin{pmatrix} 1 & \beta \\ 0 & 1 \end{pmatrix} \begin{pmatrix} a & b \\ 0 & a^{-1} \end{pmatrix} = \begin{pmatrix} 1 & \beta a^{-2} \\ 0 & 1 \end{pmatrix} \in N.$$

Thus the subgroup N is normal.

(ii) Let $\mathbb{R}_+$ be the group of real numbers under multiplication. The map from G onto $\mathbb{R}_+$ given by

$$\begin{pmatrix} a & b \\ 0 & a^{-1} \end{pmatrix} \mapsto a$$

is a *homomorphism* whose kernel is N. Consequently G/N is isomorphic to $\mathbb{R}_+$ which is isomorphic to the additive group $\mathbb{R}$.

Problem 35.   Let G be a group. Let H be a subgroup of G. For any fixed $g \in G$ one defines

$$gHg^{-1} := \{\, ghg^{-1} : h \in H \,\}.$$

The group gHg^{-1} is called a *conjugate* of the subgroup H. Show that if H is cyclic, then gHg^{-1} is cyclic.

Solution 35.   Let H be cyclic and $d \in H$ be its generator. Then all elements h of H are equal to d^p for some $p \in \mathbb{Z}$. Let $a = gdg^{-1}$, then $a \in gHg^{-1}$. Consider powers of the element a

$$\begin{aligned} a^p &= (gdg^{-1})(gdg^{-1})(gdg^{-1}) \cdots (gdg^{-1}) \\ &= gd(g^{-1}g)d(g^{-1}g)d(g^{-1} \cdots g)dg^{-1} \\ &= gd \cdots dg^{-1} \\ &= gd^p g^{-1} = ghg^{-1}. \end{aligned}$$

Thus all powers of the element a are in gHg^{-1}. Since all elements of H are of the form d^p we find that all elements of gHg^{-1} are of the form $gd^p g^{-1} = gHg^{-1}$. Thus gHg^{-1} is cyclic and its generator is gdg^{-1}.

Problem 36.   Let $G_1, G_2, \ldots, G_n$ be groups. For

$$(g_1, g_2, \ldots, g_n) \in G_1 \times G_2 \times \cdots \times G_n, \quad (h_1, h_2, \ldots, h_n) \in G_1 \times G_2 \times \cdots \times G_n$$

we define $(g_1, g_2, \ldots, g_n) \bullet (h_1, h_2, \ldots, h_n)$ to be $(g_1 h_1, g_2 h_2, \ldots, g_n h_n)$. Show that $G_1 \times G_2 \times \cdots \times G_n$ is a group under this composition (called the external direct product of the groups G_j).

Solution 36. Since $g_j, h_j \in G_j$ we have $g_j h_j \in G_j$. Thus the definition given above provides that $G_1 \times G_2 \times \cdots \times G_n$ is closed under the composition. If e_j is the identity element in G_j, then $(e_1, e_2, \ldots, e_n)$ is the identity element of $G_1 \times G_2 \times \cdots \times G_n$. The inverse of $(g_1, g_2, \ldots, g_n)$ is $(g_1^{-1}, g_2^{-1}, \ldots, g_n^{-1})$. The associative law

$$(g_1, \ldots, g_n) \bullet ((h_1, \ldots, h_n) \bullet (k_1, \ldots, k_n))$$

$$= ((g_1, \ldots, g_n) \bullet (h_1, \ldots, h_n)) \bullet (k_1, \ldots, k_n)$$

is also satisfied.

Problem 37. Let G be a finite group. A *conjugacy class* for a given element g of G is the set defined by

$$C(g) := \{ aga^{-1} : a \in G \}.$$

Show that the number of elements in a conjugacy class divides the order of G.

Solution 37. Let G be a finite group. Then for any group element x, the elements in the conjugacy class of x are in one-to-one correspondence with cosets of the *centralizer* $C_G(x)$. The centralizer $C_G(x)$ of each element x is a subgroup of group G. Thus the left and right cosets exist (for each x). When any two elements b and c belong to the same left coset, then $b = cz$ for some z in the centralizer $C_G(x)$. This implies that two conjugations of x by b and c are equal

$$bxb^{-1} = (cz)x(cz)^{-1} = cxc^{-1}.$$

Therefore the number of elements in the conjugacy class of x is the index $[G : C_G(x)]$ of the centralizer $C_G(x)$ in G. Lagrange's theorem then implies that the size of each conjugacy class is a divisor of the size of the group.

Problem 38. The 3×3 permutation matrices

$$\begin{pmatrix} 1 & 0 & 0 \\ 0 & 1 & 0 \\ 0 & 0 & 1 \end{pmatrix}, \quad \begin{pmatrix} 0 & 1 & 0 \\ 0 & 0 & 1 \\ 1 & 0 & 0 \end{pmatrix}, \quad \begin{pmatrix} 0 & 0 & 1 \\ 1 & 0 & 0 \\ 0 & 1 & 0 \end{pmatrix}$$

form a group under matrix multiplication. Find the conjugacy classes.

Solution 38.   Since the group is commutative the conjugacy classes are

$$\left\{\begin{pmatrix} 1 & 0 & 0 \\ 0 & 1 & 0 \\ 0 & 0 & 1 \end{pmatrix}\right\}, \quad \left\{\begin{pmatrix} 0 & 1 & 0 \\ 0 & 0 & 1 \\ 1 & 0 & 0 \end{pmatrix}\right\}, \quad \left\{\begin{pmatrix} 0 & 0 & 1 \\ 1 & 0 & 0 \\ 0 & 1 & 0 \end{pmatrix}\right\}.$$

Problem 39.   Let D_n be the *dihedral group* with $n \geq 3$. This is the finite nonabelian group of rigid motions of a regular n-gon. The order of the group is $2n$. Find the center

$$Z(D_n) := \{ c \in D_n : cx = xc \quad \text{for all} \quad x \in D_n \}.$$

Solution 39.   Let a, b be the generators of the group D_n, i.e.

$$D_n = \langle a, b : a^n = b^2 = e, \ ba = a^{-1}b \rangle$$

where a is a rotation by $2\pi/n$, b is a flip and e is the neutral element. It suffices to find those elements which commute with the generators a and b. Since $n \geq 3$ it follows that $a^{-1} \neq a$. Suppose $a^s b \in Z(D_n)$. Hence

$$a^{s+1}b = a(a^s b) = (a^s b)a = a^{s-1}b.$$

Thus $a^2 = e$. This is a contradiction. Hence no element of the form $a^s b$ is in the center $Z(D_n)$. Analogously, if for $1 \leq r < n$

$$a^r b = ba^r = a^{-r}b$$

then $a^{2r} = e$, which is possible only if $2r = n$. Thus a^r commutes with b if and only if $n = 2r$. If $n = 2r$, the center of D_n is $\{ e, a^r \}$. If n is odd the center is given by the neutral element $\{ e \}$.

Problem 40.   The number of elements of a group (finite or infinite) is called its *order*. The order of G is denoted by $|G|$. The order of an element g in the group G is the smallest positive integer k such that $g^k = e$, where e is the identity element of the group. It is denoted by $|g|$. If no such element exists, one says that g has infinite order. Consider the 3×3 permutation matrix

$$P = \begin{pmatrix} 0 & 1 & 0 \\ 0 & 0 & 1 \\ 1 & 0 & 0 \end{pmatrix}$$

and matrix multiplication. What is the order of P in the group of permutation matrices? What is order of the group of all 3×3 permutation matrices?

Solution 40. We have

$$P^2 = \begin{pmatrix} 0 & 0 & 1 \\ 1 & 0 & 0 \\ 0 & 1 & 0 \end{pmatrix} = P^T$$

and $P^3 = I_3$, where ^{T} denotes transpose. Thus the order of the group element P is 3. The order of the group of all 3×3 permutation matrices is $3! = 6$.

Problem 41. Consider the set of *Pauli spin matrices* σ_x, σ_y, σ_z

$$\sigma_x := \begin{pmatrix} 0 & 1 \\ 1 & 0 \end{pmatrix}, \quad \sigma_y := \begin{pmatrix} 0 & -i \\ i & 0 \end{pmatrix}, \quad \sigma_z := \begin{pmatrix} 1 & 0 \\ 0 & -1 \end{pmatrix}$$

and the 2×2 identity matrix I_2. Can we extend this set so that we obtain a group under matrix multiplication?

Solution 41. We have $\sigma_x^2 = \sigma_y^2 = \sigma_z^2 = I_2$ and

$$\sigma_x \sigma_y = i\sigma_z, \qquad \sigma_y \sigma_x = -i\sigma_z$$

$$\sigma_y \sigma_z = i\sigma_x, \qquad \sigma_z \sigma_y = -i\sigma_x$$

$$\sigma_z \sigma_x = i\sigma_y, \qquad \sigma_x \sigma_z = -i\sigma_y.$$

Furthermore $(i\sigma_z)(i\sigma_z) = -I_2$, $\sigma_z(i\sigma_z) = iI_2$, $\sigma_z(-i\sigma_z) = -iI_2$ etc. Thus we have to extend the set to the 16 elements

$$\{ I_2, \sigma_x, \sigma_y, \sigma_z, -I_2, -\sigma_x, -\sigma_y, -\sigma_z,$$

$$iI_2, i\sigma_x, i\sigma_y, i\sigma_z, -iI_2, -i\sigma_x, -i\sigma_y, -i\sigma_z \}$$

to obtain a group under matrix multiplication. Note that all matrices are unitary. A generating set is $\{ \sigma_x, \sigma_z \}$.

Problem 42. The group $SL(2, \mathbb{R})$ consists of all 2×2 matrices over $\mathbb{R}$ with determinant equal to 1. Let

$$A = \begin{pmatrix} 0 & -1 \\ 1 & 0 \end{pmatrix}, \qquad B = \begin{pmatrix} 0 & -1 \\ 1 & -1 \end{pmatrix}$$

be elements of $SL(2, \mathbb{R})$. What is the order of A, B and AB?

Solution 42. The order of the group element A is equal to 4. We have

$$A^2 = \begin{pmatrix} -1 & 0 \\ 0 & -1 \end{pmatrix} = -I_2$$

and $A^3 = -A$. Thus $A^4 = I_2$. The order of the matrix B is equal to 3. We have

$$B^2 = \begin{pmatrix} -1 & 1 \\ -1 & 0 \end{pmatrix}$$

and therefore $B^3 = I_2$. We have

$$AB = \begin{pmatrix} -1 & 1 \\ 0 & -1 \end{pmatrix}.$$

By induction we can show that

$$(AB)^n = \begin{pmatrix} (-1)^n & (-1)^{n+1}n \\ 0 & (-1)^n \end{pmatrix}.$$

Thus AB has infinite order.

Problem 43.   Consider the set $N = \{1, 2, \ldots, n\}$. The set of all permutations of N is called the *symmetric group* of degree n and is denoted by S_n. The elements of S_n have the form

$$s = \begin{pmatrix} 1 & 2 & \cdots & n \\ s(1) & s(2) & \cdots & s(n) \end{pmatrix}.$$

The order of the group is $n!$.
(i) Consider S_3. Give all the elements of this group.
(ii) Is the group commutative?
(iii) How many subgroups does S_3 have?

Solution 43.   (i) We have the six elements (e is the identity)

$$e = \begin{pmatrix} 1 & 2 & 3 \\ 1 & 2 & 3 \end{pmatrix}, \quad t = \begin{pmatrix} 1 & 2 & 3 \\ 2 & 3 & 1 \end{pmatrix}, \quad t^2 = \begin{pmatrix} 1 & 2 & 3 \\ 3 & 1 & 2 \end{pmatrix},$$

$$r = \begin{pmatrix} 1 & 2 & 3 \\ 1 & 3 & 2 \end{pmatrix}, \quad tr = \begin{pmatrix} 1 & 2 & 3 \\ 2 & 1 & 3 \end{pmatrix}, \quad t^2r = \begin{pmatrix} 1 & 2 & 3 \\ 3 & 2 & 1 \end{pmatrix}.$$

(ii) We have

$$tr = \begin{pmatrix} 1 & 2 & 3 \\ 2 & 3 & 1 \end{pmatrix} \circ \begin{pmatrix} 1 & 2 & 3 \\ 1 & 3 & 2 \end{pmatrix} = \begin{pmatrix} 1 & 2 & 3 \\ 2 & 1 & 3 \end{pmatrix}$$

and

$$rt = \begin{pmatrix} 1 & 2 & 3 \\ 1 & 3 & 2 \end{pmatrix} \circ \begin{pmatrix} 1 & 2 & 3 \\ 2 & 3 & 1 \end{pmatrix} = \begin{pmatrix} 1 & 2 & 3 \\ 3 & 2 & 1 \end{pmatrix}.$$

Thus $tr \neq rt$. Therefore the group is non-commutative.

(iii) The symmetric group S_3 has 6 subgroups. There is obviously one subgroup of order 1. There are three subgroups of order 2, namely $\{e, r\}$, $\{e, tr\}$, $\{e, t^2 r\}$. There is one subgroup of order 3, namely $\{e, t, t^2\}$. This is the cyclic permutation subgroup. There is one subgroup of order 6, namely S_3 itself.

Problem 44.   Consider the square where the corners are numbered counter-clockwise 1, 2, 3, 4. Show that the symmetry group D_4 of the square is a subgroup of the symmetric group S_4. The group S_4 is defined as the set of all permutations of $\{1, 2, 3, 4\}$. Each element of the symmetry group D_4 transforms every corner of the square to another corner. Thus it transforms every element of $\{1, 2, 3, 4\}$ (the number of the corner) to another element.

Solution 44.   We demonstrate the bijection between all elements of the group D_4 group and some subset of S_4. We prove that the product of every two elements of this subset belongs again to the subset. Then the subset is a subgroup of S_4. The characteristic property of the elements D_4 is that they preserve some geometric shapes - edges and diagonals. Two vertices having a common edge before transformation, have a common edge after transformation also, and the same for diagonals. Perform two such transformations one after another and let v_1, v_2 be two vertices, which have a common edge before the transformations. This implies that after the first transformation they also have a common edge. Therefore after the second transformation they possess the same property. This proves that the product of every two elements of D_4 belongs again to D_4. Thus it is a subgroup of the group S_4. The relation between shapes before and after transformation is a bijection. Therefore the inverse transformation in the sense of S_4 also preserves the geometric shapes. Thus when some transformation belongs to D_4, its inverse (in the sense of S_4) transformation also belongs to D_4. Therefore the symmetry group D_4 of the square is a subgroup of the symmetric group S_4. The group is generated by the two elements

$$ s = \begin{pmatrix} 1 & 2 & 3 & 4 \\ 2 & 3 & 4 & 1 \end{pmatrix}, \qquad t = \begin{pmatrix} 1 & 2 & 3 & 4 \\ 2 & 1 & 4 & 3 \end{pmatrix}. $$

The order of the group D_4 is 8.

Problem 45.   The group of all even permutations of n objects is known as the *alternating group* of degree n and is denoted by A_n. For $n > 1$, the alternating group has order $n!/2$. Find all elements of the group A_3.

Solution 45.   Since $n = 3$ we have $3!/2 = 3$ elements. The elements are

$$e = \begin{pmatrix} 1 & 2 & 3 \\ 1 & 2 & 3 \end{pmatrix}, \quad s = \begin{pmatrix} 1 & 2 & 3 \\ 2 & 3 & 1 \end{pmatrix}, \quad t = \begin{pmatrix} 1 & 2 & 3 \\ 3 & 1 & 2 \end{pmatrix}.$$

Expressed with permutation matrices (which permutes row vectors) we have

$$e \mapsto \begin{pmatrix} 1 & 0 & 0 \\ 0 & 1 & 0 \\ 0 & 0 & 1 \end{pmatrix}, \quad s \mapsto \begin{pmatrix} 0 & 0 & 1 \\ 1 & 0 & 0 \\ 0 & 1 & 0 \end{pmatrix}, \quad t \mapsto \begin{pmatrix} 0 & 1 & 0 \\ 0 & 0 & 1 \\ 1 & 0 & 0 \end{pmatrix}.$$

Problem 46.   If G is a group and $g \in G$, then $H := \{\, g^n \, : \, n \in \mathbb{Z} \,\}$ is a subgroup of G. This group is the cyclic subgroup of G generated by g. Consider the 3×3 permutation matrix

$$A = \begin{pmatrix} 0 & 1 & 0 \\ 0 & 0 & 1 \\ 1 & 0 & 0 \end{pmatrix}.$$

Show that A, A^2, A^3 form a group under matrix multiplication.

Solution 46.   From $A^3 = I_3$ we obtain $A^{-1} = A^2 = A^T$. Thus we have a group. This is the cyclic group C_3.

Problem 47.   Consider the 4×4 permutation matrix

$$A = \begin{pmatrix} 0 & 1 & 0 & 0 \\ 0 & 0 & 1 & 0 \\ 0 & 0 & 0 & 1 \\ 1 & 0 & 0 & 0 \end{pmatrix}.$$

Let I_4 be the 4×4 identity matrix. Find A^2, A^3, A^4 and thus show that we have a cyclic group.

Solution 47.   We obtain

$$A^2 = \begin{pmatrix} 0 & 0 & 1 & 0 \\ 0 & 0 & 0 & 1 \\ 1 & 0 & 0 & 0 \\ 0 & 1 & 0 & 0 \end{pmatrix}, \quad A^3 = \begin{pmatrix} 0 & 0 & 0 & 1 \\ 1 & 0 & 0 & 0 \\ 0 & 1 & 0 & 0 \\ 0 & 0 & 1 & 0 \end{pmatrix}$$

and $A^4 = I_4$. Note that $A^{-1} = A^3$, $(A^2)^{-1} = A^2$, $(A^3)^{-1} = A$.

Problem 48. (i) Consider the 4×4 matrix

$$A = \begin{pmatrix} 0 & 0 & 1 & 0 \\ 0 & 0 & 0 & 1 \\ -1 & 0 & 0 & 0 \\ 0 & -1 & 0 & 0 \end{pmatrix}.$$

Obviously rank$(A) = 4$ and therefore the inverse exists. Let I_4 be the 4×4 identity matrix. Find the inverse of A.

(ii) Does the set $\{ A, A^{-1}, I_4 \}$ form a group under matrix multiplication? If not can we find a finite extension to the set to obtain a group?

(iii) Calculate the determinants of the matrices found in (ii). Show that these numbers form a group under multiplication.

Solution 48. (i) We find

$$A^{-1} = \begin{pmatrix} 0 & 0 & -1 & 0 \\ 0 & 0 & 0 & -1 \\ 1 & 0 & 0 & 0 \\ 0 & 1 & 0 & 0 \end{pmatrix} = -A.$$

(ii) Since $A^2 = (A^{-1})^2 = -I_4$ and $A^{-1} = -A$ we find that the set

$$\{ A, -A, I_4, -I_4 \}$$

forms a group under matrix multiplication.

(iii) We obtain $+1$ for all the determinants of the matrices. Thus we have the trivial group.

Problem 49. Show that the following matrices form a group under matrix multiplication

$$s_1 = \begin{pmatrix} 1 & 0 & 0 \\ 0 & 1 & 0 \\ 0 & 0 & 1 \end{pmatrix}, \quad s_2 = \begin{pmatrix} 1 & 0 & 0 \\ 0 & 0 & 1 \\ 0 & 1 & 0 \end{pmatrix}, \quad s_3 = \begin{pmatrix} 0 & 1 & 0 \\ 1 & 0 & 0 \\ 0 & 0 & 1 \end{pmatrix},$$

$$s_4 = \begin{pmatrix} 0 & 1 & 0 \\ 0 & 0 & 1 \\ 1 & 0 & 0 \end{pmatrix}, \quad s_5 = \begin{pmatrix} 0 & 0 & 1 \\ 1 & 0 & 0 \\ 0 & 1 & 0 \end{pmatrix}, \quad s_6 = \begin{pmatrix} 0 & 0 & 1 \\ 0 & 1 & 0 \\ 1 & 0 & 0 \end{pmatrix}.$$

These matrices are the six 3×3 *permutation matrices*.

Solution 49. The multiplication table is

·	s_1	s_2	s_3	s_4	s_5	s_6
s_1	s_1	s_2	s_3	s_4	s_5	s_6
s_2	s_2	s_1	s_4	s_3	s_6	s_5
s_3	s_3	s_5	s_1	s_6	s_2	s_4
s_4	s_4	s_6	s_2	s_5	s_1	s_3
s_5	s_5	s_3	s_6	s_1	s_4	s_2
s_6	s_6	s_4	s_5	s_2	s_3	s_1

From the table we see that the element $s_1 = e$ commutes with all other elements and only s_4 and s_5 commute: $s_4 s_5 = s_5 s_4 = s_1 = e$. All other elements do not commute. This means that the group of 3×3 permutation matrices is not commutative. The inverse elements can be found from the multiplication table

$$s_1^{-1} = s_1, \quad s_2^{-1} = s_2, \quad s_3^{-1} = s_3, \quad s_4^{-1} = s_5, \quad s_5^{-1} = s_4, \quad s_6^{-1} = s_6.$$

The associative law follows the from associative law for matrix multiplication. Thus the set of the 3×3 permutation matrices forms a non-commutative group under matrix multiplication.

Problem 50. Given two manifolds M and N, a bijective map ϕ from M to N is called a *diffeomorphism* if both $\phi : M \to N$ and its inverse ϕ^{-1} are differentiable. Let $f : \mathbb{R} \to \mathbb{R}$ be given by the analytic function

$$f(x) = 4x(1-x)$$

and the analytic function $g : \mathbb{R} \to \mathbb{R}$ be given by

$$g(x) = 1 - 2x^2 \,.$$

(i) Can one find a diffeomorphism $\phi : \mathbb{R} \to \mathbb{R}$ such that

$$g = \phi \circ f \circ \phi^{-1} ?$$

(ii) Consider the diffeomorphism $\psi : \mathbb{R} \to \mathbb{R}$

$$\psi(x) = \sinh(x) \,.$$

Calculate $\psi \circ g \circ \psi^{-1}$.

Solution 50. (i) We find $\phi(x) = 2x - 1$. Thus the inverse is $\phi^{-1}(x) = (x+1)/2$. Therefore

$$(f \circ \phi^{-1})(x) = 4 \frac{x+1}{2} \left(1 - \frac{x+1}{2} \right) = 1 - x^2 \,.$$

It follows that

$$(\phi \circ f \circ \phi^{-1})(x) = 2(-x^2 + 1) - 1 = 1 - 2x^2 \,.$$

(ii) For the left composition we obtain

$$(\psi \circ g)(x) = \sinh(1 - 2x^2).$$

Then

$$(\psi \circ g \circ \psi^{-1})(x) = \sinh\left(1 - 2(\text{Arsinh}(x))^2\right).$$

Problem 51. Let G be the *dihedral group* defined by

$$D_8 := \langle a, b \,:\, a^4 = b^2 = e, \ b^{-1}ab = a^{-1} \rangle$$

where e is the identity element in the group. Define the invertible 2×2 matrices

$$A = \begin{pmatrix} 0 & 1 \\ -1 & 0 \end{pmatrix}, \qquad B = \begin{pmatrix} 1 & 0 \\ 0 & -1 \end{pmatrix}.$$

Let I_2 be the 2×2 identity matrix. Show that $A^4 = B^2 = I_2$, $B^{-1}AB = A^{-1}$ and thus show that we have a representation of the dihedral group.

Solution 51. We have $A^2 = -I_2$. Thus $A^4 = I_2$. Furthermore $B^2 = I_2$. The inverse of B is given by $B^{-1} = B$. Thus

$$B^{-1}AB = BAB = \begin{pmatrix} 1 & 0 \\ 0 & -1 \end{pmatrix} \begin{pmatrix} 0 & 1 \\ -1 & 0 \end{pmatrix} \begin{pmatrix} 1 & 0 \\ 0 & -1 \end{pmatrix} = A^{-1} \,.$$

It follows that

$$\rho : a^j b^k \to A^j B^k, \qquad j = 0, 1, 2, 3, \quad k = 0, 1$$

is a representation of D_8 over $\mathbb{R}$.

Problem 52. Consider the 2×2 matrices

$$C_3 = \begin{pmatrix} -1/2 & -\sqrt{3}/2 \\ \sqrt{3}/2 & -1/2 \end{pmatrix}, \qquad \sigma_3 = \begin{pmatrix} 1 & 0 \\ 0 & -1 \end{pmatrix}$$

together with the 2×2 identity matrix I_2. Do these matrices form a group under matrix multiplication? If not find the smallest set of matrices with C_3, σ_3 and I_2 as elements which forms a group under matrix multiplication.

Solution 52. We have $\sigma_3^2 = I_2$ and

$$C_3^2 = \begin{pmatrix} -1/2 & \sqrt{3}/2 \\ -\sqrt{3}/2 & -1/2 \end{pmatrix}, \qquad C_3^3 = I_2 \,.$$

Furthermore

$$\sigma_3 C_3 = \begin{pmatrix} -1/2 & -\sqrt{3}/2 \\ -\sqrt{3}/2 & 1/2 \end{pmatrix}, \qquad C_3 \sigma_3 = \begin{pmatrix} -1/2 & \sqrt{3}/2 \\ \sqrt{3}/2 & 1/2 \end{pmatrix}.$$

Thus $\sigma_3 C_3 \neq C_3 \sigma_3$. Since $\sigma_3 C_3 \sigma_3 = C_3^2 = C_3^{-1}$ it follows that

$$(\sigma_3 C_3)^2 = I_2, \qquad (C_3 \sigma_3)^2 = I_2.$$

Thus the elements

$$\{ I_2, \ \sigma_3, \ C_3, \ C_3^2, \ \sigma_3 C_3, \ C_3 \sigma_3 \}$$

form a non-commutative group under matrix multiplication.

Problem 53.   Let $c \in \mathbb{R}$ and $c \neq 0$. Show that the 2×2 matrices

$$A(c) = \begin{pmatrix} c & c \\ c & c \end{pmatrix}$$

form a group under matrix multiplication. Notice that the matrices $A(c)$ are not invertible in the usual sense, i.e. $\det(A(c)) = 0$. Find $\exp(A(c))$.

Solution 53.   Multiplication of such two matrices yields

$$\begin{pmatrix} c_1 & c_1 \\ c_1 & c_1 \end{pmatrix} \begin{pmatrix} c_2 & c_2 \\ c_2 & c_2 \end{pmatrix} = \begin{pmatrix} 2c_1 c_2 & 2c_1 c_2 \\ 2c_1 c_2 & 2c_1 c_2 \end{pmatrix}.$$

Thus the right-hand side is an element of the set again. The neutral element is the matrix

$$\begin{pmatrix} 1/2 & 1/2 \\ 1/2 & 1/2 \end{pmatrix}.$$

The inverse element is the matrix

$$\begin{pmatrix} 1/(4c) & 1/(4c) \\ 1/(4c) & 1/(4c) \end{pmatrix}.$$

We obtain

$$\exp(A(c)) = I_2 + \begin{pmatrix} 1 & 1 \\ 1 & 1 \end{pmatrix} \sum_{j=1}^{\infty} \frac{c^j}{j!} 2^{j-1}.$$

Problem 54.   (i) Find all 2×2 matrices A over $\mathbb{R}$ with

$$\det(A) = a_{11} a_{22} - a_{12} a_{21} = 1 \tag{1}$$

and

$$A\frac{1}{\sqrt{2}}\begin{pmatrix}1\\1\end{pmatrix} = \frac{1}{\sqrt{2}}\begin{pmatrix}1\\1\end{pmatrix}. \tag{2}$$

(ii) Do these matrices form a group under matrix multiplication?

Solution 54. (i) From (2) we find the system of linear equations

$$a_{11} + a_{12} = 1, \qquad a_{21} + a_{22} = 1.$$

From $\det(A) = 1$ we have $a_{11}a_{22} - a_{12}a_{21} = 1$. Eliminating a_{22} we obtain

$$a_{11} + a_{12} = 1, \qquad a_{11}(1 - a_{21}) - a_{12}a_{21} = 1.$$

Eliminating a_{12} we obtain $a_{11} - a_{21} = 1$ with $a_{12} = 1 - a_{11}$ and $a_{22} = 2 - a_{11}$. Thus

$$A = \begin{pmatrix} a_{11} & 1 - a_{11} \\ a_{11} - 1 & 2 - a_{11} \end{pmatrix}$$

with a_{11} arbitrary.
(ii) Consider the matrices

$$\begin{pmatrix} \alpha & 1 - \alpha \\ \alpha - 1 & 2 - \alpha \end{pmatrix}$$

where $\alpha \in \mathbb{R}$. For $\alpha = 1$ one has the identity matrix I_2. Then

$$\begin{pmatrix} \alpha & 1 - \alpha \\ \alpha - 1 & 2 - \alpha \end{pmatrix}\begin{pmatrix} \beta & 1 - \beta \\ \beta - 1 & 2 - \beta \end{pmatrix} = \begin{pmatrix} \gamma & 1 - \gamma \\ \gamma - 1 & 2 - \gamma \end{pmatrix}$$

where $\gamma = \alpha + \beta - 1$. The inverse of the matrix exists and is given by

$$\begin{pmatrix} 2 - \alpha & \alpha - 1 \\ 1 - \alpha & \alpha \end{pmatrix}.$$

Thus we have a group under matrix multiplication.

Problem 55.   Let G be a group. If a subset H of G itself is a group under the operation of G, then H is called a subgroup of G. Consider the group of the 3×3 permutation matrices and matrix multiplication. Find a subgroup of order 3.

Solution 55.   We find the abelian group

$$\begin{pmatrix} 1 & 0 & 0 \\ 0 & 1 & 0 \\ 0 & 0 & 1 \end{pmatrix}, \quad \begin{pmatrix} 0 & 1 & 0 \\ 0 & 0 & 1 \\ 1 & 0 & 0 \end{pmatrix}, \quad \begin{pmatrix} 0 & 0 & 1 \\ 1 & 0 & 0 \\ 0 & 1 & 0 \end{pmatrix}.$$

Problem 56.   Consider finite groups. Then *Lagrange's theorem* tells us that the order of a subgroup of a finite group is always a divisor of the order of the group. Show that we cannot conclude that if G is a finite group and if m is a divisor of the order of G that G will contain a subgroup of order m. Consider the subgroup A_4 with determinant $+1$ of the group of all 4×4 permutation matrices. The order of this group is $4!/2 = 12$.

Solution 56.   We show that there is no subgroup of order 6 of the group A_4. The group A_4 has order 12. According to Lagrange's theorem the order of a subgroup of a finite group is always a divisor of the order of the group. Thus a subgroup of A_4 can have order 6. However the group A_4 has no subgroup of order 6. A_4 is the smallest group demonstrating that a group does not need to not have a subgroup of every order that divides the group's order. We list all elements of the group A_4 in lexicographical order

$$g_1 = \begin{pmatrix} 1 & 0 & 0 & 0 \\ 0 & 1 & 0 & 0 \\ 0 & 0 & 1 & 0 \\ 0 & 0 & 0 & 1 \end{pmatrix}, \quad g_2 = \begin{pmatrix} 1 & 0 & 0 & 0 \\ 0 & 0 & 1 & 0 \\ 0 & 0 & 0 & 1 \\ 0 & 1 & 0 & 0 \end{pmatrix}, \quad g_3 = \begin{pmatrix} 1 & 0 & 0 & 0 \\ 0 & 0 & 0 & 1 \\ 0 & 1 & 0 & 0 \\ 0 & 0 & 1 & 0 \end{pmatrix},$$

$$g_4 = \begin{pmatrix} 0 & 1 & 0 & 0 \\ 1 & 0 & 0 & 0 \\ 0 & 0 & 0 & 1 \\ 0 & 0 & 1 & 0 \end{pmatrix}, \quad g_5 = \begin{pmatrix} 0 & 1 & 0 & 0 \\ 0 & 0 & 1 & 0 \\ 1 & 0 & 0 & 0 \\ 0 & 0 & 0 & 1 \end{pmatrix}, \quad g_6 = \begin{pmatrix} 0 & 1 & 0 & 0 \\ 0 & 0 & 0 & 1 \\ 0 & 0 & 1 & 0 \\ 1 & 0 & 0 & 0 \end{pmatrix},$$

$$g_7 = \begin{pmatrix} 0 & 0 & 1 & 0 \\ 1 & 0 & 0 & 0 \\ 0 & 1 & 0 & 0 \\ 0 & 0 & 0 & 1 \end{pmatrix}, \quad g_8 = \begin{pmatrix} 0 & 0 & 1 & 0 \\ 0 & 1 & 0 & 0 \\ 0 & 0 & 0 & 1 \\ 1 & 0 & 0 & 0 \end{pmatrix}, \quad g_9 = \begin{pmatrix} 0 & 0 & 1 & 0 \\ 0 & 0 & 0 & 1 \\ 1 & 0 & 0 & 0 \\ 0 & 1 & 0 & 0 \end{pmatrix},$$

$$g_{10} = \begin{pmatrix} 0 & 0 & 0 & 1 \\ 1 & 0 & 0 & 0 \\ 0 & 0 & 1 & 0 \\ 0 & 1 & 0 & 0 \end{pmatrix}, \quad g_{11} = \begin{pmatrix} 0 & 0 & 0 & 1 \\ 0 & 1 & 0 & 0 \\ 1 & 0 & 0 & 0 \\ 0 & 0 & 1 & 0 \end{pmatrix}, \quad g_{12} = \begin{pmatrix} 0 & 0 & 0 & 1 \\ 0 & 0 & 1 & 0 \\ 0 & 1 & 0 & 0 \\ 1 & 0 & 0 & 0 \end{pmatrix}.$$

The multiplication table of the group A_4 is given by

·	g_1	g_2	g_3	g_4	g_5	g_6	g_7	g_8	g_9	g_{10}	g_{11}	g_{12}
g_1	g_1	g_2	g_3	g_4	g_5	g_6	g_7	g_8	g_9	g_{10}	g_{11}	g_{12}
g_2	g_2	g_3	g_1	g_6	g_4	g_5	g_8	g_9	g_7	g_{12}	g_{10}	g_{11}
g_3	g_3	g_1	g_2	g_5	g_6	g_4	g_9	g_7	g_8	g_{11}	g_{12}	g_{10}
g_4	g_4	g_7	g_{10}	g_1	g_8	g_{11}	g_2	g_5	g_{12}	g_3	g_6	g_9
g_5	g_5	g_9	g_{11}	g_3	g_7	g_{12}	g_1	g_6	g_{10}	g_2	g_4	g_8
g_6	g_6	g_8	g_{12}	g_2	g_9	g_{10}	g_3	g_4	g_{11}	g_1	g_5	g_7
g_7	g_7	g_{10}	g_4	g_{11}	g_1	g_8	g_5	g_{12}	g_2	g_9	g_3	g_6
g_8	g_8	g_{12}	g_6	g_{10}	g_2	g_9	g_4	g_{11}	g_3	g_7	g_1	g_5
g_9	g_9	g_{11}	g_5	g_{12}	g_3	g_7	g_6	g_{10}	g_1	g_8	g_2	g_4
g_{10}	g_{10}	g_4	g_7	g_8	g_{11}	g_1	g_{12}	g_2	g_5	g_6	g_9	g_3
g_{11}	g_{11}	g_5	g_9	g_7	g_{12}	g_3	g_{10}	g_1	g_6	g_4	g_8	g_2
g_{12}	g_{12}	g_6	g_8	g_9	g_{10}	g_2	g_{11}	g_3	g_4	g_5	g_7	g_1

We calculate orders of all elements. The result is given in the table below. The first column is the group element, which we use to generate the cyclic group. The second column is the order of the generator. In the following columns sequential powers of the generator are presented.

Generator	Order	g_1	g_2	g_3
g_1	1	g_1		
g_2	3	g_2	g_3	g_1
g_3	3	g_3	g_2	g_1
g_4	2	g_4	g_1	
g_5	3	g_5	g_7	g_1
g_6	3	g_6	g_{10}	g_1
g_7	3	g_7	g_5	g_1
g_8	3	g_8	g_{11}	g_1
g_9	2	g_9	g_1	
g_{10}	3	g_{10}	g_6	g_1
g_{11}	3	g_{11}	g_8	g_1
g_{12}	2	g_{12}	g_1	

From this table we see that A_4 has cyclic subgroups of order 2 and 3. Suppose that some element g of a group G is in its subgroup H. Then all powers of g are also in H. All cyclic subgroups generated by G are in H and the orders of element g in G and H are equal. 2 and 3 are prime divisors of 6. Thus the subgroup of A_4 of order 6 according to *Cauchy's theorem* can have cyclic subgroups of orders 2 and 3. These cyclic subgroups must exist in A_4. The subgroups share the identity element g_1. A subgroup containing a cyclic subgroup of order m and a different cyclic subgroup of order n has order of at least $1 + (m - 1) + (n - 1) = m + n - 1$.

Thus to find subgroup of order 6 we use the following algorithm:
- we choose an arbitrary cyclic subgroup of order 2 and an arbitrary cyclic subgroup of order 3;

- we calculate all products of elements of the first cyclic subgroup by elements of the second cyclic subgroup;
- we calculate all products of elements of the second cyclic subgroup by elements of the first cyclic subgroup;
- we count all distinct elements in the result.

If the sum is more then 6 there is not a subgroup of order 6, containing both cyclic subgroups. If the sum is equal to 6 additional considerations are necessary. There are no such cyclic subgroups.

The results can be summarized in the following table.

e_1	e_1^2	e_2	e_2^2	e_2^3	Count
g_4	g_1	g_2	g_3	g_1	8
g_4	g_1	g_5	g_7	g_1	8
g_4	g_1	g_6	g_{10}	g_1	8
g_4	g_1	g_8	g_{11}	g_1	8
g_9	g_1	g_2	g_3	g_1	8
g_9	g_1	g_5	g_7	g_1	8
g_9	g_1	g_6	g_{10}	g_1	8
g_9	g_1	g_8	g_{11}	g_1	8
g_{12}	g_1	g_2	g_3	g_1	8
g_{12}	g_1	g_5	g_7	g_1	8
g_{12}	g_1	g_6	g_{10}	g_1	8
g_{12}	g_1	g_8	g_{11}	g_1	8

In the first two columns the elements of the first cyclic subgroup of order 2 are presented. In the next three columns the elements of the second cyclic subgroup of order 3 are presented. The last column contains the count of distinct products of elements of the first cyclic subgroup by elements of the second cyclic subgroup. In all the cases the count is more than 6. Thus the group A_4 of order 12 does not have a subgroup of order 6.

Problem 57.   Do the twelve 2×2 orthogonal matrices

$$\begin{pmatrix} 1 & 0 \\ 0 & 1 \end{pmatrix}, \quad \begin{pmatrix} 1 & 0 \\ 0 & -1 \end{pmatrix}, \quad \begin{pmatrix} -1 & 0 \\ 0 & 1 \end{pmatrix}, \quad \begin{pmatrix} -1 & 0 \\ 0 & -1 \end{pmatrix},$$

$$\begin{pmatrix} 0 & 1 \\ 1 & 0 \end{pmatrix}, \quad \begin{pmatrix} 0 & 1 \\ -1 & 0 \end{pmatrix}, \quad \begin{pmatrix} 0 & -1 \\ 1 & 0 \end{pmatrix}, \quad \begin{pmatrix} 0 & -1 \\ -1 & 0 \end{pmatrix},$$

$$\frac{1}{\sqrt{2}}\begin{pmatrix} 1 & 1 \\ 1 & -1 \end{pmatrix}, \quad \frac{1}{\sqrt{2}}\begin{pmatrix} 1 & -1 \\ 1 & 1 \end{pmatrix}, \quad \frac{1}{\sqrt{2}}\begin{pmatrix} -1 & 1 \\ 1 & 1 \end{pmatrix}, \quad \frac{1}{\sqrt{2}}\begin{pmatrix} 1 & 1 \\ -1 & 1 \end{pmatrix}$$

form a group under matrix multiplication? If not add the matrices so that one has a group.

Solution 57. We have to add the orthogonal matrices

$$\frac{1}{\sqrt{2}}\begin{pmatrix} 1 & -1 \\ -1 & -1 \end{pmatrix}, \quad \frac{1}{\sqrt{2}}\begin{pmatrix} -1 & 1 \\ -1 & -1 \end{pmatrix}, \quad \frac{1}{\sqrt{2}}\begin{pmatrix} -1 & -1 \\ 1 & -1 \end{pmatrix}, \quad \frac{1}{\sqrt{2}}\begin{pmatrix} -1 & -1 \\ -1 & 1 \end{pmatrix}$$

to the set. Then we have a non-commutative group of order 16.

Problem 58. Consider the group $GL(n, \mathbb{R})$ of all invertible $n \times n$ matrices over $\mathbb{R}$. Show that the $n \times n$ matrices with integer entries and determinant $+1$ or -1 form a *subgroup* of $GL(n, \mathbb{R})$.

Solution 58. The set of all integers is a subset of $\mathbb{R}$. All $n \times n$ matrices with integer entries and determinant $+1$ or -1 are a subset of $GL(n, \mathbb{R})$ of all invertible $n \times n$ matrices over $\mathbb{R}$. If the determinant of the matrix is equal to ± 1, the matrix is invertible. The identity element of the group $GL(n, \mathbb{R})$ is the identity matrix I_n. All its entries are integers and the determinant is $+1$. Consider two $n \times n$ matrices with integer entries and determinant $+1$ or -1. Their product is another $n \times n$ matrix with integer entries and determinant $+1$ or -1. The determinant of the product is equal to product of the determinants. The associative law follows from the same law for matrix multiplication. The inverse of $n \times n$ matrix with integer entries and determinant $+1$ or -1 is also $n \times n$ matrix with integer entries and the same determinant. The formula for components of the inverse is

$$(A^{-1})_{ij} = (-1)^{i+j} \det(M_{ji}) / \det(A)$$

where $\det(M_{ji})$ is the determinant of the $(n-1) \times (n-1)$ matrix that results from deleting row j and column i of the matrix A. When all entries of A are integers and $\det(A) = \pm 1$, all entries of $(A^{-1})_{ij}$ are also integers. The determinant of the inverse matrix is equal to $1/\det(A)$. This implies

$$\det(A) = \pm 1 \Rightarrow \det(A^{-1}) = \pm 1.$$

Thus the $n \times n$ matrices with integer entries and determinant $+1$ or -1 form a subgroup of the group $GL(n, \mathbb{R})$.

Problem 59. Consider the group $G = \{a, b, c\}$ with group operation $\cdot : G \times G \to G$ defined by the table

$\cdot$	a	b	c
a	a	b	c
b	b	c	a
c	c	a	b

Let $M(n)$ denote the vector space of $n \times n$ matrices over $\mathbb{C}$.
(i) Show that $f : G \to M(3)$ is a faithful representation for $(G, \cdot)$ where

$$f(a) := \begin{pmatrix} 1 & 0 & 0 \\ 0 & 1 & 0 \\ 0 & 0 & 1 \end{pmatrix}$$

$$f(b) := \begin{pmatrix} \frac{\sqrt{3}}{2}i - \frac{1}{2} & 0 & 0 \\ 0 & -\frac{1}{2} & \frac{\sqrt{3}}{2} \\ 0 & -\frac{\sqrt{3}}{2} & -\frac{1}{2} \end{pmatrix}$$

$$f(c) := \begin{pmatrix} -\frac{\sqrt{3}}{2}i - \frac{1}{2} & 0 & 0 \\ 0 & -\frac{1}{2} & -\frac{\sqrt{3}}{2} \\ 0 & \frac{\sqrt{3}}{2} & -\frac{1}{2} \end{pmatrix}.$$

(ii) Is the representation in (i) reducible? Prove or disprove.

Solution 59.   (i) We have

$$f(a)f(a) = \begin{pmatrix} 1 & 0 & 0 \\ 0 & 1 & 0 \\ 0 & 0 & 1 \end{pmatrix} = f(a)$$

$$f(a)f(b) = \begin{pmatrix} \frac{\sqrt{3}}{2}i - \frac{1}{2} & 0 & 0 \\ 0 & -\frac{1}{2} & \frac{\sqrt{3}}{2} \\ 0 & -\frac{\sqrt{3}}{2} & -\frac{1}{2} \end{pmatrix} = f(b)$$

$$f(a)f(c) = \begin{pmatrix} -\frac{\sqrt{3}}{2}i - \frac{1}{2} & 0 & 0 \\ 0 & -\frac{1}{2} & -\frac{\sqrt{3}}{2} \\ 0 & \frac{\sqrt{3}}{2} & -\frac{1}{2} \end{pmatrix} = f(c)$$

$$f(b)f(a) = \begin{pmatrix} \frac{\sqrt{3}}{2}i - \frac{1}{2} & 0 & 0 \\ 0 & -\frac{1}{2} & \frac{\sqrt{3}}{2} \\ 0 & -\frac{\sqrt{3}}{2} & -\frac{1}{2} \end{pmatrix} = f(b)$$

$$f(b)f(b) = \begin{pmatrix} -\frac{\sqrt{3}}{2}i - \frac{1}{2} & 0 & 0 \\ 0 & -\frac{1}{2} & -\frac{\sqrt{3}}{2} \\ 0 & \frac{\sqrt{3}}{2} & -\frac{1}{2} \end{pmatrix} = f(c)$$

$$f(b)f(c) = \begin{pmatrix} 1 & 0 & 0 \\ 0 & 1 & 0 \\ 0 & 0 & 1 \end{pmatrix} = f(a)$$

$$f(c)f(a) = \begin{pmatrix} -\frac{\sqrt{3}}{2}i - \frac{1}{2} & 0 & 0 \\ 0 & -\frac{1}{2} & -\frac{\sqrt{3}}{2} \\ 0 & \frac{\sqrt{3}}{2} & -\frac{1}{2} \end{pmatrix} = f(c)$$

$$f(c)f(b) = \begin{pmatrix} 1 & 0 & 0 \\ 0 & 1 & 0 \\ 0 & 0 & 1 \end{pmatrix} = f(a)$$

$$f(c)f(c) = \begin{pmatrix} \frac{\sqrt{3}}{2}i - \frac{1}{2} & 0 & 0 \\ 0 & -\frac{1}{2} & \frac{\sqrt{3}}{2} \\ 0 & -\frac{\sqrt{3}}{2} & -\frac{1}{2} \end{pmatrix} = f(b).$$

Thus f is a representation of the group.
(ii) Let $\oplus$ be the direct sum. Since

$$f(a) = \begin{pmatrix} 1 & 0 & 0 \\ 0 & 1 & 0 \\ 0 & 0 & 1 \end{pmatrix} = (1) \oplus \begin{pmatrix} 1 & 0 \\ 0 & 1 \end{pmatrix} = (1) \oplus (1) \oplus (1)$$

$$f(b) = \begin{pmatrix} \frac{\sqrt{3}}{2}i - \frac{1}{2} & 0 & 0 \\ 0 & -\frac{1}{2} & \frac{\sqrt{3}}{2} \\ 0 & -\frac{\sqrt{3}}{2} & -\frac{1}{2} \end{pmatrix} = \left(\frac{\sqrt{3}}{2}i - \frac{1}{2} \right) \oplus \begin{pmatrix} -\frac{1}{2} & \frac{\sqrt{3}}{2} \\ -\frac{\sqrt{3}}{2} & -\frac{1}{2} \end{pmatrix}$$

$$= (e^{\frac{2\pi i}{3}}) \oplus \begin{pmatrix} \cos\left(\frac{2\pi i}{3}\right) & \sin\left(\frac{2\pi i}{3}\right) \\ -\sin\left(\frac{2\pi i}{3}\right) & \cos\left(\frac{2\pi i}{3}\right) \end{pmatrix}$$

$$f(c) = \begin{pmatrix} -\frac{\sqrt{3}}{2}i - \frac{1}{2} & 0 & 0 \\ 0 & -\frac{1}{2} & -\frac{\sqrt{3}}{2} \\ 0 & \frac{\sqrt{3}}{2} & -\frac{1}{2} \end{pmatrix} = \left(-\frac{\sqrt{3}}{2}i - \frac{1}{2} \right) \oplus \begin{pmatrix} -\frac{1}{2} & -\frac{\sqrt{3}}{2} \\ \frac{\sqrt{3}}{2} & -\frac{1}{2} \end{pmatrix}$$

$$= (e^{\frac{4\pi i}{3}}) \oplus \begin{pmatrix} \cos\left(\frac{4\pi i}{3}\right) & \sin\left(\frac{4\pi i}{3}\right) \\ -\sin\left(\frac{4\pi i}{3}\right) & \cos\left(\frac{4\pi i}{3}\right) \end{pmatrix}$$

we find that the representation is reducible.

Problem 60. The *direct sum* of the 2×2 matrices A, B is defined as

$$A \oplus B := \begin{pmatrix} A & 0_2 \\ 0_2 & B \end{pmatrix}$$

where 0_2 is the 2×2 zero matrix. The 2×2 matrices

$$I_2 = \begin{pmatrix} 1 & 0 \\ 0 & 1 \end{pmatrix}, \qquad N = \begin{pmatrix} 0 & 1 \\ 1 & 0 \end{pmatrix}$$

form a group under matrix multiplication. Show that the 4×4 matrices

$$I_2 \oplus I_2, \quad I_2 \oplus N, \quad N \oplus I_2, \quad N \oplus N$$

form a group under matrix multiplication.

Solution 60.   For the direct sum we find the permutation matrices

$$
I_2 \oplus I_2 = \begin{pmatrix} 1 & 0 & 0 & 0 \\ 0 & 1 & 0 & 0 \\ 0 & 0 & 1 & 0 \\ 0 & 0 & 0 & 1 \end{pmatrix} = I_4, \quad A = I_2 \oplus N = \begin{pmatrix} 1 & 0 & 0 & 0 \\ 0 & 1 & 0 & 0 \\ 0 & 0 & 0 & 1 \\ 0 & 0 & 1 & 0 \end{pmatrix},
$$

$$
B = N \oplus I_2 = \begin{pmatrix} 0 & 1 & 0 & 0 \\ 1 & 0 & 0 & 0 \\ 0 & 0 & 1 & 0 \\ 0 & 0 & 0 & 1 \end{pmatrix}, \quad C = N \oplus N = \begin{pmatrix} 0 & 1 & 0 & 0 \\ 1 & 0 & 0 & 0 \\ 0 & 0 & 0 & 1 \\ 0 & 0 & 1 & 0 \end{pmatrix}.
$$

The multiplication table is

$\cdot$	I_4	A	B	C
I_4	I_4	A	B	C
A	A	I_4	C	B
B	B	C	I_4	A
C	C	B	A	I_4

For the inverse elements we find $A^{-1} = A$, $B^{-1} = B$, $C^{-1} = C$. Thus the set of direct sums of matrices forms a group under matrix multiplication.

Problem 61.   Let A, B be 2×2 matrices. We define the composition

$$
A \star B := \begin{pmatrix} a_{11} & 0 & 0 & a_{12} \\ 0 & b_{11} & b_{12} & 0 \\ 0 & b_{21} & b_{22} & 0 \\ a_{21} & 0 & 0 & a_{22} \end{pmatrix}.
$$

Let

$$
E = \begin{pmatrix} 1 & 0 \\ 0 & 1 \end{pmatrix}, \quad N = \begin{pmatrix} 0 & 1 \\ 1 & 0 \end{pmatrix}.
$$

Show that the 4×4 matrices $E \star E$, $E \star N$, $N \star E$, $N \star N$ form a group under matrix multiplication.

Solution 61.   We obtain the permutation matrices

$$
E \star E = \begin{pmatrix} 1 & 0 & 0 & 0 \\ 0 & 1 & 0 & 0 \\ 0 & 0 & 1 & 0 \\ 0 & 0 & 0 & 1 \end{pmatrix}, \quad E \star N = \begin{pmatrix} 1 & 0 & 0 & 0 \\ 0 & 0 & 1 & 0 \\ 0 & 1 & 0 & 0 \\ 0 & 0 & 0 & 1 \end{pmatrix},
$$

$$
N \star E = \begin{pmatrix} 0 & 0 & 0 & 1 \\ 0 & 1 & 0 & 0 \\ 0 & 0 & 1 & 0 \\ 1 & 0 & 0 & 0 \end{pmatrix}, \quad N \star N = \begin{pmatrix} 0 & 0 & 0 & 1 \\ 0 & 0 & 1 & 0 \\ 0 & 1 & 0 & 0 \\ 1 & 0 & 0 & 0 \end{pmatrix}.
$$

Matrix multiplication shows that these four matrices form a group.

Problem 62. The *Kronecker product* of the 2×2 matrices A, B is defined as the 4×4 matrix

$$A \otimes B := \begin{pmatrix} a_{11}B & a_{12}B \\ a_{21}B & a_{22}B \end{pmatrix}.$$

Let

$$N = \begin{pmatrix} 0 & 1 \\ 1 & 0 \end{pmatrix}.$$

Show that the 4×4 matrices $I_2 \otimes I_2$, $I_2 \otimes N$, $N \otimes I_2$, $N \otimes N$ form a group under matrix multiplication.

Solution 62. For the Kronecker product we obtain the permutation matrices

$$g_1 = I_2 \otimes I_2 = \begin{pmatrix} 1 & 0 & 0 & 0 \\ 0 & 1 & 0 & 0 \\ 0 & 0 & 1 & 0 \\ 0 & 0 & 0 & 1 \end{pmatrix} = I_4, \quad g_2 = I_2 \otimes N = \begin{pmatrix} 0 & 1 & 0 & 0 \\ 1 & 0 & 0 & 0 \\ 0 & 0 & 0 & 1 \\ 0 & 0 & 1 & 0 \end{pmatrix},$$

$$g_3 = N \otimes I_2 = \begin{pmatrix} 0 & 0 & 1 & 0 \\ 0 & 0 & 0 & 1 \\ 1 & 0 & 0 & 0 \\ 0 & 1 & 0 & 0 \end{pmatrix}, \quad g_4 = N \otimes N = \begin{pmatrix} 0 & 0 & 0 & 1 \\ 0 & 0 & 1 & 0 \\ 0 & 1 & 0 & 0 \\ 1 & 0 & 0 & 0 \end{pmatrix}.$$

The multiplication table is

$\cdot$	g_1	g_2	g_3	g_4
g_1	g_1	g_2	g_3	g_4
g_2	g_2	g_1	g_4	g_3
g_3	g_3	g_4	g_1	g_2
g_4	g_4	g_3	g_2	g_1

The inverse elements are given by $g_2^{-1} = g_2$, $g_3^{-1} = g_3$, $g_4^{-1} = g_4$. Thus the set $\{\, g_1, g_2, g_3, g_4 \,\}$ forms a group under matrix multiplication.

Problem 63. Which of the 4×4 permutation matrices can be written as Kronecker products of 2×2 permutation matrices?

Solution 63. There are only two 2×2 permutation matrices, namely

$$I_2 = \begin{pmatrix} 1 & 0 \\ 0 & 1 \end{pmatrix}, \quad N = \begin{pmatrix} 0 & 1 \\ 1 & 0 \end{pmatrix}.$$

Thus we find four 4×4 permutation matrices

$$I_2 \otimes I_2 = \begin{pmatrix} 1 & 0 & 0 & 0 \\ 0 & 1 & 0 & 0 \\ 0 & 0 & 1 & 0 \\ 0 & 0 & 0 & 1 \end{pmatrix}, \qquad I_2 \otimes N = \begin{pmatrix} 0 & 1 & 0 & 0 \\ 1 & 0 & 0 & 0 \\ 0 & 0 & 0 & 1 \\ 0 & 0 & 1 & 0 \end{pmatrix},$$

$$N \otimes I_2 = \begin{pmatrix} 0 & 0 & 1 & 0 \\ 0 & 0 & 0 & 1 \\ 1 & 0 & 0 & 0 \\ 0 & 1 & 0 & 0 \end{pmatrix}, \qquad N \otimes N = \begin{pmatrix} 0 & 0 & 0 & 1 \\ 0 & 0 & 1 & 0 \\ 0 & 1 & 0 & 0 \\ 1 & 0 & 0 & 0 \end{pmatrix}.$$

Problem 64. Show that the four 2×2 matrices

$$A = \begin{pmatrix} 1 & 0 \\ 0 & 1 \end{pmatrix}, \quad B = \begin{pmatrix} -1 & 0 \\ 0 & -1 \end{pmatrix}, \quad C = \begin{pmatrix} 0 & 1 \\ 1 & 0 \end{pmatrix}, \quad D = \begin{pmatrix} 0 & -1 \\ -1 & 0 \end{pmatrix}$$

form a group under matrix multiplication. Is the group abelian?

Solution 64. The group table is

·	A	B	C	D
A	A	B	C	D
B	B	A	D	C
C	C	D	A	B
D	D	C	B	A

From the table we find that the matrix product of any two matrices is again an element of the set $\{A, B, C, D\}$. The inverse elements are

$$A^{-1} = A, \quad B^{-1} = B, \quad C^{-1} = C, \quad D^{-1} = D.$$

Thus the set of four matrices forms a group under matrix multiplication. From the group table we see that the group is abelian.

Problem 65. Consider the set of all 2×2 matrices over the set of integers $\mathbb{Z}$ with determinant equal to 1. Show that these matrices form a group under matrix multiplication. This group is called $SL(2, \mathbb{Z})$.

Solution 65. Let $A, B \in SL(2, \mathbb{Z})$. We have

$$AB = \begin{pmatrix} a_{11} & a_{12} \\ a_{21} & a_{22} \end{pmatrix} \begin{pmatrix} b_{11} & b_{12} \\ b_{21} & b_{22} \end{pmatrix} = \begin{pmatrix} a_{11}b_{11} + a_{12}b_{21} & a_{11}b_{12} + a_{12}b_{22} \\ a_{21}b_{11} + a_{22}b_{21} & a_{21}b_{12} + a_{22}b_{22} \end{pmatrix}.$$

Since $a_{ij}b_{kl} \in \mathbb{Z}$, and

$$\det A = a_{11}a_{22} - a_{12}a_{21} = 1, \quad \det B = b_{11}b_{22} - b_{12}b_{21} = 1$$

with $\det(AB) = \det(A)\det(B) = 1$ we obtain that $AB \in SL(2, \mathbb{Z})$. The neutral element (identity) is the 2×2 identity matrix. The inverse of A is given by

$$A^{-1} = \begin{pmatrix} a_{22} & -a_{12} \\ -a_{21} & a_{11} \end{pmatrix}$$

with $\det(A^{-1}) = a_{11}a_{22} - a_{12}a_{21} = 1$ and $-a_{12}, -a_{21} \in \mathbb{Z}$. Matrix multiplication is associative. Thus we have a group.

Problem 66. Consider the group of all 4×4 permutation matrices

$$A = \begin{pmatrix} 1 & 0 & 0 & 0 \\ 0 & 1 & 0 & 0 \\ 0 & 0 & 1 & 0 \\ 0 & 0 & 0 & 1 \end{pmatrix}, \quad B = \begin{pmatrix} 1 & 0 & 0 & 0 \\ 0 & 1 & 0 & 0 \\ 0 & 0 & 0 & 1 \\ 0 & 0 & 1 & 0 \end{pmatrix}, \quad C = \begin{pmatrix} 1 & 0 & 0 & 0 \\ 0 & 0 & 1 & 0 \\ 0 & 1 & 0 & 0 \\ 0 & 0 & 0 & 1 \end{pmatrix},$$

$$D = \begin{pmatrix} 1 & 0 & 0 & 0 \\ 0 & 0 & 1 & 0 \\ 0 & 0 & 0 & 1 \\ 0 & 1 & 0 & 0 \end{pmatrix}, \quad E = \begin{pmatrix} 1 & 0 & 0 & 0 \\ 0 & 0 & 0 & 1 \\ 0 & 1 & 0 & 0 \\ 0 & 0 & 1 & 0 \end{pmatrix}, \quad F = \begin{pmatrix} 1 & 0 & 0 & 0 \\ 0 & 0 & 0 & 1 \\ 0 & 0 & 1 & 0 \\ 0 & 1 & 0 & 0 \end{pmatrix},$$

$$G = \begin{pmatrix} 0 & 1 & 0 & 0 \\ 1 & 0 & 0 & 0 \\ 0 & 0 & 1 & 0 \\ 0 & 0 & 0 & 1 \end{pmatrix}, \quad H = \begin{pmatrix} 0 & 1 & 0 & 0 \\ 1 & 0 & 0 & 0 \\ 0 & 0 & 0 & 1 \\ 0 & 0 & 1 & 0 \end{pmatrix}, \quad I = \begin{pmatrix} 0 & 1 & 0 & 0 \\ 0 & 0 & 1 & 0 \\ 1 & 0 & 0 & 0 \\ 0 & 0 & 0 & 1 \end{pmatrix},$$

$$J = \begin{pmatrix} 0 & 1 & 0 & 0 \\ 0 & 0 & 1 & 0 \\ 0 & 0 & 0 & 1 \\ 1 & 0 & 0 & 0 \end{pmatrix}, \quad K = \begin{pmatrix} 0 & 1 & 0 & 0 \\ 0 & 0 & 0 & 1 \\ 1 & 0 & 0 & 0 \\ 0 & 0 & 1 & 0 \end{pmatrix}, \quad L = \begin{pmatrix} 0 & 1 & 0 & 0 \\ 0 & 0 & 0 & 1 \\ 0 & 0 & 1 & 0 \\ 1 & 0 & 0 & 0 \end{pmatrix},$$

$$M = \begin{pmatrix} 0 & 0 & 1 & 0 \\ 1 & 0 & 0 & 0 \\ 0 & 1 & 0 & 0 \\ 0 & 0 & 0 & 1 \end{pmatrix}, \quad N = \begin{pmatrix} 0 & 0 & 1 & 0 \\ 1 & 0 & 0 & 0 \\ 0 & 0 & 0 & 1 \\ 0 & 1 & 0 & 0 \end{pmatrix}, \quad O = \begin{pmatrix} 0 & 0 & 1 & 0 \\ 0 & 1 & 0 & 0 \\ 1 & 0 & 0 & 0 \\ 0 & 0 & 0 & 1 \end{pmatrix},$$

$$P = \begin{pmatrix} 0 & 0 & 1 & 0 \\ 0 & 1 & 0 & 0 \\ 0 & 0 & 0 & 1 \\ 1 & 0 & 0 & 0 \end{pmatrix}, \quad Q = \begin{pmatrix} 0 & 0 & 1 & 0 \\ 0 & 0 & 0 & 1 \\ 1 & 0 & 0 & 0 \\ 0 & 1 & 0 & 0 \end{pmatrix}, \quad R = \begin{pmatrix} 0 & 0 & 1 & 0 \\ 0 & 0 & 0 & 1 \\ 0 & 1 & 0 & 0 \\ 1 & 0 & 0 & 0 \end{pmatrix},$$

$$S = \begin{pmatrix} 0 & 0 & 0 & 1 \\ 1 & 0 & 0 & 0 \\ 0 & 1 & 0 & 0 \\ 0 & 0 & 1 & 0 \end{pmatrix}, \quad T = \begin{pmatrix} 0 & 0 & 0 & 1 \\ 1 & 0 & 0 & 0 \\ 0 & 0 & 1 & 0 \\ 0 & 1 & 0 & 0 \end{pmatrix}, \quad U = \begin{pmatrix} 0 & 0 & 0 & 1 \\ 0 & 1 & 0 & 0 \\ 1 & 0 & 0 & 0 \\ 0 & 0 & 1 & 0 \end{pmatrix},$$

$$V = \begin{pmatrix} 0 & 0 & 0 & 1 \\ 0 & 1 & 0 & 0 \\ 0 & 0 & 1 & 0 \\ 1 & 0 & 0 & 0 \end{pmatrix}, \quad W = \begin{pmatrix} 0 & 0 & 0 & 1 \\ 0 & 0 & 1 & 0 \\ 1 & 0 & 0 & 0 \\ 0 & 1 & 0 & 0 \end{pmatrix}, \quad X = \begin{pmatrix} 0 & 0 & 0 & 1 \\ 0 & 0 & 1 & 0 \\ 0 & 1 & 0 & 0 \\ 1 & 0 & 0 & 0 \end{pmatrix}.$$

Find all subgroups and discuss whether they are commutative or not.

Solution 66.   Note that

$$B^2 = C^2 = F^2 = G^2 = H^2 = O^2 = Q^2 = V^2 = X^2 = A.$$

Applying the *Lagrange's theorem* we find that the proper subgroups must have order 12, 8, 6, 4, 3, 2, 1. The subgroup with order 12 is given by the set

$$\{ A, D, E, H, I, L, M, P, Q, T, U, X \}.$$

The group is not commutative since, for example, $BX \neq XB$.
There are three subgroups with order 8 given by the sets

$$\{ A, B, G, H, Q, R, W, X \}, \quad \{ A, C, H, K, N, Q, V, X \},$$

$$\{ A, F, H, J, O, Q, S, X \}.$$

There are four subgroups with order 6 given by the sets

$$\{ A, B, C, D, E, F \}, \quad \{ A, B, O, P, U, V \},$$

$$\{ A, C, G, I, M, O \}, \quad \{ A, F, G, L, T, V \}.$$

There are seven subgroups with order 4 given by the sets

$$\{ A, B, G, H \}, \quad \{ A, C, V, X \}, \quad \{ A, F, O, Q \}, \quad \{ A, H, Q, X \},$$

$$\{ A, H, R, W \}, \quad \{ A, J, Q, S \}, \quad \{ A, K, N, X \}.$$

There are four subgroups with order 3 given by the sets

$$\{ A, D, E \}, \quad \{ A, I, M \}, \quad \{ A, L, T \}, \quad \{ A, P, U \}.$$

There are nine subgroups with order 2 given by the sets

$$\{ A, B \}, \quad \{ A, C \}, \quad \{ A, F \}, \quad \{ A, G \}, \quad \{ A, H \},$$

$$\{ A, O \}, \quad \{ A, Q \}, \quad \{ A, V \}, \quad \{ A, X \}.$$

The subgroup with order 1 is obviously given by the set $\{ A \}$.

Problem 67.   The *Heisenberg group* is the group of 3×3 upper triangular matrices of the form

$$\begin{pmatrix} 1 & a_{12} & a_{13} \\ 0 & 1 & a_{23} \\ 0 & 0 & 1 \end{pmatrix}$$

where $a_{jk} \in \mathbb{F}$ with $\mathbb{F}$ an arbitrary field. In the following we consider $\mathbb{F} = \mathbb{C}$.
(i) Is the Heisenberg group commutative?
(ii) Let A, B be elements of the Heisenberg group. Does the *braid like relation* $ABBA = BAAB$ hold?

Solution 67. (i) The Heisenberg group is not commutative. We have

$$AB = \begin{pmatrix} 1 & a_{12} + b_{12} & a_{13} + a_{12}b_{23} + b_{13} \\ 0 & 1 & a_{23} + b_{23} \\ 0 & 0 & 1 \end{pmatrix}$$

and

$$BA = \begin{pmatrix} 1 & a_{12} + b_{12} & a_{13} + a_{23}b_{12} + b_{13} \\ 0 & 1 & a_{23} + b_{23} \\ 0 & 0 & 1 \end{pmatrix}.$$

(ii) From (i) we find that $ABBA = BAAB$. Thus the elements of the Heisenberg group satisfy the braid like relation.

Problem 68. The group $SL(2, \mathbb{Z})$ is generated by

$$T = \begin{pmatrix} 1 & 1 \\ 0 & 1 \end{pmatrix}, \qquad S = \begin{pmatrix} 0 & 1 \\ -1 & 0 \end{pmatrix}.$$

Find $A \in SL(2, \mathbb{Z})$ in terms of T and S with

$$A = \begin{pmatrix} -2 & 1 \\ -1 & 0 \end{pmatrix}.$$

Solution 68. We obtain

$$A = TTS = \begin{pmatrix} 1 & 1 \\ 0 & 1 \end{pmatrix} \begin{pmatrix} 1 & 1 \\ 0 & 1 \end{pmatrix} \begin{pmatrix} 0 & 1 \\ -1 & 0 \end{pmatrix} = \begin{pmatrix} -2 & 1 \\ -1 & 0 \end{pmatrix}.$$

Problem 69. Consider the symmetric group S_3 of the 3×3 permutation matrices. Do the group elements which satisfy the condition $P_j^2 = I_3$ form a subgroup of S_3?

Solution 69. No, they do not. Consider the matrices

$$P_1 = \begin{pmatrix} 1 & 0 & 0 \\ 0 & 0 & 1 \\ 0 & 1 & 0 \end{pmatrix}, \qquad P_2 = \begin{pmatrix} 0 & 1 & 0 \\ 1 & 0 & 0 \\ 0 & 0 & 1 \end{pmatrix}.$$

Then $P_1^2 = P_2^2 = I_3$. Now

$$P_1 P_2 = \begin{pmatrix} 0 & 1 & 0 \\ 0 & 0 & 1 \\ 1 & 0 & 0 \end{pmatrix}$$

with $(P_1 P_2)^2 \neq I_3$.

Problem 70.   (i) Consider the group G of all 3×3 permutation matrices. Show that

$$\Pi = \frac{1}{|G|} \sum_{g \in G} g$$

is a *projection matrix*, i.e. $\Pi = \Pi^*$ and $\Pi^2 = \Pi$. Here $|G|$ denotes the number of elements in the group.
(ii) Consider the subgroup H of G given by the matrices

$$\begin{pmatrix} 1 & 0 & 0 \\ 0 & 1 & 0 \\ 0 & 0 & 1 \end{pmatrix}, \quad \begin{pmatrix} 0 & 0 & 1 \\ 0 & 1 & 0 \\ 1 & 0 & 0 \end{pmatrix}.$$

Show that

$$\Pi = \frac{1}{|H|} \sum_{g \in H} g$$

is a projection matrix.

Solution 70.   (i) We have $|G| = 6$. Thus

$$\Pi = \frac{1}{|G|} \sum_{g \in G} g = \frac{1}{3} \begin{pmatrix} 1 & 1 & 1 \\ 1 & 1 & 1 \\ 1 & 1 & 1 \end{pmatrix}.$$

We find $\Pi = \Pi^*$ (i.e. Π is hermitian) and $\Pi^2 = \Pi$. Hence Π is a projection matrix.
(ii) We have $|H| = 2$. Thus

$$\frac{1}{|H|} \sum_{g \in H} g = \frac{1}{2} \begin{pmatrix} 1 & 0 & 1 \\ 0 & 2 & 0 \\ 1 & 0 & 1 \end{pmatrix}.$$

This symmetric matrix over $\mathbb{R}$ is a projection matrix.

Problem 71.   Consider the permutation group S_n. Let $\sigma \in S_n$. Is $\varphi(\sigma) = \mathrm{sgn}(\sigma)$ a *homomorphism*? Here $\mathrm{sgn}(\sigma)$ denotes the sign of a permutation σ.

Solution 71. Yes, $\phi(\sigma) = \text{sgn}(\sigma)$ is a homomorphism. To prove this we need to show that the sign of a product of two permutations is the product of signs of these permutations. The simplest way to define the sign of permutation is to say that it is equal to the determinant of the permutation matrix. The result from the fact that the determinant of the product is equal to the product of determinants.

Problem 72. (i) Consider the permutation group S_n. The *alternating group* A_n consists of all the even permutations of S_n. Show that they form a group under composition.

(ii) Consider S_3 with the matrix representation by the 3×3 permutation matrices

$$
\begin{pmatrix} 1 & 0 & 0 \\ 0 & 1 & 0 \\ 0 & 0 & 1 \end{pmatrix}, \quad
\begin{pmatrix} 1 & 0 & 0 \\ 0 & 0 & 1 \\ 0 & 1 & 0 \end{pmatrix}, \quad
\begin{pmatrix} 0 & 1 & 0 \\ 1 & 0 & 0 \\ 0 & 0 & 1 \end{pmatrix},
$$

$$
\begin{pmatrix} 0 & 1 & 0 \\ 0 & 0 & 1 \\ 1 & 0 & 0 \end{pmatrix}, \quad
\begin{pmatrix} 0 & 0 & 1 \\ 1 & 0 & 0 \\ 0 & 1 & 0 \end{pmatrix}, \quad
\begin{pmatrix} 0 & 0 & 1 \\ 0 & 1 & 0 \\ 1 & 0 & 0 \end{pmatrix}.
$$

Find the matrix representation of A_3 using the 3×3 permutation matrices.

Solution 72. (i) The product of even permutations of S_n is again an even permutation. The order of A_n is half of S_n, i.e. $n!/2$. The simplest way to define the sign of a permutation is to say that it is equal to the determinant of the permutation matrix. The result follows from the fact that the determinant of a matrix product is equal to product of the determinants. If the determinants of two matrices are equal to 1, the determinant of their product is also equal to 1. Thus the set of even permutations is closed under group multiplication. If the determinant of a matrix is equal to 1, the inverse matrix exists and its determinant is also equal to 1. Thus the set of even permutations is closed under inversion. Thus A_n is a subgroup of S_n.

(ii) The three matrices with determinant 1 form the matrix representation of A_3. They are given by

$$
I_3 = \begin{pmatrix} 1 & 0 & 0 \\ 0 & 1 & 0 \\ 0 & 0 & 1 \end{pmatrix}, \quad
A = \begin{pmatrix} 0 & 1 & 0 \\ 0 & 0 & 1 \\ 1 & 0 & 0 \end{pmatrix}, \quad
B = \begin{pmatrix} 0 & 0 & 1 \\ 1 & 0 & 0 \\ 0 & 1 & 0 \end{pmatrix}.
$$

We see that this is a cyclic group with generator A. We have

$$
A^2 = B, \qquad A^3 = I_3.
$$

Problem 73.   Let $x \in \mathbb{R}$ and $z \in \mathbb{C}$. Show that the triangular 3×3 matrices

$$g(z,x) = \begin{pmatrix} 1 & \bar{z} & \frac{1}{2}\bar{z}z + ix \\ 0 & 1 & z \\ 0 & 0 & 1 \end{pmatrix}$$

form a group under matrix multiplication.

Solution 73.   The neutral element is the 3×3 identity matrix, where $x = 0$ and $z = 0$. Multiplication of two such matrices yields

$$g(z,x)g(w,y) = \begin{pmatrix} 1 & \bar{z} & \frac{1}{2}\bar{z}z + ix \\ 0 & 1 & z \\ 0 & 0 & 1 \end{pmatrix} \begin{pmatrix} 1 & \bar{w} & \frac{1}{2}\bar{w}w + iy \\ 0 & 1 & w \\ 0 & 0 & 1 \end{pmatrix}$$

$$= \begin{pmatrix} 1 & \bar{w} + \bar{z} & \bar{w}w/2 + iy + \bar{z}w + \bar{z}z/2 + ix \\ 0 & 1 & w + z \\ 0 & 0 & 1 \end{pmatrix}.$$

The inverse of $g(z,x)$ is given by $g^{-1}(z,x) = g(-z,-x)$. Thus $\bar{z} \to -\bar{z}$ and

$$g^{-1}(z,x) = \begin{pmatrix} 1 & -\bar{z} & \frac{1}{2}\bar{z}z - ix \\ 0 & 1 & -z \\ 0 & 0 & 1 \end{pmatrix}.$$

The associative law holds for matrix multiplication.

Problem 74.   Let $x \in \mathbb{R}$ and $z_1, z_2 \in \mathbb{C}$. We set $\mathbf{z} = (z_1, z_2)^T$. Show that the triangular 4×4 matrices

$$g(\mathbf{z},x) = \begin{pmatrix} 1 & \bar{z}_1 & \bar{z}_2 & \frac{1}{2}(\mathbf{z}|\mathbf{z}) + ix \\ 0 & 1 & 0 & z_1 \\ 0 & 0 & 1 & z_2 \\ 0 & 0 & 0 & 1 \end{pmatrix}$$

form a group under matrix multiplication, where $(\mathbf{z}|\mathbf{z})$ is the scalar product

$$(\mathbf{z}|\mathbf{z}) \equiv \bar{\mathbf{z}}\mathbf{z} := |z_1|^2 + |z_2|^2 .$$

Solution 74.   The neutral element is the 4×4 identity matrix, where $x = 0$ and $z_1 = z_2 = 0$. Multiplication of two matrices yields

$$g(\mathbf{z},x)g(\mathbf{w},y)$$

$$= \begin{pmatrix} 1 & \bar{w}_1 + \bar{z}_1 & \bar{w}_2 + \bar{z}_2 & \bar{w}w/2 + iy + \bar{z}_1 w_1 + \bar{z}_2 w_2 + \bar{z}z/2 + ix \\ 0 & 1 & 0 & w_1 + z_1 \\ 0 & 0 & 1 & w_2 + z_2 \\ 0 & 0 & 0 & 1 \end{pmatrix}.$$

The inverse of $g(\mathbf{z}, x)$ is given by $g^{-1}(\mathbf{z}, x) = g(-\mathbf{z}, -x)$. Thus $\bar{\mathbf{z}} \to -\bar{\mathbf{z}}$ and

$$g^{-1}(\mathbf{z}, x) = \begin{pmatrix} 1 & -\bar{z}_1 & -\bar{z}_2 & \frac{1}{2}\bar{z}z - ix \\ 0 & 1 & 0 & -z_1 \\ 0 & 0 & 1 & -z_2 \\ 0 & 0 & 0 & 1 \end{pmatrix}.$$

Problem 75.   All invertible $n \times n$ matrices form a group under matrix multiplication. Let $a, b, c \in \mathbb{R}$. Consider the 4×4 matrix

$$M = \frac{i}{4} \begin{pmatrix} -2c & -a+ib & -a+ib & 0 \\ -2a-2ib & c-a & c+a & 2b-2ic \\ -2a-2ib & c+a & c-a & -2b+2ic \\ 0 & b+ic & -b-ic & 2a \end{pmatrix}.$$

Let T be the invertible 4×4 matrix

$$T = \begin{pmatrix} 1 & 0 & 0 & 0 \\ 0 & 1/2 & 1/2 & 0 \\ 0 & -1/2 & 1/2 & 0 \\ 0 & 0 & 0 & 1 \end{pmatrix}.$$

Show that the 4×4 matrix TMT^{-1} can be written as a direct sum of two 2×2 matrices.

Solution 75.   The inverse of T is given by

$$T^{-1} = \begin{pmatrix} 1 & 0 & 0 & 0 \\ 0 & 1 & -1 & 0 \\ 0 & 1 & 1 & 0 \\ 0 & 0 & 0 & 1 \end{pmatrix}.$$

Then we find

$$TMT^{-1} = -\frac{i}{2} \begin{pmatrix} c & a-ib & 0 & 0 \\ a+ib & -c & 0 & 0 \\ 0 & 0 & a & b-ic \\ 0 & 0 & b+ic & -a \end{pmatrix}$$

$$= -\frac{i}{2}\left(\begin{pmatrix} c & a-ib \\ a+ib & -c \end{pmatrix} \oplus \begin{pmatrix} a & b-ic \\ b+ic & -a \end{pmatrix} \right).$$

Problem 76.   (i) Show that the cyclic group C_3 is generated by

$$A = \begin{pmatrix} 0 & 1 \\ -1 & 0 \end{pmatrix}$$

under matrix multiplication. Are the matrices linearly independent?
(ii) Let $r_0, r_1, r_2, r_3 \in \mathbb{R}$. Consider the set of all linear combinations of the form

$$r_0 I_2 + r_1 A + r_2 A^2 + r_3 A^3 .$$

Show that this set is a *ring*.

Solution 76.   (i) We have

$$A^2 = \begin{pmatrix} -1 & 0 \\ 0 & -1 \end{pmatrix}, \qquad A^3 = \begin{pmatrix} 0 & -1 \\ 1 & 0 \end{pmatrix}$$

and $A^4 = I_2$. The inverse of A is A^3. The inverse of A^2 is A^2. I_2 and A^2 are linearly dependent and A and A^3 are linearly dependent.
(ii) Consider two elements

$$E_1 = r_0 I_2 + r_1 A + r_2 A^2 + r_3 A^3, \qquad E_2 = \rho_0 I_2 + \rho_1 A + \rho_2 A^2 + \rho_3 A^3.$$

The set S is closed under addition. We have

$$E_1 + E_2 = (r_0 + \rho_0) I_2 + (r_1 + \rho_1) A + (r_2 + \rho_2) A^2 + (r_3 + \rho_3) A^3 \Rightarrow (E_1 + E_2) \in S.$$

The coefficients $r_0, r_1, r_2, r_3 \in \mathbb{R}$ and therefore the associative law follows from associativity law for real number addition.
- Zero element 0_2 exists. This element corresponds to zero values of all coefficients.
- Unique inverse element $-E$ exists for all nonzero elements. One sets $\rho_i = -r_i$.
Since $r_0, r_1, r_2, r_3 \in \mathbb{R}$ the commutative law follows from commutativity law for real number addition.
Properties of multiplication: The set S is closed under multiplication. We have

$$\begin{aligned} E_1 E_2 = {} & (r_0 \rho_0) I_2 + (r_0 \rho_1 + r_1 \rho_0) A + (r_0 \rho_2 + r_1 \rho_1 + r_2 \rho_0) A^2 \\ & + (r_0 \rho_3 + r_1 \rho_2 + r_2 \rho_1 + r_3 \rho_0) A^3 + (r_1 \rho_3 + r_2 \rho_2 + r_3 \rho_1) A^4 \\ & + (r_2 \rho_3 + r_3 \rho_2) A^5 + (r_3 \rho_3) A^6 . \end{aligned}$$

Since $A^4 = I_2$, $A^5 = A$, $A^6 = A^2$ it follows that

$$\begin{aligned} E_1 E_2 = {} & (r_0 \rho_0 + (r_1 \rho_3 + r_2 \rho_2 + r_3 \rho_1)) I_2 + (r_0 \rho_1 + r_1 \rho_0 + r_2 \rho_3 + r_3 \rho_2) A \\ & + (r_0 \rho_2 + r_1 \rho_1 + r_2 \rho_0 + r_3 \rho_3) A^2 + (r_0 \rho_3 + r_1 \rho_2 + r_2 \rho_1 + r_3 \rho_0) A^3 . \end{aligned}$$

It follows that $(E_1 E_2) \in S$. We also have

$$(E_1 E_2) E_3 = E_1 (E_2 E_3) .$$

- Existence of multiplicative identity. The neutral element is the identity matrix I_2. This corresponds to $r_0 = 1$ and $r_j = 0$, $j > 0$.
The left distributive law follows from the matrix addition and multiplication distributive laws

$$E_1(E_2 + E_3) = (E_1E_2) + (E_1E_3).$$

The right distributive law follows from the matrix addition and multiplication distributive laws

$$(E_2 + E_3)E_1 = (E_2E_1) + (E_3E_1).$$

Thus all requirements are satisfied and the set S is a ring.

Problem 77.   Let L be the invertible matrix

$$L = \begin{pmatrix} 1 & 1 \\ 0 & 1 \end{pmatrix}.$$

(i) Find all invertible 2×2 matrices M such that $M^{-1}LM = L$.
(ii) Find all invertible 2×2 matrices S such that $S^{-1}LS = L^{-1}$.
(iii) Calculate the matrix $(MS)^{-1}L(MS)$. Discuss.

Solution 77.   (i) Let

$$M = \begin{pmatrix} a & b \\ c & d \end{pmatrix}.$$

Then

$$ML = \begin{pmatrix} a & a+b \\ c & c+d \end{pmatrix}, \qquad LM = \begin{pmatrix} a+c & b+d \\ c & d \end{pmatrix}.$$

From $M^{-1}LM = L$ it follows that $ML = LM$. Therefore $c = 0$, $a = d \neq 0$. Finally

$$M = \begin{pmatrix} \alpha & \beta \\ 0 & \alpha \end{pmatrix}, \qquad \alpha \neq 0.$$

(ii) The inverse of L is given by

$$L^{-1} = \begin{pmatrix} 1 & -1 \\ 0 & 1 \end{pmatrix}.$$

Let S be an invertible matrix

$$S = \begin{pmatrix} a & b \\ c & d \end{pmatrix}.$$

From $S^{-1}LS = L^{-1}$ it follows that $LS = SL^{-1}$. Now

$$SL^{-1} = \begin{pmatrix} a & b-a \\ c & d-c \end{pmatrix}, \qquad LS = \begin{pmatrix} a+c & b+d \\ c & d \end{pmatrix}.$$

Thus $c = 0$, $a = -d \neq 0$ and

$$S = \begin{pmatrix} \gamma & \delta \\ 0 & -\gamma \end{pmatrix} \quad \gamma \neq 0.$$

(iii) We have

$$(MS)^{-1} L (MS) = S^{-1} (M^{-1} L M) S = S^{-1} L S = L^{-1}.$$

Thus we obtain the inverse of L.

Problem 78.   Consider the square. Find all symmetry operations which leave the square invariant. Use permutation matrices.

Solution 78.   The D_4 group contains two different sets of transformations:
- four rotations R_i (including identical transformation);
- four reflections S_i.
We enumerate the vertices of the square in the following way. We place the first vertex in the point $(1,0)$, the second vertex in the point $(0,1)$, the third vertex in the point $(-1,0)$ and, finally, the fourth vertex in the point $(0,-1)$.
Rotations are:
- by the angle 0:

$$R_0 = I_4 = \begin{pmatrix} 1 & 0 & 0 & 0 \\ 0 & 1 & 0 & 0 \\ 0 & 0 & 1 & 0 \\ 0 & 0 & 0 & 1 \end{pmatrix}$$

- by the angle $\pi/2$:

$$R_1 = \begin{pmatrix} 0 & 0 & 0 & 1 \\ 1 & 0 & 0 & 0 \\ 0 & 1 & 0 & 0 \\ 0 & 0 & 1 & 0 \end{pmatrix}$$

By this transformation each vertex goes to the vertex with the next number. We express this fact using the permutation matrix R_i, which sends each k-th row of arbitrary matrix M to $(k+i)$-th row of result by multiplication $R_i M$.
- by the angle π:

$$R_2 = \begin{pmatrix} 0 & 0 & 1 & 0 \\ 0 & 0 & 0 & 1 \\ 1 & 0 & 0 & 0 \\ 0 & 1 & 0 & 0 \end{pmatrix}$$

- by the angle $3\pi/2$:

$$R_3 = \begin{pmatrix} 0 & 1 & 0 & 0 \\ 0 & 0 & 1 & 0 \\ 0 & 0 & 0 & 1 \\ 1 & 0 & 0 & 0 \end{pmatrix}$$

Reflections are:
- reflection in X axis:

$$S_0 = \begin{pmatrix} 1 & 0 & 0 & 0 \\ 0 & 0 & 0 & 1 \\ 0 & 0 & 1 & 0 \\ 0 & 1 & 0 & 0 \end{pmatrix}$$

(points 2 and 4 exchange their places).
- reflection in $x = y$ axis:

$$S_1 = \begin{pmatrix} 0 & 1 & 0 & 0 \\ 1 & 0 & 0 & 0 \\ 0 & 0 & 0 & 1 \\ 0 & 0 & 1 & 0 \end{pmatrix}$$

(points (1,2) and (3,4) exchange their places).
- reflection in Y axis:

$$S_2 = \begin{pmatrix} 0 & 0 & 1 & 0 \\ 0 & 1 & 0 & 0 \\ 1 & 0 & 0 & 0 \\ 0 & 0 & 0 & 1 \end{pmatrix}$$

(points 1 and 3 exchange their places).
- reflection in $x = -y$ axis:

$$S_3 = \begin{pmatrix} 0 & 0 & 0 & 1 \\ 0 & 0 & 1 & 0 \\ 0 & 1 & 0 & 0 \\ 1 & 0 & 0 & 0 \end{pmatrix}$$

(points 1,4 and 2,3 exchange their places). The multiplication table of D_4 is

$\cdot$	R_0	R_1	R_2	R_3	S_0	S_1	S_2	S_3
R_0	R_0	R_1	R_2	R_3	S_0	S_1	S_2	S_3
R_1	R_1	R_2	R_3	R_0	S_1	S_2	S_3	S_0
R_2	R_2	R_3	R_0	R_1	S_2	S_3	S_0	S_1
R_3	R_3	R_0	R_1	R_2	S_3	S_0	S_1	S_2
S_0	S_0	S_3	S_2	S_1	R_0	R_3	R_2	R_1
S_1	S_1	S_0	S_3	S_2	R_1	R_0	R_3	R_2
S_2	S_2	S_1	S_0	S_3	R_2	R_1	R_0	R_3
S_3	S_3	S_2	S_1	S_0	R_3	R_2	R_1	R_0

Problem 79. Consider the group $GL(2, \mathbb{F})$, where the underlying field $\mathbb{F}$ could be $\mathbb{Q}$, $\mathbb{R}$, $\mathbb{C}$ or $\mathbb{Z}_p$, where p is prime. Consider the 2×2 matrix

$$A = \begin{pmatrix} 3 & 2 \\ 1 & 5 \end{pmatrix}.$$

(i) Find the inverse of A if the underlying field is $\mathbb{Q}$.
(ii) Find the inverse of A if the underlying field is $\mathbb{Z}_5$.

Solution 79. (i) The inverse of a 2×2 matrix

$$M = \begin{pmatrix} a & b \\ c & d \end{pmatrix}$$

if it exists is given by

$$M^{-1} = \frac{1}{ad - cb} \begin{pmatrix} d & -b \\ -c & a \end{pmatrix}$$

where $ad - cb$ is the determinant of M. We obtain $\det(A) = 13$. Thus

$$A^{-1} = \frac{1}{13} \begin{pmatrix} 5 & -2 \\ -1 & 3 \end{pmatrix}.$$

(ii) We have

$$(3 \cdot 5 - 1 \cdot 2) \bmod 5 = 13 \bmod 5 = 3 \bmod 5.$$

Now the inverse of 3 is 2 since $(3 \cdot 2) \bmod 5 = 1 \bmod 5$. Thus the inverse is

$$A^{-1} = \begin{pmatrix} 5 \cdot 2 & -2 \cdot 2 \\ -1 \cdot 2 & 3 \cdot 2 \end{pmatrix} \bmod 5 = \begin{pmatrix} 0 & 1 \\ 3 & 1 \end{pmatrix}.$$

Problem 80. Show that $SL(2, \mathbb{R})$ is a normal subgroup of $GL(2, \mathbb{R})$. The group $SL(2, \mathbb{R})$ consists of all 2×2 matrices over $\mathbb{R}$ with determinant 1. Use the following theorem.

A subgroup H of G is normal in G if and only if $gHg^{-1} \subseteq H$ for all $g \in G$.

Solution 80. Let $g \in GL(2, \mathbb{R})$ and $h \in SL(2, \mathbb{R})$. Now

$$\det(ghg^{-1}) = \det(g) \det(h) \det(g^{-1})$$
$$= \det(g) \det(h) \det(g)^{-1}$$
$$= \det(g) \det(g)^{-1}$$
$$= 1.$$

Thus $ghg^{-1} \in H$ and therefore $gHg^{-1} \subseteq H$.

Problem 81. Consider the group $G = GL(2, \mathbb{R})$ and the normal subgroup $H = SL(2, \mathbb{R})$. Let $g \in G$ with $\det(g) = 2$. Show that the set gH is the set of all 2×2 matrices in G with determinant 2.

Solution 81. Let $k \in gH$. Thus $k = gh$ for some $h \in H$. It follows that

$$\det(k) = \det(gh) = \det(g)\det(h) = \det(g) = 2.$$

Problem 82. Let G be a group and let $g \in G$. Show that

$$H := \{\, g^n \; : \; n \in \mathbb{Z} \,\}$$

is a subgroup of G and is the smallest subgroup of G that contains the element g.

Solution 82. We have to check the conditions to be a subgroup. Since $g^r g^s = g^{r+s}$ for $r, s \in \mathbb{Z}$, we find that the product in G of two elements of H is again in H. Thus the set H is closed under the group operation of G. Furthermore $g^0 = e$, where e is the neutral element. Thus $e \in H$ and for $g^r \in H$, $g^{-r} \in H$ we have $g^{-r}g^r = e$. Thus all the conditions for a subgroup are satisfied. Let S be a subgroup of G with $g \in S$. Then $g^n \in S$. Consequently $H \subseteq S$ so that H is the smallest subgroup containing g.

Problem 83. The *Klein four-group* is an abelian group of order 4. It is the smallest non-cyclic group. The Klein four-group consists of the identity element e and the elements i, j, k with the composition

$$i^2 = j^2 = k^2 = ijk = e.$$

Give a matrix representations with 2×2 matrices.

Solution 83. We have

$$e \mapsto \begin{pmatrix} 1 & 0 \\ 0 & 1 \end{pmatrix}, \quad i \mapsto \begin{pmatrix} 1 & 0 \\ 0 & -1 \end{pmatrix}, \quad j \mapsto \begin{pmatrix} -1 & 0 \\ 0 & 1 \end{pmatrix}, \quad k \mapsto \begin{pmatrix} -1 & 0 \\ 0 & -1 \end{pmatrix}.$$

Note that all the matrices can be written as direct sums of 1×1 matrices.

Problem 84. Let $\omega = \exp(i2\pi/3)$, $\bar{\omega} = \exp(-i2\pi/3)$. Show that the unitary matrix

$$U = \frac{1}{\sqrt{3}} \begin{pmatrix} 1 & \omega & \bar{\omega} \\ 1 & \bar{\omega} & \omega \\ 1 & 1 & 1 \end{pmatrix}$$

reduces the natural representation of C_3

$$
\begin{pmatrix} 1 & 0 & 0 \\ 0 & 1 & 0 \\ 0 & 0 & 1 \end{pmatrix}, \quad
\begin{pmatrix} 0 & 1 & 0 \\ 0 & 0 & 1 \\ 1 & 0 & 0 \end{pmatrix}, \quad
\begin{pmatrix} 0 & 0 & 1 \\ 1 & 0 & 0 \\ 0 & 1 & 0 \end{pmatrix}
$$

to its irreducible form.

Solution 84.   Since $\omega\bar\omega = 1$ the inverse of U is given by

$$
U^{-1} = U^* = \frac{1}{\sqrt{3}} \begin{pmatrix} 1 & 1 & 1 \\ \bar\omega & \omega & 1 \\ \omega & \bar\omega & 1 \end{pmatrix}.
$$

Thus we find

$$
U \begin{pmatrix} 1 & 0 & 0 \\ 0 & 1 & 0 \\ 0 & 0 & 1 \end{pmatrix} U^* = \begin{pmatrix} 1 & 0 & 0 \\ 0 & 1 & 0 \\ 0 & 0 & 1 \end{pmatrix} = (1) \oplus (1) \oplus (1)
$$

$$
U \begin{pmatrix} 0 & 1 & 0 \\ 0 & 0 & 1 \\ 1 & 0 & 0 \end{pmatrix} U^* = \begin{pmatrix} 1 & 0 & 0 \\ 0 & w & 0 \\ 0 & 0 & \bar w \end{pmatrix} = (1) \oplus (\omega) \oplus (\bar\omega)
$$

$$
U \begin{pmatrix} 0 & 0 & 1 \\ 1 & 0 & 0 \\ 0 & 1 & 0 \end{pmatrix} U^* = \begin{pmatrix} 1 & 0 & 0 \\ 0 & \bar w & 0 \\ 0 & 0 & w \end{pmatrix} = (1) \oplus (\bar\omega) \oplus (\omega).
$$

Problem 85.   Let $a, b \in \mathbb{R}$. What are the conditions on a, b such that the matrices

$$
\begin{pmatrix} a & b & 0 \\ b & a & b \\ 0 & b & a \end{pmatrix}
$$

form a group under matrix multiplication?

Solution 85.   Note that

$$
\det \begin{pmatrix} a & b & 0 \\ b & a & b \\ 0 & b & a \end{pmatrix} = a(a^2 - 2b^2).
$$

Let

$$
A = \begin{pmatrix} a_1 & b_1 & 0 \\ b_1 & a_1 & b_1 \\ 0 & b_1 & a_1 \end{pmatrix}, \quad
B = \begin{pmatrix} a_2 & b_2 & 0 \\ b_2 & a_2 & b_2 \\ 0 & b_2 & a_2 \end{pmatrix}.
$$

Then matrix multiplication yields

$$AB = \begin{pmatrix} a_1a_2 + b_1b_2 & a_1b_2 + b_1a_2 & b_1b_2 \\ b_1a_2 + a_1b_2 & b_1b_2 + a_1a_2 + b_1b_2 & a_1b_2 + b_1a_2 \\ b_1b_2 & b_1a_2 + a_1b_2 & b_1b_2 + a_1a_2 \end{pmatrix}.$$

We see that the product AB is a matrix of the same kind if

$$b_1b_2 = 0 \Rightarrow b_1 = 0 \vee b_2 = 0.$$

Since A^2 must be a group element we find $b_1 = 0$. In this case the matrices A and B are proportional to the identity matrix

$$A = a_1 I_3, \qquad B = a_2 I_3.$$

These matrices form a group under matrix multiplication, when $a_1 \neq 0$, $a_2 \neq 0$. Otherwise no inverse element exists.

Problem 86. Let $a, b, c \in \mathbb{R}$. Show that the matrices

$$\begin{pmatrix} 1 & a & c \\ 0 & 1 & b \\ 0 & 0 & 1 \end{pmatrix}$$

form a non-commutative group under matrix multiplication. Find the *center* of the group G.

Solution 86. Let

$$A = \begin{pmatrix} 1 & a_{12} & a_{13} \\ 0 & 1 & a_{23} \\ 0 & 0 & 1 \end{pmatrix}, \qquad B = \begin{pmatrix} 1 & b_{12} & b_{13} \\ 0 & 1 & b_{23} \\ 0 & 0 & 1 \end{pmatrix}.$$

Then

$$AB = \begin{pmatrix} 1 & a_{12} + b_{12} & a_{13} + b_{13} + a_{12}b_{23} \\ 0 & 1 & a_{23} + b_{23} \\ 0 & 0 & 1 \end{pmatrix}$$

and

$$BA = \begin{pmatrix} 1 & b_{12} + a_{12} & b_{13} + a_{13} + b_{12}a_{23} \\ 0 & 1 & b_{23} + a_{23} \\ 0 & 0 & 1 \end{pmatrix}.$$

Thus the matrix product of these two matrices is again a matrix of the same kind. Associativity law follows from the same law for matrix multiplication. The inverse matrix is

$$A^{-1} = \begin{pmatrix} 1 & -a_{12} & -a_{13} + a_{12}a_{23} \\ 0 & 1 & -a_{23} \\ 0 & 0 & 1 \end{pmatrix}.$$

Thus the set of matrices is closed under multiplication and inversion and forms a group. As $a_{12}b_{23} \neq b_{12}a_{23}$, the group is not commutative. If $B \in Z(G)$, then

$$a_{12}b_{23} - b_{12}a_{23} = 0$$

for all values of a_{12} and a_{23}. It follows that $b_{23} = b_{12} = 0$. Thus the center of the group $Z(G)$ is the set of elements

$$B = \begin{pmatrix} 1 & 0 & b_{13} \\ 0 & 1 & 0 \\ 0 & 0 & 1 \end{pmatrix}.$$

Problem 87.   Let A_n be the *alternating group*. The order of A_n is $n!/2$. Is A_n a normal subgroup of S_n?

Solution 87.   Yes, A_n is a normal subgroup of S_n. The characteristic property of A_n is that all its elements are even. In particular this means that the determinant of the permutation matrix, which represents this element, is equal to 1. Consider the conjugation gag^{-1} of some element $a \in A_n$ by an arbitrary element of S_n. We have

$$\det(gag^{-1}) = \det(g)\det(a)/\det(g) = \det(a) = 1 \Rightarrow gag^{-1} \in A_n.$$

Thus A_n is a normal subgroup of S_n.

Problem 88.   Let S be a set and G be a group. An *action* of G on S is a map

$$\varphi : G \times S \to S, \qquad \varphi(g, x) = gx$$

such that
1. $ex = x$ for the identity element e of G and for all $x \in S$
2. $(g_1 g_2)x = g_1(g_2 x)$ for all $x \in S$ and all $g_1, g_2 \in G$.

Under these conditions, S is a G-set. Give an example for G, S and the action.

Solution 88.   Consider the matrix group

$$\begin{pmatrix} 1 & 0 \\ 0 & 1 \end{pmatrix}, \qquad \begin{pmatrix} 0 & 1 \\ 1 & 0 \end{pmatrix}$$

under matrix multiplication and S is the vector space $\mathbb{R}^2$.

Problem 89. (i) Show that the set of matrices

$$G_0 = \left\{ \begin{pmatrix} a & b \\ 0 & 1 \end{pmatrix} : a \neq 0 \right\} \subset GL(2, \mathbb{C})$$

form a group under matrix multiplication.
(ii) Is the group commutative?
(iii) Find the *center* of the group.

Solution 89. (i) We have

$$\begin{pmatrix} a & b \\ 0 & 1 \end{pmatrix} \begin{pmatrix} c & d \\ 0 & 1 \end{pmatrix} = \begin{pmatrix} ac & ad + b \\ 0 & 1 \end{pmatrix}.$$

The neutral element is the 2×2 identity matrix. The inverse element is

$$\begin{pmatrix} 1/a & -b/a \\ 0 & 1 \end{pmatrix}.$$

Finally matrix multiplication is associative.
(ii) The group is not commutative since

$$\begin{pmatrix} a & b \\ 0 & 1 \end{pmatrix} \begin{pmatrix} c & d \\ 0 & 1 \end{pmatrix} \neq \begin{pmatrix} c & d \\ 0 & 1 \end{pmatrix} \begin{pmatrix} a & b \\ 0 & 1 \end{pmatrix}$$

in general.
(iii) To find the group center, we denote its element by

$$z = \begin{pmatrix} x & y \\ 0 & 1 \end{pmatrix}.$$

By direct calculation we have

$$\begin{pmatrix} a & b \\ 0 & 1 \end{pmatrix} \begin{pmatrix} x & y \\ 0 & 1 \end{pmatrix} = \begin{pmatrix} ax & ay + b \\ 0 & 1 \end{pmatrix}$$

$$\begin{pmatrix} x & y \\ 0 & 1 \end{pmatrix} \begin{pmatrix} a & b \\ 0 & 1 \end{pmatrix} = \begin{pmatrix} xa & xb + y \\ 0 & 1 \end{pmatrix}.$$

We obtain the condition $ay + b = xb + y$. We find $y = 0$, $x = 1$. Thus the group center consists of the identity matrix I_2.

Problem 90. The group $SL(2, \mathbb{R})$ is given by the matrices

$$\begin{pmatrix} a & b \\ c & d \end{pmatrix}, \quad a, b, c, d \in \mathbb{R}, \quad ad - bc = 1.$$

Find the *center* of $SL(2, \mathbb{R})$.

Solution 90.   Denote the arbitrary element of $Z(SL(2, \mathbb{R}))$ as

$$z = \begin{pmatrix} p & q \\ r & s \end{pmatrix} \in Z(SL(2, \mathbb{R}))$$

and an arbitrary element of $SL(2, \mathbb{R})$ as

$$g = \begin{pmatrix} a & b \\ c & d \end{pmatrix} \in G.$$

From the condition $zg - gz = 0_2$ we find $q = r = 0$, $p = s \neq 0$. From the condition $\det(z) = 1$ follows that $ps = 1$. There are two solutions: $p = s = 1$, $p = s = -1$. Thus the center of $SL(2, \mathbb{R})$ contains only the two matrices I_2 and $-I_2$.

Problem 91.   Let A be an $n \times n$ matrix. Then the *vec operation* is defined by

$$\text{vec}(A) := (a_{11}, \ldots, a_{n1}, a_{12}, \ldots, a_{1n}, \ldots, a_{nn})^T$$

i.e. we stack the columns under each other. Consider the Pauli matrices σ_x and σ_z

$$\sigma_x = \begin{pmatrix} 0 & 1 \\ 1 & 0 \end{pmatrix}, \qquad \sigma_z = \begin{pmatrix} 1 & 0 \\ 0 & -1 \end{pmatrix}.$$

Then both $\{\sigma_x, I_2\}$ and $\{\sigma_z, I_2\}$ form a group under matrix multiplication.
(i) Let $\otimes$ be the Kronecker product. Show that $\{\sigma_z \otimes \sigma_x, I_2 \otimes I_2\}$ form a group under matrix multiplication.
(ii) Show that $\text{vec}(\sigma_x) = (\sigma_z \otimes \sigma_x)\text{vec}(\sigma_z)$.

Solution 91.   (i) Let us denote

$$A = \sigma_z \otimes \sigma_x = \begin{pmatrix} 0 & 1 & 0 & 0 \\ 1 & 0 & 0 & 0 \\ 0 & 0 & 0 & -1 \\ 0 & 0 & -1 & 0 \end{pmatrix}.$$

Thus

$$A^2 = (\sigma_z \otimes \sigma_x)(\sigma_z \otimes \sigma_x) = (\sigma_z \sigma_z) \otimes (\sigma_x \sigma_x) = I_2 \otimes I_2 = I_4.$$

(ii) Direct calculation yields

$$(\sigma_z \otimes \sigma_x)\text{vec}(\sigma_z) = \begin{pmatrix} 0 & 1 & 0 & 0 \\ 1 & 0 & 0 & 0 \\ 0 & 0 & 0 & -1 \\ 0 & 0 & -1 & 0 \end{pmatrix} \begin{pmatrix} 1 \\ 0 \\ 0 \\ -1 \end{pmatrix} = \begin{pmatrix} 0 \\ 1 \\ 1 \\ 0 \end{pmatrix} = \text{vec}(\sigma_x).$$

Problem 92. (i) Consider the group given by the matrices

$$I_2 = \begin{pmatrix} 1 & 0 \\ 0 & 1 \end{pmatrix}, \qquad U = \begin{pmatrix} 0 & 1 \\ 1 & 0 \end{pmatrix}$$

and matrix multiplication. Let $x, y \in \mathbb{R}$. Find the solutions of the equation

$$(xI_2 + yU)^2 = xI_2 + yU.$$

(ii) Using the results from (i) show that the matrices $xI_2 + yU$ are projection matrices.

Solution 92. (i) Since

$$(xI_2 + yU)^2 = x^2 I_2 + y^2 I_2 + 2xyU = (x^2 + y^2)I_2 + 2xyU$$

we find that the condition $(xI_2 + yU)^2 = xI_2 + yU$ yields the system of equations $x^2 + y^2 = x$, $2xy = y$. We have to do a case study. Case 1. If $y = 0$, then we obtain $x^2 = x$ with the solutions $x = 0$ and $x = 1$. Case 2. If $y \neq 0$, then $x = 1/2$ and $y = \pm 1/2$.
(ii) For the case 1 we find the 2×2 zero matrix and the 2×2 identity matrix. Both are projection matrices. For the case 2 we find

$$xI_2 + yU = \frac{1}{2}I_2 + \frac{1}{2}U = \frac{1}{2}\begin{pmatrix} 1 & 1 \\ 1 & 1 \end{pmatrix}$$

and

$$xI_2 + yU = \frac{1}{2}I_2 - \frac{1}{2}U = \frac{1}{2}\begin{pmatrix} 1 & -1 \\ -1 & 1 \end{pmatrix}.$$

Both are projection matrices.

Problem 93. The *tetrahedral group* T is of order 12. It can be generated by taking powers of products of the 3×3 matrices

$$g_1 = \begin{pmatrix} 1 & 0 & 0 \\ 0 & -1 & 0 \\ 0 & 0 & -1 \end{pmatrix}, \qquad g_2 = \begin{pmatrix} 0 & 0 & 1 \\ 1 & 0 & 0 \\ 0 & 1 & 0 \end{pmatrix}.$$

Note that g_2 is a permutation matrix. Calculate the orders of g_1 and g_2.

Solution 93. Note that $g_1^2 = I_3$ and $g_2^3 = I_3$. Thus g_1, g_1^2 and g_2, g_2^2, $g_2^3 = I_3$ form subgroups of T. Thus the orders are 2 and 3.

Problem 94. Let G be a finite group. For a finite group G the number of conjugacy classes is equivalent to the number of non-equivalent irreducible matrix representations. Let $\{C_1, \ldots, C_r\}$ be the set of conjugacy classes, where one sets C_1 to be the conjugacy class of 1, the neutral element of the finite group G. Thus $C_1 = \{1\}$. Let $\{\chi_1, \ldots, \chi_r\}$ be the set of distinct irreducible characters of G. One chooses χ_1 to be the trivial character of G, i.e. $\chi_1(g) = 1$ for all $g \in G$. Then the *character table* is the $r \times r$ table $X(G) = (a_{jk})$ defined by $a_{jk} = \chi_j(c_k)$ for all $j, k = 1, \ldots, r$, where c_k is any element of C_k. The value of a character is constant on each conjugacy class.

(i) Let $\omega = e^{2\pi i/3}$. Show that $1, \omega, \omega^2$ form a group under multiplication.

(ii) Find the character table of the *cyclic group* C_3.

Solution 94. (i) The key fact is that $\omega^3 = 1$. Thus the elements $1, \omega, \omega^2$ form a cyclic group of order 3 with generator ω under multiplication. We have the group table

$\cdot$	1	ω	ω^2
1	1	ω	ω^2
ω	ω	ω^2	1
ω^2	ω^2	1	ω

(ii) Since the group is commutative each element of the group forms a conjugacy class. Thus there are three irreducible representations of the group. Let e be the neutral element. The character table is

C_3	e	c	c^2
$D^{(1)}$	1	1	1
$D^{(2)}$	1	ω	ω^2
$D^{(3)}$	1	ω^2	ω

We have

$$1^2 + 1^2 + 1^2 = 3, \quad 1 + \omega + \omega^2 = \frac{1 - \omega^3}{1 - \omega} = 0.$$

Problem 95. Consider the six 3×3 permutation matrices

$$P_{123} = \begin{pmatrix} 1 & 0 & 0 \\ 0 & 1 & 0 \\ 0 & 0 & 1 \end{pmatrix}, \quad P_{132} = \begin{pmatrix} 1 & 0 & 0 \\ 0 & 0 & 1 \\ 0 & 1 & 0 \end{pmatrix}, \quad P_{213} = \begin{pmatrix} 0 & 1 & 0 \\ 1 & 0 & 0 \\ 0 & 0 & 1 \end{pmatrix}$$

$$P_{231} = \begin{pmatrix} 0 & 1 & 0 \\ 0 & 0 & 1 \\ 1 & 0 & 0 \end{pmatrix}, \quad P_{312} = \begin{pmatrix} 0 & 0 & 1 \\ 1 & 0 & 0 \\ 0 & 1 & 0 \end{pmatrix}, \quad P_{321} = \begin{pmatrix} 0 & 0 & 1 \\ 0 & 1 & 0 \\ 1 & 0 & 0 \end{pmatrix}$$

which form a group under matrix multiplication.

(i) Find the conjugacy classes.

(ii) Find the irreducible representations.
(iii) Write down the character table.

Solution 95.   (i) The inverse matrices are given by

$$P_{123}^{-1} = P_{123}, \qquad P_{132}^{-1} = P_{132}, \qquad P_{213}^{-1} = P_{213}$$

$$P_{231}^{-1} = P_{312}, \qquad P_{312}^{-1} = P_{231}, \qquad P_{321}^{-1} = P_{321}.$$

Obviously, $C_1 = \{ P_{123} = I_3 \}$ forms it own class. Now

$$P_{123}P_{132}P_{123} = P_{132}, \quad P_{132}P_{132}P_{132}^{-1} = P_{132}, \quad P_{213}P_{132}P_{213}^{-1} = P_{321},$$

$$P_{231}P_{132}P_{231}^{-1} = P_{213}, \quad P_{312}P_{132}P_{312}^{-1} = P_{321}, \quad P_{321}P_{132}P_{321}^{-1} = P_{213}.$$

Thus we have the class with three elements

$$C_2 = \{ P_{132}, \ P_{213}, \ P_{321} \}.$$

The determinant of these three matrices is equal to -1. Analogously we find the class with two elements

$$C_3 = \{ P_{231}, \ P_{312} \}.$$

The determinant of these two matrices is $+1$.
(ii) The number of conjugacy classes is equal to the number of irreducible representations. We find two one-dimensional representations. The trivial is $P_{jk\ell} \to +1$ for all $jk\ell$. The non-trivial one is using the result from (i) about the determinant

$$P_{123} \to +1, \quad P_{132} \to -1, \quad P_{213} \to -1,$$

$$P_{321} \to -1, \quad P_{231} \to +1, \quad P_{312} \to +1.$$

The faithful two-dimensional representation is given by $P_{123} \to I_2$ and

$$P_{231} \to \begin{pmatrix} -1/2 & \sqrt{3}/2 \\ -\sqrt{3}/2 & -1/2 \end{pmatrix}, \quad P_{312} \to \begin{pmatrix} -1/2 & -\sqrt{3}/2 \\ \sqrt{3}/2 & -1/2 \end{pmatrix}$$

$$P_{213} \to \begin{pmatrix} -1/2 & \sqrt{3}/2 \\ \sqrt{3}/2 & 1/2 \end{pmatrix}, \quad P_{132} \to \begin{pmatrix} 1 & 0 \\ 0 & -1 \end{pmatrix}, \quad P_{321} \to \begin{pmatrix} -1/2 & -\sqrt{3}/2 \\ -\sqrt{3}/2 & 1/2 \end{pmatrix}.$$

(iii) For the character table we only need the trace of the faithful two-dimensional representation given in (i). Using the result from (i) the character table is

S_3	C_1	C_2	C_3
A_1	1	1	1
A_2	1	-1	1
E	2	0	-1

Problem 96.   The *dihedral group* D_3 is a non-commutative group of order 6. The group table is given by

•	I	A	B	C	D	E
I	I	A	B	C	D	E
A	A	B	I	D	E	C
B	B	I	A	E	C	D
C	C	E	D	I	B	A
D	D	C	E	A	I	B
E	E	D	C	B	A	I

where I is the neutral element.
(i) Find a faithful representation by 2×2 matrices starting from the matrices

$$\begin{pmatrix} \cos\phi & \sin\phi \\ -\sin\phi & \cos\phi \end{pmatrix} \quad \text{and} \quad \begin{pmatrix} \cos\phi & \sin\phi \\ \sin\phi & -\cos\phi \end{pmatrix}.$$

The determinant of the matrix on the left-hand side is $+1$ (which includes the identity matrix) and the determinant on the right-hand side is -1.
(ii) Find the conjugacy classes.

Solution 96.   (i) We have $A^3 = B^3 = I_2$. The elements $\{I, A, B\}$ form a commutative subgroup. We find the faithful representation

$$I \to \begin{pmatrix} 1 & 0 \\ 0 & 1 \end{pmatrix}, \quad A \to \begin{pmatrix} -1/2 & -\sqrt{3}/2 \\ \sqrt{3}/2 & -1/2 \end{pmatrix}, \quad B \to \begin{pmatrix} -1/2 & \sqrt{3}/2 \\ -\sqrt{3}/2 & -1/2 \end{pmatrix},$$

$$C \to \begin{pmatrix} 1/2 & \sqrt{3}/2 \\ \sqrt{3}/2 & -1/2 \end{pmatrix}, \quad D \to \begin{pmatrix} -1 & 0 \\ 0 & 1 \end{pmatrix}, \quad E \to \begin{pmatrix} 1/2 & -\sqrt{3}/2 \\ -\sqrt{3}/2 & -1/2 \end{pmatrix}$$

with $\det(I) = \det(A) = \det(B) = 1$, $\det(C) = \det(D) = \det(E) = -1$ and $\operatorname{tr}(I) = 2$, $\operatorname{tr}(A) = \operatorname{tr}(B) = -1$, $\operatorname{tr}(C) = \operatorname{tr}(D) = \operatorname{tr}(E) = 0$.
(ii) The three conjugacy classes are $\{I\}$, $\{A, B\}$, $\{C, D, E\}$. Thus there are three irreducible representations. The two dimensional is given above.

Problem 97.   Consider the six-dimensional vector space V consisting of polynomials of degree 2 in two real variables x_1, x_2

$$p(x_1, x_2) = c_{20}x_1^2 + c_{11}x_1x_2 + c_{02}x_2^2 + c_{10}x_1 + c_{01}x_2 + c_{00}$$

where the c_{jk}'s are real constants. Consider the faithful representation of the dihedral group given by 2×2 matrices in the previous problem. Find a six-dimensional representation of D_3 in V. Let

$$\mathbf{x} = \begin{pmatrix} x_1 \\ x_2 \end{pmatrix} \in \mathbb{R}^2$$

and g be one of the 2×2 matrices of the representation. One defines

$$O_g f(\mathbf{x}) := f(g^{-1}\mathbf{x}).$$

Solution 97. We have the six linear transformations

$$\begin{pmatrix} x_1' \\ x_2' \end{pmatrix} = I_2 \begin{pmatrix} x_1 \\ x_2 \end{pmatrix} = \begin{pmatrix} 1 & 0 \\ 0 & 1 \end{pmatrix} \begin{pmatrix} x_1 \\ x_2 \end{pmatrix}$$

$$\begin{pmatrix} x_1' \\ x_2' \end{pmatrix} = A \begin{pmatrix} x_1 \\ x_2 \end{pmatrix} = \begin{pmatrix} -1/2 & -\sqrt{3}/2 \\ \sqrt{3}/2 & -1/2 \end{pmatrix} \begin{pmatrix} x_1 \\ x_2 \end{pmatrix}$$

$$\begin{pmatrix} x_1' \\ x_2' \end{pmatrix} = B \begin{pmatrix} x_1 \\ x_2 \end{pmatrix} = \begin{pmatrix} -1/2 & \sqrt{3}/2 \\ -\sqrt{3}/2 & -1/2 \end{pmatrix} \begin{pmatrix} x_1 \\ x_2 \end{pmatrix}$$

$$\begin{pmatrix} x_1' \\ x_2' \end{pmatrix} = C \begin{pmatrix} x_1 \\ x_2 \end{pmatrix} = \begin{pmatrix} 1/2 & \sqrt{3}/2 \\ \sqrt{3}/2 & -1/2 \end{pmatrix} \begin{pmatrix} x_1 \\ x_2 \end{pmatrix}$$

$$\begin{pmatrix} x_1' \\ x_2' \end{pmatrix} = D \begin{pmatrix} x_1 \\ x_2 \end{pmatrix} = \begin{pmatrix} -1 & 0 \\ 0 & 1 \end{pmatrix} \begin{pmatrix} x_1 \\ x_2 \end{pmatrix}$$

$$\begin{pmatrix} x_1' \\ x_2' \end{pmatrix} = E \begin{pmatrix} x_1 \\ x_2 \end{pmatrix} = \begin{pmatrix} 1/2 & -\sqrt{3}/2 \\ -\sqrt{3}/2 & -1/2 \end{pmatrix} \begin{pmatrix} x_1 \\ x_2 \end{pmatrix}.$$

These linear transformations provide $x_1'^2 + x_2'^2 = x_1^2 + x_2^2$. Consider now the polynomial

$$p(x_1, x_2) = c_{20}x_1^2 + c_{11}x_1x_2 + c_{02}x_2^2 + c_{10}x_1 + c_{01}x_2 + c_{00}.$$

The identity transformation yields $p(x_1, x_2)$ again. The ordering of the coefficients is $c_{20}, c_{11}, c_{02}, c_{10}, c_{01}, c_{00}$. Thus I induces the identity transformation of the coefficients c_{ij}, i.e. $I \to I_6$, where I_6 is the 6×6 identity matrix. Consider now the rotation by $2\pi/3$. The inverse of the matrix A is given by

$$A^{-1} = \begin{pmatrix} -1/2 & \sqrt{3}/2 \\ -\sqrt{3}/2 & -1/2 \end{pmatrix}.$$

Applying this matrix to the vector $\mathbf{x} \in \mathbb{R}^2$ yields

$$\begin{pmatrix} -1/2 & \sqrt{3}/2 \\ -\sqrt{3}/2 & -1/2 \end{pmatrix} \begin{pmatrix} x_1 \\ x_2 \end{pmatrix} = \begin{pmatrix} -x_1/2 + \sqrt{3}x_2/2 \\ -\sqrt{3}x_1/2 - x_2/2 \end{pmatrix}.$$

Thus the polynomial takes the form

$$x_1^2\left(\frac{c_{20}}{4} + \frac{\sqrt{3}c_{11}}{4} + \frac{3c_{02}}{4}\right) + x_1x_2\left(-\frac{\sqrt{3}c_{20}}{2} - \frac{c_{11}}{2} + \frac{\sqrt{3}c_{02}}{2}\right) + x_2^2\left(\frac{3c_{20}}{4} - \frac{\sqrt{3}c_{11}}{4} + \frac{c_{02}}{4}\right)$$

$$+ x_1\left(-\frac{c_{10}}{2} - \frac{\sqrt{3}c_{01}}{2}\right) + x_2\left(\frac{\sqrt{3}c_{10}}{2} - \frac{c_{01}}{2}\right) + c_{00}.$$

This induces the transformation of the coefficients c_{ij}

$$A \to \begin{pmatrix} 1/4 & \sqrt{3}/4 & 3/4 \\ -\sqrt{3}/2 & -1/2 & \sqrt{3}/2 \\ 3/4 & -\sqrt{3}/4 & 1/4 \end{pmatrix} \oplus \begin{pmatrix} -1/2 & -\sqrt{3}/2 \\ \sqrt{3}/2 & -1/2 \end{pmatrix} \oplus (1).$$

Analogously we find

$$B \to \begin{pmatrix} 1/4 & -\sqrt{3}/4 & 3/4 \\ \sqrt{3}/2 & -1/2 & -\sqrt{3}/2 \\ 3/4 & \sqrt{3}/4 & 1/4 \end{pmatrix} \oplus \begin{pmatrix} -1/2 & \sqrt{3}/2 \\ -\sqrt{3}/2 & -1/2 \end{pmatrix} \oplus (1)$$

$$C \to \begin{pmatrix} 1/4 & \sqrt{3}/4 & 3/4 \\ \sqrt{3}/2 & 1/2 & -\sqrt{3}/2 \\ 3/4 & -\sqrt{3}/4 & 1/4 \end{pmatrix} \oplus \begin{pmatrix} 1/2 & \sqrt{3}/2 \\ \sqrt{3}/2 & -1/2 \end{pmatrix} \oplus (1)$$

$$D \to \begin{pmatrix} 1 & 0 & 0 \\ 0 & -1 & 0 \\ 0 & 0 & 1 \end{pmatrix} \oplus \begin{pmatrix} -1 & 0 \\ 0 & 1 \end{pmatrix} \oplus (1)$$

$$E \to \begin{pmatrix} 1/4 & -\sqrt{3}/4 & 3/4 \\ -\sqrt{3}/2 & 1/2 & \sqrt{3}/2 \\ 3/4 & \sqrt{3}/4 & 1/4 \end{pmatrix} \oplus \begin{pmatrix} 1/2 & -\sqrt{3}/2 \\ -\sqrt{3}/2 & -1/2 \end{pmatrix} \oplus (1).$$

The tedious calculations to find the group table and the representations should be done using Computer Algebra. A *Maxima* implementation is given below

```
/* dihedral.mac */

I: matrix([1,0],[0,1])$
A: matrix([-1,-sqrt(3)],[ sqrt(3),-1])/2$
B: matrix([-1, sqrt(3)],[-sqrt(3),-1])/2$
D: matrix([-1,0],[0,1])$
C: matrix([ 1, sqrt(3)],[ sqrt(3),-1])/2$
E: matrix([ 1,-sqrt(3)],[-sqrt(3),-1])/2$

for i in [ I, A, B, C, D, E ] do block (
  [x],
  x: [],
  for j in [ I, A, B, C, D, E ] do
   x: endcons(subst([I='I,A='A,B='B,C='C,D='D,E='E], i . j), x),
  print(x)
)$


p(x,y):= c20*x^2+c11*x*y+c02*y^2+c10*x+c01*y+c00$

for i in [ I, A, B, C, D, E ] do block (
  [x, y, xp, yp, z, v, pp, k, R],
```

```
R: genmatrix(R, 6, 6),
v: matrix([x],[y]),
v: invert(i) . v,
xp: matrix([1, 0]). v,
yp: matrix([0, 1]). v,
pp: expand(p(xp,yp)),
print("p -> ", pp),
k: expand(subst([x=0, y=0], diff(pp, x, 2)/2)),
R[1,1]: coeff(k,c20),R[1,2]: coeff(k,c11), R[1,3]: coeff(k,c02),
R[1,4]: coeff(k,c10),R[1,5]: coeff(k,c01), R[1,6]: coeff(k,c00),
k: subst([x=0, y=0], diff(pp, x, 1, y, 1)),
R[2,1]: coeff(k,c20),R[2,2]: coeff(k,c11),R[2,3]: coeff(k,c02),
R[2,4]: coeff(k,c10),R[2,5]: coeff(k,c01),R[2,6]: coeff(k,c00),
k: expand(subst([x=0,y=0],diff(pp,y,2)/2)),
R[3,1]: coeff(k,c20),R[3,2]: coeff(k,c11),R[3,3]: coeff(k,c02),
R[3,4]: coeff(k,c10),R[3,5]: coeff(k,c01),R[3,6]: coeff(k,c00),
k: subst([x=0,y=0],diff(pp,x,1)),
R[4,1]: coeff(k,c20),R[4,2]: coeff(k,c11),R[4,3]: coeff(k,c02),
R[4,4]: coeff(k,c10),R[4,5]: coeff(k,c01),R[4,6]: coeff(k,c00),
k: subst([x=0,y=0],diff(pp,y,1)),
R[5,1]: coeff(k,c20),R[5,2]: coeff(k,c11),R[5,3]: coeff(k,c02),
R[5,4]: coeff(k,c10),R[5,5]: coeff(k,c01),R[5,6]: coeff(k,c00),
k: subst([x=0,y=0],pp),
R[6,1]: coeff(k,c20),R[6,2]: coeff(k,c11),R[6,3]: coeff(k,c02),
R[6,4]: coeff(k,c10),R[6,5]: coeff(k,c01),R[6,6]: coeff(k,c00),
print(R)
)$
```

Problem 98. Let $A = (a_{ij})$ be a $2n \times 2n$ skew-symmetric matrix. The *Pfaffian* is defined as

$$\mathrm{Pf}(A) := \frac{1}{2^n n!} \sum_{\sigma \in S_{2n}} \mathrm{sgn}(\sigma) \prod_{j=1}^{n} a_{\sigma(2j-1),\sigma(2j)}$$

where S_{2n} is the symmetric group and $\mathrm{sgn}(\sigma)$ is the signature of permutation σ. Consider the case with $n = 2$, i.e.

$$A = \begin{pmatrix} 0 & a_{12} & a_{13} & a_{14} \\ -a_{12} & 0 & a_{23} & a_{24} \\ -a_{13} & -a_{23} & 0 & a_{34} \\ -a_{14} & -a_{24} & -a_{34} & 0 \end{pmatrix}.$$

Calculate $\mathrm{Pf}(A)$.

Solution 98. We have $n = 2$ and $4! = 24$ permutations. Thus

$$\mathrm{Pf}(A) = \frac{1}{8} \sum_{\sigma \in S_4} \mathrm{sgn}(\sigma) \prod_{j=1}^{2} a_{\sigma(2j-1),\sigma(2j)}$$

$$= \frac{1}{8} \sum_{\sigma \in S_4} \mathrm{sgn}(\sigma) a_{\sigma(1)\sigma(2)} a_{\sigma(3)\sigma(4)}$$

$$= a_{12}a_{34} - a_{13}a_{24} + a_{23}a_{14}$$

where we have taken into account that for $j > k$ we have $a_{jk} = -a_{kj}$. The summation over all permutations can be avoided. Let Π be the set of all partitions of the set $\{1, 2, \ldots, 2n\}$ into pairs without regard to order. There are $2n - 1$ such partitions. An element $\alpha \in \Pi$ can be written as

$$\alpha = \{(i_1, j_1), (i_2, j_2), \ldots, (i_n, j_n)\}$$

with $i_k < j_k$ and $i_1 < i_2 < \cdots < i_n$. Let

$$\pi = \begin{pmatrix} 1 & 2 & 3 & 4 & \ldots & 2n \\ i_1 & j_1 & i_2 & j_2 & \ldots & j_n \end{pmatrix}$$

be a corresponding permutation. Given a partition α we define the number

$$A_\alpha := \mathrm{sgn}(\pi) a_{i_1 j_1} a_{i_2 j_2} \cdots a_{i_n j_n} .$$

The Pfaffian of A is then given by

$$\mathrm{Pf}(A) = \sum_{\alpha \in \Pi} A_\alpha .$$

For $n = 2$ we have $(1, 2)(3, 4)$, $(1, 3)(2, 4)$, $(1, 4)(2, 3)$, i.e. we have 3 partitions.

Problem 99. The *action* or realization of a group G on a set M is defined as a map φ_g

$$\varphi_g : \mathbf{x} \mapsto \mathbf{x}' = \varphi_g(\mathbf{x})$$

where $\mathbf{x}, \mathbf{x}' \in M$ and $g \in G$. With each $g \in G$ a function is identified and

$$\varphi_e(\mathbf{x}) = x, \quad \varphi_{g_1} \circ \varphi_{g_2} = \varphi_{g_2 g_1}$$

for all $\mathbf{x} \in M$ and g_1, g_2, e (identity element) $\in G$. Let $x_1 \sim x_2$ if and only if there exists $g \in G$ such that $\varphi_g(x_1) = x_2$. Show that $\sim$ defines an *equivalence relation*.

Solution 99. For each $x \in M$ we have $\varphi_e(x) = x$. Thus $x \sim x$ and $\sim$ is reflexive. Suppose that $x_1 \sim x_2$, i.e. $\varphi_g(x_1) = x_2$ for some $g \in G$. Then

$$\varphi_{g^{-1}}(x_2) = (\varphi_{g^{-1}} \circ \varphi_g)(x_1) = \varphi_{g^{-1}g}(x_1) = \varphi_e(x_1) = x_1 .$$

Thus $\sim$ is symmetric. If $x_1 \sim x_2$ and $x_2 \sim x_3$ then

$$\varphi_{g_2}(x_2) = x_3, \qquad \varphi_{g_1}(x_1) = x_2 .$$

Thus

$$(\varphi_{g_2} \circ \varphi_{g_1})(x_1) = (\varphi_{g_2}(\varphi_{g_1}(x_1)) = \varphi_{g_2}(x_2) = x_3 .$$

Thus $x_1 \sim x_3$ and $\sim$ is transitive.

Problem 100. Let $n \geq 3$ and let $\sigma_1, \dots, \sigma_{n-1}$ be the generators of the braid group $\mathcal{B}_n$. The *braid group* $\mathcal{B}_n$ on n-strings where $n \geq 3$ has a finite presentation of B_n given by

$$\langle \sigma_1, \dots, \sigma_{n-1} : \sigma_i \sigma_j = \sigma_j \sigma_i, \quad \sigma_{i+1} \sigma_i \sigma_{i+1} = \sigma_i \sigma_{i+1} \sigma_i \rangle$$

where $1 \leq i, j < n - 1$, $|i - j| > 1$ or $j = n - 1$. Here $\sigma_i \sigma_j = \sigma_j \sigma_i$ and $\sigma_i \sigma_{i+1} \sigma_i = \sigma_{i+1} \sigma_i \sigma_{i+1}$ are called the braid relations. The second one is also called the Yang-Baxter equation.
(i) Consider B_3, $a = \sigma_1 \sigma_2 \sigma_1$ and $b = \sigma_1 \sigma_2$. Show that $a^2 = b^3$.
(ii) Consider B_3. The cosets $[\sigma_1]$ of σ_1 and $[\sigma_2]$ of σ_2 map to the 2×2 matrices

$$[\sigma_1] \mapsto R = \begin{pmatrix} 1 & 1 \\ 0 & 1 \end{pmatrix}, \qquad [\sigma_2] \mapsto L^{-1} = \begin{pmatrix} 1 & 0 \\ -1 & 1 \end{pmatrix}$$

where $L, R \in SL(2, \mathbb{Z})$. Thus $L^{-1}, R^{-1} \in SL(2, \mathbb{Z})$. Show that

$$RL^{-1}R = L^{-1}RL^{-1} .$$

Solution 100. (i) Using the *Yang-Baxter relation* $\sigma_1 \sigma_2 \sigma_1 = \sigma_2 \sigma_1 \sigma_2$ we have

$$a^2 = \sigma_1 \sigma_2 \sigma_1 \sigma_1 \sigma_2 \sigma_1 = \sigma_1 \sigma_2 \sigma_1 \sigma_2 \sigma_1 \sigma_2 = b^3 .$$

(ii) We have

$$RL^{-1}R = \begin{pmatrix} 1 & 1 \\ 0 & 1 \end{pmatrix} \begin{pmatrix} 1 & 0 \\ -1 & 1 \end{pmatrix} \begin{pmatrix} 1 & 1 \\ 0 & 1 \end{pmatrix} = \begin{pmatrix} 0 & 1 \\ -1 & 0 \end{pmatrix}$$

and

$$L^{-1}RL^{-1} = \begin{pmatrix} 1 & 0 \\ -1 & 1 \end{pmatrix} \begin{pmatrix} 1 & 1 \\ 0 & 1 \end{pmatrix} \begin{pmatrix} 1 & 0 \\ -1 & 1 \end{pmatrix} = \begin{pmatrix} 0 & 1 \\ -1 & 0 \end{pmatrix} .$$

Taking the inverse it also follows that

$$R^{-1}LR^{-1} = LR^{-1}L .$$

Note that L and R are the standard left and right moves on the Stern-Brocot tree.

Programming Problems

Problem 101. Give an efficient implementation of a group in C++ and LISP. For C++ use `map<pair<G,G>,G>` or `map<G,map<G,G> >` from the Standard Template Library. The `map` class in C++ implements a many to one relationship between keys and values. The `map` class is a template class `map<keytype,valuetype>` which can be used in similar way to arrays, i.e. `m[key]=value`. Consider for example the simplest non-trivial group $\{a, b\}$ with group operation $\cdot$ given by $a \cdot a = b \cdot b = a$ and $a \cdot b = b \cdot a = b$.

Solution 101. An efficient solution implements an efficient lookup table for the group operation from the group table. We give an interactive example using string concatenation instead of `pair`.

```
// group.cpp

#include <iostream>
#include <map>
#include <string>
using namespace std;

int main(void)
{
map<string,string> group;
string a = "a", b = "b", c;
group[ a + "*" + a ] = a;   group[ a + "*" + b ] = b;
group[ b + "*" + a ] = b;   group[ b + "*" + b ] = a;
cin >> c;
cout << group[c] << endl;
return 0;
}
```

The following LISP program uses associative lists for the group table.

```
; group.lisp

(defvar g1 '( ((a a) a) ((a b) b)
              ((b a) b) ((b b) a) ) )

(defun group (g) (lambda (x y)
                (cadr (assoc (list x y) g :test #'equal)) ) )

(labels ( (* (x y)
            (funcall (group g1) x y)) )
          (format t "~A~%~A~%"    ; ~A : value, ~% : newline
            (* 'a 'b)
            (* 'b 'b) ) )
```

The following LISP program uses a *hash table* for the group table.

```
; group2.lisp

(setf g1 (make-hash-table :test #'equal))
(setf (gethash '(a a) g1) 'a)   ; a*a = a
(setf (gethash '(a b) g1) 'b)   ; a*b = b
(setf (gethash '(b a) g1) 'b)   ; b*a = b
(setf (gethash '(b b) g1) 'a)   ; b*b = a

(defun group (g) (lambda (x y) (gethash (list x y) g)))

(labels ( (* (x y)
             (funcall (group g1) x y)) )
        (format t "~A~%~A~%"      ; ~A : value, ~% : newline
             (* 'a 'b)
             (* 'b 'b) ) )
```

Problem 102. Consider the permutation group S_3. Write a C++ program for the composition of the group elements using the `map` class of the Standard Template Library. Then implement the inverse of each group element. Finally we determine the conjugacy classes. The group consists of six elements which we denote by $a[0]$, $a[1]$, ..., $a[5]$. The neutral (identity) element is denoted by $a[0]$. Thus we have

$$a[0] * a[j] = a[j] * a[0] = a[j]$$

for $j = 0, 1, \ldots, 5$.

Solution 102. The group is nonabelian.

```
// group.cpp

#include <iostream>
#include <sstream>
#include <map>
#include <string>
#include <vector>
using namespace std;

map<string,map<string,string> > group;

string operator*(const string &s1,const string &s2)
{ return group[s1][s2]; }

int main(void)
```

```
{
 int i, j, k, n = 6;
 string res;
 vector<string> a(n), g(n), cl1(n), cl2(n);

 for(i=0;i<n;i++)
 {
  ostringstream name;
  name << "a" << "[" << i << "]";
  a[i] = name.str();
 }

 // a[0] is the neutral element
 for(i=0;i<n;i++)
  group[a[0]][a[i]] = group[a[i]][a[0]] = a[i];

 group[a[1]][a[1]] = a[0];
 group[a[1]][a[2]] = a[3]; group[a[2]][a[1]] = a[4];
 group[a[1]][a[3]] = a[2]; group[a[3]][a[1]] = a[5];
 group[a[1]][a[4]] = a[5]; group[a[4]][a[1]] = a[2];
 group[a[1]][a[5]] = a[4]; group[a[5]][a[1]] = a[3];
 group[a[2]][a[2]] = a[0];
 group[a[2]][a[3]] = a[5]; group[a[3]][a[2]] = a[1];
 group[a[2]][a[4]] = a[1]; group[a[4]][a[2]] = a[5];
 group[a[2]][a[5]] = a[3]; group[a[5]][a[2]] = a[4];
 group[a[3]][a[3]] = a[4];
 group[a[3]][a[4]] = a[0]; group[a[4]][a[3]] = a[0];
 group[a[3]][a[5]] = a[2]; group[a[5]][a[3]] = a[1];
 group[a[4]][a[4]] = a[3];
 group[a[4]][a[5]] = a[1]; group[a[5]][a[4]] = a[2];
 group[a[5]][a[5]] = a[0];

 res = a[0]*a[1]*a[2]*a[3]*a[4]*a[5];
 cout << "res = " << res << endl << endl;

 // find the inverse
 g[0] = a[0];
 for(j=0;j<n;j++)
  for(k=0;k<n;k++)
   if(a[j]*a[k]==a[0] && a[k]*a[j]==a[0]) { g[j] = a[k]; k = n; }

 for(j=0;j<n;j++) cout << "g[" << j << "] = " << g[j] << endl;
 cout << endl;

 // conjugacy class of the group element a[1]
 for(j=0;j<n;j++) cl1[j] = a[j]*a[1]*g[j];
  for(j=0;j<n;j++)
```

```
   cout << "cl1[" << j << "] = " << cl1[j] << endl;
  cout << endl;

  // conjugacy class of the group element a[3]
  for(j=0;j<n;j++) cl2[j] = a[j]*a[3]*g[j];
   for(j=0;j<n;j++)
    cout << "cl2[" << j << "] = " << cl2[j] << endl;
  return 0;
}
```

The output is

```
res = a[2]

g[0] = a[0]
g[1] = a[1]
g[2] = a[2]
g[3] = a[4]
g[4] = a[3]
g[5] = a[5]

cl1[0] = a[1]
cl1[1] = a[1]
cl1[2] = a[5]
cl1[3] = a[2]
cl1[4] = a[5]
cl1[5] = a[2]
cl2[0] = a[3]
cl2[1] = a[4]
cl2[2] = a[4]
cl2[3] = a[3]
cl2[4] = a[3]
cl2[5] = a[4]
```

Supplementary Problems

Problem 103.   Let A, B be 3×3 matrices over $\mathbb{R}$. We define the composition
$$A \bullet B := A + B - 2AB.$$
Does this composition define a group? The neutral element is 0_3.

Problem 104.   Consider the set
$$G = \{(a, b) \in \mathbb{R}^2 : -\infty < a < \infty, \ 0 \le b < 2\pi\}.$$
We define the composition $(a, b) \bullet (c, d) := (ac, (b + d) \bmod 2\pi)$. Show that this composition defines a group.

Problem 105.   Let $m \in \mathbb{Z}$, $m \ge 2$, and let $G(m)$ be the group of 2×2 matrices of determinant ± 1 and with entries in the ring $\mathbb{Z}/m\mathbb{Z}$, i.e.

$$G(m) := \left\{ \begin{pmatrix} a_{11} & a_{12} \\ a_{21} & a_{22} \end{pmatrix} : a_{11}, a_{12}, a_{21}, a_{22} \in \mathbb{Z}/m\mathbb{Z}, \ |a_{11}a_{22} - a_{12}a_{21}| = 1 \right\}.$$

Show that the set $G(m)$ forms a finite group under matrix multiplication.

Problem 106.   Show that there is only one group of order three. Is the group commutative? Is the group isomorphic to a subgroup of S_3?

Problem 107.   Let $\phi : G \to H$ be a homomorphism between the groups G and H. Show that $\ker(\phi)$ is a normal subgroup of G.

Problem 108.   Let $\alpha, \beta, \phi \in \mathbb{R}$ and $\alpha, \beta \ne 0$. Consider the matrices
$$A(\alpha, \beta, \phi) = \begin{pmatrix} \alpha \cos \phi & -\beta \sin \phi \\ \beta^{-1} \sin \phi & \alpha^{-1} \cos \phi \end{pmatrix}.$$
Do the matrices form a group under matrix multiplication?

Problem 109.   Let S_n be the group of all permutations of n objects. Show that every finite group is isomorphic to a group of permutations.

Problem 110.   Let G be a group. An *automorphism* of G is an isomorphism sending G onto itself. Show that the set $\text{Aut}(G)$ of automorphisms of G is a group with respect to the operation of composition of automorphisms.

Chapter 2

Lie Groups

Let $\mathbb{R}$ be the field of real numbers. Let

$$\mathbb{R}^m := \{\, \mathbf{x} = (x_1, \ldots, x_m) \,:\, x_j \in \mathbb{R} \quad 1 \le j \le m \,\}$$

that is the set of all ordered m-tuples of real numbers. The real number x_j is called the j-th coordinate of the point $\mathbf{x} \in \mathbb{R}^m$. For any $\mathbf{x}, \mathbf{y} \in \mathbb{R}^m$, $c \in \mathbb{R}$ we can define addition and scalar multiplication in $\mathbb{R}$ making $\mathbb{R}^m$ an m-dimensional vector space over $\mathbb{R}$. To introduce a topological structure we define the metric

$$d(\mathbf{x}, \mathbf{y}) := \sqrt{\sum_{j=1}^{m} (x_j - y_j)^2}$$

where $\mathbf{x}, \mathbf{y} \in \mathbb{R}^m$. The m-dimensional vector space $\mathbb{R}^m$ with this metric is called the m-dimensional Euclidean space.

A finite dimensional differentiable manifold M is a connected topological space with the properties:
(i) M is locally homeomorphic to $\mathbb{R}^m$ for some $m < \infty$. Thus for every point $p \in M$ there exists an open neighborhood U and a homeomorphism ϕ from U into an open ball in $\mathbb{R}^m$.
(ii) If (U_1, ϕ_1) and (U_2, ϕ_2) are two such coordinate charts then the overlap functions

$$\phi_2 \circ \phi_1^{-1} \,:\, \phi_1(U_1 \cap U_2) \to \phi_2(U_1 \cap U_2)$$

are smooth (i.e., infinitely differentiable, or C^∞).

In some cases one considers C^ω (analytic functions) instead of C^∞. If f can be expressed as convergent series in a neighborhood of any point of U, then $f \in C^\omega(U)$.

A Lie group can be defined as follows. Let G be a nonempty set. If G is a group (whose operation is denoted by multiplication $\cdot$), G is an r-dimensional smooth manifold and the inverse map $\tau : G \to G$ such that $\tau(g) = g^{-1}$ and the multiplication map $\phi : G \times G \to G$ such that $\phi(g_1 \cdot g_2) = g_1 \cdot g_2$ are both smooth maps.

Instead of C^r we can also consider the vector spaces C^∞ and C^ω.

For C^ω we can summarize the definition as follows. Let a group G be also an analytic manifold. We call the group a Lie group if the mapping

$$G \times G \ni (g_1, g_2) \to g_1 \cdot g_2^{-1} \in G$$

is analytic.

A Lie group G possesses two sets of diffeomorphisms - the right and left translation. For $g \in G$, the *right translation* by g on G is $R_g : G \to G$ such that

$$R_g(x) = x \cdot g$$

and the *left translation* is $L_g : G \to G$ such that

$$L_g(x) = g \cdot x .$$

Both R_g and L_g are diffeomorphisms from G to itself.

A group G is called a *Lie transformation group* of a differentiable manifold M if there is a differentiable map

$$\varphi : G \times M \to M, \qquad \varphi(g, \mathbf{x}) = g\mathbf{x}$$

such that $(g_1 \cdot g_2)\mathbf{x} = g_1 \cdot (g_2\mathbf{x})$ for $\mathbf{x} \in M$ and $g_1, g_2 \in G$ and $e\mathbf{x} = \mathbf{x}$ for the identity element e of G and $\mathbf{x} \in M$ are satisfied. Thus $\mathbf{x} \in M$ is transformed to $g\mathbf{x}$ by the transformation φ. This is known as the group action on $\mathbf{x}$. The classical groups as well as the affine groups are examples of Lie transformation groups. In most cases we have $M = \mathbb{R}^n$ or $M = \mathbb{C}^n$.

Every Lie group G has a Lie algebra L (the tangent space to the Lie group manifold at the identity element of the group) which is a vector space with a skew-symmetric product, the Lie bracket $[,]$.

Problem 1. A (global) group of $n \times n$ matrices is compact if it is a bounded, closed subset of the set of all $n \times n$ matrices. A set U of $n \times n$ matrices is bounded if there exists a constant $K > 0$ such that $|A_{ik}| \leq K$ for $1 \leq i, k \leq n$ and all $A \in U$. The set U is closed provided every Cauchy sequence in U converges to a matrix in U. A sequence of $n \times n$ matrices $\{A^{(p)}\}$ is a *Cauchy sequence* if each of the sequences of matrix elements $\{A_{ik}^{(p)}\}$, $1 \leq i, k \leq n$ is Cauchy. The orthogonal group $O(3, \mathbb{R})$ consists of all 3×3 matrices A over $\mathbb{R}$ with $AA^T = I_3$. Show that the orthogonal group $O(3, \mathbb{R})$ is compact.

Solution 1. If $A \in O(3, \mathbb{R})$ then $A^T A = I_3$, i.e.

$$\sum_{j=1}^{3} A_{j\ell} A_{jk} = \delta_{\ell k}.$$

Setting $\ell = k$ we obtain

$$\sum_{j=1}^{3} (A_{jk})^2 = 1.$$

Thus $|A_{jk}| \leq 1$ for all j, k. Thus the matrix elements are bounded. Let $\{A^{(p)}\}$ be a Cauchy sequence in $O(3, \mathbb{R})$ with limit A. Then

$$I_3 = \lim_{p \to \infty} (A^{(p)})^T A^{(p)} = A^T A$$

so $A \in O(3, \mathbb{R})$ and the Lie group $O(3, \mathbb{R})$ is compact.

Problem 2. The Lie group $SU(1, 1)$ consist of the set of all 2×2 pseudo-unitary matrices (of determinant 1) preserving the quadratic form

$$|z_1|^2 - |z_2|^2, \quad (z_1, z_2 \in \mathbb{C}).$$

(i) Show that the 2×2 matrix

$$U(\tau, \alpha, \beta) = \begin{pmatrix} \cosh(\tau/2) e^{-i\alpha} & \sinh(\tau/2) e^{-i\beta} \\ \sinh(\tau/2) e^{i\beta} & \cosh(\tau/2) e^{i\alpha} \end{pmatrix} \tag{1}$$

preserves the quadratic form $|z_1|^2 - |z_2|^2$ $(z_1, z_2 \in \mathbb{C})$, where $\tau, \alpha, \beta \in \mathbb{R}$.
(ii) Show that $\det U(\tau, \alpha, \beta) = 1$.
(iii) Give the inverse of $U(\tau, \alpha, \beta)$.
(iv) Show that $U_1(\tau_1, \alpha_1, \beta_1) U_2(\tau_2, \alpha_2, \beta_2)$ with $\alpha_1 + \alpha_2 = \beta_1 - \beta_2$ is again a matrix of the form (1).

Solution 2. (i) We have

$$\begin{pmatrix} \tilde{z}_1 \\ \tilde{z}_2 \end{pmatrix} = U \begin{pmatrix} z_1 \\ z_2 \end{pmatrix} = \begin{pmatrix} z_1 \cosh(\tau/2) e^{-i\alpha} + z_2 \sinh(\tau/2) e^{-i\beta} \\ z_1 \sinh(\tau/2) e^{i\beta} + z_2 \cosh(\tau/2) e^{i\alpha} \end{pmatrix}.$$

Using that $\cosh^2(\tau/2) - \sinh^2(\tau/2) = 1$ we obtain

$$
\begin{aligned}
|\tilde{z}_1|^2 - |\tilde{z}_2|^2 &= z_1 z_1^* \cosh^2(\tau/2) + z_2 z_2^* \sinh^2(\tau/2) \\
&\quad - z_1 z_1^* \sinh^2(\tau/2) - z_2 z_2^* \cosh^2(\tau/2) \\
&= z_1 z_1^* - z_2 z_2^* \\
&= |z_1|^2 - |z_2|^2 .
\end{aligned}
$$

(ii) Using that $\cosh^2(\tau/2) - \sinh^2(\tau/2) = 1$ we obtain $\det U = 1$.

(iii) The inverse of U is given by the replacements $\tau \to -\tau$, $\alpha \to -\alpha$, $\beta \to \beta$. Thus

$$
U^{-1} = \begin{pmatrix} \cosh(\tau/2)e^{i\alpha} & -\sinh(\tau/2)e^{-i\beta} \\ -\sinh(\tau/2)e^{i\beta} & \cosh(\tau/2)e^{-i\alpha} \end{pmatrix} .
$$

(iv) Using the identities

$$
\begin{aligned}
\cosh(x+y) &\equiv \cosh(x)\cosh(y) + \sinh(x)\sinh(y) \\
\sinh(x+y) &\equiv \sinh(x)\cosh(y) + \cosh(x)\sinh(y)
\end{aligned}
$$

and $\alpha_1 + \alpha_2 = \beta_1 - \beta_2$ we obtain

$$
U_1 U_2 = \begin{pmatrix} \cosh((\tau_1+\tau_2)/2)e^{-i(\alpha_1+\alpha_2)} & \sinh((\tau_1+\tau_2)/2)e^{-i(\beta_1-\alpha_2)} \\ \sinh((\tau_1+\tau_2)/2)e^{i(\beta_1-\alpha_2)} & \cosh((\tau_1+\tau_2)/2)e^{i(\alpha_1+\alpha_2)} \end{pmatrix} .
$$

Problem 3. The underlying field is the real numbers $\mathbb{R}$. The elements of the Lie group $SO(n)$ satisfy $A^T A = I_n$ and $\det A = 1$. Given any rotation matrix, $R \in SO(n)$. If R does not admit -1 as an eigenvalue, then there is a unique skew symmetric matrix, S, $(S^T = -S)$ so that

$$
R = (I_n - S)(I_n + S)^{-1} .
$$

The matrix R is called the *Cayley transform* of S. Let $n = 2$ and

$$
S = \begin{pmatrix} 0 & 1 \\ -1 & 0 \end{pmatrix} .
$$

Find R.

Solution 3. We have

$$
I_2 - S = \begin{pmatrix} 1 & -1 \\ 1 & 1 \end{pmatrix} , \qquad I_2 + S = \begin{pmatrix} 1 & 1 \\ -1 & 1 \end{pmatrix}
$$

and

$$
(I_2 + S)^{-1} = \frac{1}{2} \begin{pmatrix} 1 & -1 \\ 1 & 1 \end{pmatrix} .
$$

It follows that

$$(I_2 - S)(I_2 + S)^{-1} = \begin{pmatrix} 1 & -1 \\ 1 & 1 \end{pmatrix} \frac{1}{2} \begin{pmatrix} 1 & -1 \\ 1 & 1 \end{pmatrix} = \begin{pmatrix} 0 & -1 \\ 1 & 0 \end{pmatrix} = R.$$

Thus $R = S^T = -S$.

Problem 4. The *group commutator* $[A, B]$ of two elements A and B of a group G is defined by

$$[A, B] := A^{-1}B^{-1}AB.$$

The commutator is a kind of measure as to how near A and B come to commuting. $[A, B]$ being the identity element of G if and only if $AB = BA$.
(i) Consider a matrix group with matrix multiplication as composition. Calculate $\det([A, B])$.
(ii) The Lie group $U(2)$ consists of all 2×2 unitary matrices U, i.e. $UU^* = I_2$. Now

$$A = \begin{pmatrix} 0 & 1 \\ 1 & 0 \end{pmatrix}, \qquad B = \frac{1}{\sqrt{2}} \begin{pmatrix} 1 & 1 \\ 1 & -1 \end{pmatrix}$$

are elements of $U(2)$. Calculate $[A, B]$.

Solution 4. (i) Obviously $\det([A, B]) = 1$.
(ii) Since $A^{-1} = A$, $B = B^{-1}$ we have

$$[A, B] = A^{-1}B^{-1}AB = ABAB = (AB)^2.$$

Thus

$$[A, B] = \begin{pmatrix} 0 & -1 \\ 1 & 0 \end{pmatrix}.$$

Whereas A and B have determinant -1 the commutator has determinant $+1$ and is an element of $SU(2)$.

Problem 5. The rotation matrix

$$A(\phi) = \begin{pmatrix} \cos\phi & \sin\phi \\ -\sin\phi & \cos\phi \end{pmatrix}$$

is an element of $SO(2)$, i.e. $A(\phi)A^T(\phi) = I_2$ and $\det(A(\phi)) = 1$. Is the matrix

$$B(\phi) = \begin{pmatrix} \cos\phi & 0 & 0 & \sin\phi \\ 0 & \cos\phi & \sin\phi & 0 \\ 0 & -\sin\phi & \cos\phi & 0 \\ -\sin\phi & 0 & 0 & \cos\phi \end{pmatrix}$$

an element of $SO(4)$?

Solution 5.   Yes, since $B(\phi)B^T(\phi) = I_4$ and $\det(B(\phi)) = 1$.

Problem 6.   Let $\epsilon \in \mathbb{R}$. The matrix

$$B(\epsilon) = \begin{pmatrix} 1 & 0 & \epsilon \\ 0 & 1 & 0 \\ 0 & 0 & 1 \end{pmatrix}$$

is an element of the Lie group $SL(3, \mathbb{R})$. Find a matrix A such that $\exp(\epsilon A) = B(\epsilon)$.

Solution 6.   We find

$$A = \begin{pmatrix} 0 & 0 & 1 \\ 0 & 0 & 0 \\ 0 & 0 & 0 \end{pmatrix}$$

since $A^2 = 0_3$.

Problem 7.   Consider the 2×2 matrix

$$g = \begin{pmatrix} a + ib & c + id \\ -c + id & a - ib \end{pmatrix}$$

where $a, b, c, d \in \mathbb{R}$ and $a^2 + b^2 + c^2 + d^2 = 1$. Thus $g \in SU(2)$. Let σ_1, σ_2, σ_3 be the Pauli spin matrices. We define $\tau_j = \frac{1}{2}\sigma_j$ for $j = 1, 2, 3$.
(i) Calculate

$$g^{-1}\tau_1 g, \qquad g^{-1}\tau_2 g, \qquad g^{-1}\tau_3 g.$$

(ii) Find the 3×3 matrix G from the relation

$$(g^{-1}\tau_1 g, g^{-1}\tau_2 g, g^{-1}\tau_3 g) = (\tau_1, \tau_2, \tau_3)G$$

where $(\tau_1, \tau_2, \tau_3)G$ is defined as

$$(\tau_1, \tau_2, \tau_3)G = (\tau_1 G_{11} + \tau_2 G_{21} + \tau_3 G_{31}, \tau_1 G_{12} + \tau_2 G_{22} + \tau_3 G_{32}, \tau_1 G_{13} + \tau_2 G_{23} + \tau_3 G_{33}.$$

(iii) Let

$$M = \begin{pmatrix} 0 & -b & c \\ b & 0 & -d \\ -c & d & 0 \end{pmatrix}.$$

Show that G can be written as

$$G = I_3 + 2aM + 2M^2. \tag{1}$$

Solution 7. (i) Since the inverse of g is given

$$g^{-1} = \begin{pmatrix} a - ib & -c - id \\ c - id & a + ib \end{pmatrix}$$

we obtain

$$g^{-1}\tau_1 g = (a^2 - b^2 - c^2 + d^2)\tau_1 + 2(ab + cd)\tau_2 - 2(ac - bd)\tau_3$$
$$g^{-1}\tau_2 g = -2(ab - cd)\tau_1 + (a^2 - b^2 + c^2 - d^2)\tau_2 + 2(ad + bc)\tau_3$$
$$g^{-1}\tau_3 g = 2(ac + bd)\tau_1 - 2(ad - bc)\tau_2 + (a^2 + b^2 - c^2 - d^2)\tau_3.$$

(ii) We find

$$G = \begin{pmatrix} a^2 - b^2 - c^2 + d^2 & -2(ab - cd) & 2(ac + bd) \\ 2(ab + cd) & a^2 - b^2 + c^2 - d^2 & -2(ad - bc) \\ -2(ac - bd) & 2(ad + bc) & a^2 + b^2 - c^2 - d^2 \end{pmatrix}.$$

Using $a^2 + b^2 + c^2 + d^2 = 1$ we can simplify the matrix G to

$$G = \begin{pmatrix} 1 - 2(b^2 + c^2) & -2(ab - cd) & 2(ac + bd) \\ 2(ab + cd) & 1 - 2(b^2 + d^2) & -2(ad - bc) \\ -2(ac - bd) & 2(ad + bc) & 1 - 2(c^2 + d^2) \end{pmatrix}.$$

(iii) From (ii) we see that G can be written as

$$G = I_3 + 2a \begin{pmatrix} 0 & -b & c \\ b & 0 & -d \\ -c & d & 0 \end{pmatrix} + 2 \begin{pmatrix} -(b^2 + c^2) & cd & bd \\ cd & -(b^2 + d^2) & bc \\ bd & bc & -(c^2 + d^2) \end{pmatrix}.$$

Thus (1) follows.

Problem 8. (i) Consider the Lie group $GL(2, \mathbb{R})$. Let

$$A = \begin{pmatrix} a & b \\ c & d \end{pmatrix} \in GL(2, \mathbb{R}).$$

Thus $\det A \neq 0$. Show that the so called *Möbius transformation*

$$z \to \frac{az + b}{cz + d}$$

of the extended complex plane $\mathbb{C} \cup \{\infty\}$ forms a group under function composition.

(ii) Show that this group is *isomorphic* to the quotient of $GL(2, \mathbb{R})$ by its center.

(iii) Show that the Möbius group is isomorphic to $SL(2, \mathbb{R})$.

Solution 8.   (i) Consider the set of functions

$$g(x) = \frac{ax + b}{cx + d}, \qquad a, b, c, d \in \mathbb{R}$$

where $ad - bc \neq 0$. These are linear fractional functions. The Möbius transformation is a subset of linear fractional functions. Linear fractional functions form a group under function composition. The group multiplication law is

$$g_2(g_1(x)) = \frac{a_2 g_1(x) + b_2}{c_2 g_1(x) + d_2} = \frac{a_2(a_1 x + b_1) + b_2(c_1 x + d_1)}{c_2(a_1 x + b_1) + d_2(c_1 x + d_1)}$$
$$= \frac{(a_2 a_1 + b_2 c_1)x + (a_2 b_1 + b_2 d_1)}{(c_2 a_1 + d_2 c_1)x + (c_2 b_1 + d_2 d_1)}.$$

Thus the composition of two linear fractional functions is again a linear fractional function. The unit element is the function $g(x) = x$, i.e. $a = d = 1$, $b = c = 0$. The inverse element is the function

$$g^{-1}(x) = \frac{dx - b}{-cx + a}.$$

We have

$$g(g^{-1}(x)) = g^{-1}(g(x)) = x.$$

The associativity law is true for function composition. Thus the linear fractional functions form a group under functions composition. The Möbius transformation forms a subgroup of the linear fractional functions, because the set $\{g(x) : a, b, c, d \in \mathbb{R}\}$ is closed under multiplication and inversion.
(ii) We find first the center of $GL(2, \mathbb{R})$. Denote an arbitrary element z of $Z(GL(2, \mathbb{R}))$ as

$$z = \begin{pmatrix} p & q \\ r & s \end{pmatrix} \in Z(GL(2, \mathbb{R}))$$

and an arbitrary element of $GL(2, \mathbb{R})$ as

$$g = \begin{pmatrix} a & b \\ c & d \end{pmatrix} \in G.$$

Then the commutator of z and g is

$$zg - gz = \begin{pmatrix} qc - br & (pb + qd) - (aq + bs) \\ (ra + sc) - (cp + dr) & rb - cq \end{pmatrix}.$$

It follows that the entries of z must satisfy the conditions $q = r = 0$, $p = s \neq 0$. Thus $Z(GL(2, \mathbb{R}))$ is a set of matrices proportional to the identity matrix I_2

$$z = \begin{pmatrix} p & 0 \\ 0 & p \end{pmatrix}, \qquad p \neq 0.$$

The quotient of $GL(2, \mathbb{R})$ by the center is the group of 2×2 matrices $SL(2, \mathbb{R})$. Each element of $GL(2, \mathbb{R})$ can be represented as a product of an element of $SL(2, \mathbb{R})$ element and a matrix z.

(iii) The group multiplication law is the same for the Möbius group and $SL(2, \mathbb{R})$. We multiply the coefficients a, b, c, d of by the same nonzero multiplier μ. This transformation does not change the function g. Let us choose μ in such a way that $ad - bc = 1$ and consider a, b, c, d as matrix entries. This matrix belongs to $SL(2, \mathbb{R})$. This mapping preserves the group operation. Thus the Möbius group is isomorphic to $SL(2, \mathbb{R})$.

Problem 9. (i) Let $A \in SL(2, \mathbb{R})$, i.e. $\det(A) = 1$. Show that

$$A^2 - \operatorname{tr}(A)A + I_2 = 0_2 \,.$$

(ii) Let $A \in SL(2, \mathbb{R})$. Show that $\operatorname{tr}(A) = \operatorname{tr}(A^{-1})$.

(iii) $A, B \in SL(2, \mathbb{R})$. Show that $\operatorname{tr}(AB) = \operatorname{tr}A \operatorname{tr}B - \operatorname{tr}(AB^{-1})$.

Solution 9. (i) We show that $A^2 - \operatorname{tr}(A)A = -I_2$. Since

$$\det(A) = a_{11}a_{22} - a_{12}a_{21} = 1, \quad \operatorname{tr}A = a_{11} + a_{22}$$

we obtain

$$A^2 - \operatorname{tr}(A)A = \begin{pmatrix} -a_{11}a_{22} + a_{12}a_{21} & 0 \\ 0 & -a_{11}a_{22} + a_{12}a_{21} \end{pmatrix} = \begin{pmatrix} -1 & 0 \\ 0 & -1 \end{pmatrix} \,.$$

(ii) Let

$$A = \begin{pmatrix} a_{11} & a_{12} \\ a_{21} & a_{22} \end{pmatrix}$$

with $\det A = a_{11}a_{22} - a_{12}a_{21} = 1$. Then the inverse is given by

$$A^{-1} = \begin{pmatrix} a_{22} & -a_{12} \\ -a_{21} & a_{11} \end{pmatrix} \,.$$

Thus $\operatorname{tr}(A) = \operatorname{tr}(A^{-1}) = a_{11} + a_{22}$.

(iii) For the left-hand side we find

$$\operatorname{tr}(AB) = a_{11}b_{11} + a_{12}b_{21} + a_{21}b_{12} + a_{22}b_{22} \,.$$

For the right-hand side we have

$$\operatorname{tr}(A)\operatorname{tr}(B) - \operatorname{tr}(AB^{-1}) = a_{11}b_{11} + a_{22}b_{22} + a_{12}b_{21} + a_{21}b_{12} \,.$$

Problem 10. (i) Let $A \in SL(2, \mathbb{C})$. Show that

$$A^2 = \operatorname{tr}(A)A - I_2, \qquad A + A^{-1} = \operatorname{tr}(A)I_2 \,.$$

(ii) Show that
$$A^n = U_{n-1}(x)A - U_{n-2}I_2$$

where $x := \frac{1}{2}\mathrm{tr}A$ and $U_n(x)$ are *Chebyshev's polynomial* of the second kind defined by

$$U_{-1}(x) := 0, \qquad U_0(x) := 1$$
$$U_{n+1}(x) := 2xU_n(x) - U_{n-1}(x), \quad n = 1, 2, \dots.$$

Apply the *Cayley-Hamilton theorem*. The Cayley-Hamilton theorem states that every $n \times n$ matrix over a commutative ring (such as $\mathbb{R}$, $\mathbb{C}$, char$\mathbb{F} = 2$) satisfies its own characteristic equation.

Solution 10.   (i) Consider the *characteristic equation* of $A \in SL(2, \mathbb{C})$

$$p(\lambda) = \det(A - \lambda I_2) = 0.$$

Thus $p(\lambda) = \lambda^2 - \mathrm{tr}(A)\lambda + \det(A) = 0$ and

$$p(A) = A^2 - A\mathrm{tr}(A) + \det(A)I_2 = 0_2.$$

Now $A \in SL(2, \mathbb{C})$ implies $\det(A) = 1$. Therefore $A^2 = A\mathrm{tr}(A) - I_2$ follows. To find the second identity we multiply this equation by A^{-1}

$$A + A^{-1} = \mathrm{tr}(A)I_2.$$

(ii) To find A^n we start from $A^3 = A^2\mathrm{tr}(A) - A$. Inserting A^2 yields

$$A^3 = (A\mathrm{tr}(A) - I_2)\mathrm{tr}(A) - A = A(\mathrm{tr}(A)^2 - 1) - \mathrm{tr}(A)I_2.$$

Repeating this step yields $A^n = P_n A - Q_n I_2$, $n = 1, 2, \dots$. To obtain recurrences for P_n and Q_n we calculate

$$A^{n+1} = P_n A^2 - Q_n A = P_n(A\mathrm{tr}(A) - I_2) - Q_n A = A(P_n\mathrm{tr}(A) - Q_n) - P_n I_2.$$

Comparing terms with I_2 we find $Q_n = P_{n-1}$. Consequently

$$P_{n+1} = P_n\mathrm{tr}(A) - P_{n-1}.$$

Since $x := \frac{1}{2}\mathrm{tr}(A)$ it follows that

$$P_{n+1} = 2xP_n - P_{n-1}$$

with the initial conditions $P_2 = \mathrm{tr}(A) = 2x$, $P_1 = 1$. The recurrence relation with these initial conditions has a unique solution. The Chebyshev's polynomial of the second kind satisfies it. We have $P_n(x) = U_{n-1}(x)$.

Problem 11. Consider the Lie group $SL(2, \mathbb{R})$ and its Lie algebra $s\ell(2, \mathbb{R})$. The Lie algebra $s\ell(2, \mathbb{R})$ consists of all 2×2 matrices over $\mathbb{R}$ with trace equal to 0. Let

$$A = \begin{pmatrix} -1 & 1 \\ 0 & -1 \end{pmatrix} \in SL(2, \mathbb{R})$$

with eigenvalues -1 (twice). Note that the matrix A is not normal, i.e. $AA^T \neq A^T A$. Show that $A \notin \exp(s\ell(2, \mathbb{R}))$. This shows that the map $\exp : s\ell(2, \mathbb{R}) \to SL(2, \mathbb{R})$ is not surjective although the Lie group $SL(2, \mathbb{R})$ is connected.

Solution 11. Consider a 2×2 matrix of the form

$$\begin{pmatrix} \alpha & \beta \\ 0 & \gamma \end{pmatrix} \Rightarrow \begin{pmatrix} \alpha & \beta \\ 0 & \gamma \end{pmatrix}^n = \begin{pmatrix} \alpha^n & * \\ 0 & \gamma^n \end{pmatrix}.$$

Then the exponential has the same form. Assume that there is an $X \in s\ell(2, \mathbb{R})$ such that $A = \exp(X)$. Since A is not diagonalizable X is also not diagonalizable. Therefore X has a double eigenvalue. Since $\text{tr}(X) = 0$ the two eigenvalues are 0. Consequently $A = e^X$ has eigenvalue 1 which is a contradiction. There is a matrix M with $\text{tr}(M) = 2i\pi$ such that $A = \exp(M)$, namely

$$M = \begin{pmatrix} i\pi & -1 \\ 0 & i\pi \end{pmatrix}.$$

Problem 12. Given the invertible 4×4 matrix

$$G = \begin{pmatrix} -1 & 0 & 0 & 0 \\ 0 & 1 & 0 & 0 \\ 0 & 0 & 1 & 0 \\ 0 & 0 & 0 & 1 \end{pmatrix}.$$

Find two non-diagonal matrices L over $\mathbb{R}$ such that $L^T G L = G$. Such transformations are known as *Lorentz transformations*.

Solution 12. Using the identities

$$\cosh^2(\beta) - \sinh^2(\beta) = 1, \quad \cos^2(\theta) + \sin^2(\theta) = 1$$

we find

$$L_1 = \begin{pmatrix} \cosh(\beta) & \sinh(\beta) & 0 & 0 \\ \sinh(\beta) & \cosh(\beta) & 0 & 0 \\ 0 & 0 & 1 & 0 \\ 0 & 0 & 0 & 1 \end{pmatrix}, \quad L_2 = \begin{pmatrix} 1 & 0 & 0 & 0 \\ 0 & 1 & 0 & 0 \\ 0 & 0 & \cos(\theta) & -\sin(\theta) \\ 0 & 0 & \sin(\theta) & \cos(\theta) \end{pmatrix}.$$

Problem 13.   The group $Sp(2n, \mathbb{R})$ consists of all real $2n \times 2n$ matrices S which obey the condition

$$S^T J S = J$$

where J is the $2n \times 2n$ skew-symmetric matrix

$$J := \begin{pmatrix} 0_n & I_n \\ -I_n & 0_n \end{pmatrix}$$

with I_n the $n \times n$ identity matrix and 0_n the $n \times n$ zero matrix. Let V be a $2n \times 2n$ real symmetric positive definite matrix. Show that there exists an $S \in Sp(2n, \mathbb{R})$ such that

$$S^T V S = D^2 > 0, \qquad D^2 = \text{diag}(\kappa_1, \kappa_2, \ldots, \kappa_n, \kappa_1, \kappa_2, \ldots, \kappa_n).$$

Solution 13.   Since $J^T = -J$, it follows that $V^{-1/2} J V^{-1/2}$ is antisymmetric. Hence there exists a $2n \times 2n$ matrix $R \in SO(2n)$ such that

$$R^T V^{-1/2} J V^{-1/2} R = \begin{pmatrix} 0_n & \Omega \\ -\Omega & 0_n \end{pmatrix}, \qquad \Omega = \text{diagonal} > 0.$$

We define a diagonal positive definite matrix

$$D = \begin{pmatrix} \Omega^{-1/2} & 0_n \\ 0_n & \Omega^{-1/2} \end{pmatrix}.$$

Then we have

$$D R^T V^{-1/2} J V^{-1/2} R D = J.$$

Now we define $S := V^{-1/2} R D$. Then $S^T J S = J$ and $S^T V S = D^2$, where D is a diagonal matrix with $d_{jj} > 0$ for $j = 1, 2, \ldots, 2n$.

Problem 14.   Consider the trace-less matrices

$$A = \begin{pmatrix} 0 & 1 \\ 0 & 0 \end{pmatrix}, \quad B = \begin{pmatrix} 0 & 0 \\ 1 & 0 \end{pmatrix}, \quad C = \begin{pmatrix} 1 & 0 \\ 0 & -1 \end{pmatrix}.$$

Let $t \in \mathbb{R}$. Show that the matrices

$$e^{tA}, \quad e^{tB}, \quad e^{tC}$$

are elements of the Lie group $SL(2, \mathbb{R})$.

Solution 14.   Since $A^2 = B^2 = 0_2$ we obtain

$$e^{tA} = \begin{pmatrix} 1 & t \\ 0 & 1 \end{pmatrix}, \quad e^{tB} = \begin{pmatrix} 1 & 0 \\ t & 1 \end{pmatrix}, \quad e^{tC} = \begin{pmatrix} e^t & 0 \\ 0 & e^{-t} \end{pmatrix}$$

which all have determinant equal to 1. We could use the identity for any $n \times n$ matrix M

$$e^{\,\mathrm{tr}(M)} \equiv \det(e^M).$$

If $\mathrm{tr}(M) = 0$, then we have $\det(e^M) = 1$.

Problem 15. The group of complex rotations $O(n, \mathbb{C})$ is defined as the group of all $n \times n$ complex matrices O, such that $OO^T = O^T O = I_n$, where ^{T} means transpose. These transformations preserve the real scalar product

$$\mathbf{x} \cdot \mathbf{y} = \sum_{j=1}^{n} x_j y_j$$

so that $(O\mathbf{x}) \cdot O\mathbf{y} = \mathbf{x} \cdot \mathbf{y}$, where $\mathbf{x}$ and $\mathbf{y}$ are complex vectors in general, i.e. $x_j, y_j \in \mathbb{C}$.
(i) Show that the 2×2 matrix ($\alpha \in \mathbb{R}$)

$$O = \begin{pmatrix} \cosh \alpha & i \sinh \alpha \\ -i \sinh \alpha & \cosh \alpha \end{pmatrix}$$

is an element of $O(2, \mathbb{C})$.
(ii) Find the partial derivatives under complex orthogonal transformations $O \in O(n, \mathbb{C})$

$$w_j(\mathbf{x}) := (O\mathbf{x})_j = \sum_{k=1}^{n} O_{jk} x_k, \qquad j = 1, 2, \dots, n$$

i.e. $\partial/\partial w_j$ with $j = 1, 2, \dots, n$.

Solution 15. (i) We have

$$O^T = \begin{pmatrix} \cosh \alpha & -i \sinh \alpha \\ i \sinh \alpha & \cosh \alpha \end{pmatrix}.$$

Thus $O^T O = OO^T = I_2$.
(ii) Using the *chain rule* we have

$$\frac{\partial}{\partial w_j} = \sum_{k=1}^{n} \frac{\partial x_k}{\partial w_j} \frac{\partial}{\partial x_k} = \sum_{k=1}^{n} (O^{-1})_{kj} \frac{\partial}{\partial x_k} = \sum_{k=1}^{n} O_{jk} \frac{\partial}{\partial x_k}.$$

Thus the partial derivatives transform exactly as the coordinates, since $O^{-1} = O^T$.

Problem 16. (i) The non-compact Lie group $SL(2, \mathbb{C})$ consists of all 2×2 matrices over $\mathbb{C}$ with determinant equal to 1. Give a 2×2 matrix

A which is an element of $SL(2,\mathbb{C})$ but not an element of the compact Lie group $SU(2)$.

(ii) The maximal compact Lie subgroup of $SL(3,\mathbb{R})$ is the Lie group $SO(3)$. Give an element which is in $SL(3,\mathbb{R})$ but not in $SO(3)$.

(iii) Consider the Lie group $SL(2,\mathbb{C})$. Give a compact and non-compact subgroup of $SL(2,\mathbb{C})$.

Solution 16.   (i) An example is

$$A = \begin{pmatrix} 2i & 0 \\ 0 & -i/2 \end{pmatrix}$$

with $\det(A) = 1$ and

$$AA^* = \begin{pmatrix} 4 & 0 \\ 0 & 1/4 \end{pmatrix} \neq I_2 \,.$$

(ii) An example is

$$A = \begin{pmatrix} \alpha^2 & 0 & 0 \\ 0 & 1/\alpha & 0 \\ 0 & 0 & 1/\alpha \end{pmatrix}, \qquad \alpha \in \mathbb{R}, \ \alpha > 0$$

with $\det(A) = 1$.

(iii) A compact subgroup is $SU(2)$. A non-compact subgroup is $SU(1,1)$.

Problem 17.   Consider the Lie group $SL(2,\mathbb{R})$, i.e. the set of all real 2×2 matrices with determinant equal to 1. A dynamical system in $SL(2,\mathbb{R})$ can be defined by

$$M_{k+2} = M_k M_{k+1}, \qquad k = 0, 1, 2, \dots \tag{1}$$

with the initial matrices $M_0, M_1 \in SL(2,\mathbb{R})$. Let $F_k := \mathrm{tr} M_k$. Show that

$$F_{k+3} = F_{k+2} F_{k+1} - F_k, \qquad k = 0, 1, 2, \dots \,. \tag{2}$$

Hint. Use that property that for any 2×2 matrix A we have

$$A^2 - A\mathrm{tr}(A) + I_2 \det(A) = 0 \,. \tag{3}$$

Solution 17.   From (1) it follows that

$$M_{k+3} = M_{k+1} M_{k+2} = M_{k+1} M_k M_{k+1} \,.$$

Taking the trace of this equation and *cyclic invariance* provides

$$\mathrm{tr}(M_{k+3}) = \mathrm{tr}(M_{k+1} M_k M_{k+1}) = \mathrm{tr}(M_{k+1}^2 M_k) \,.$$

Using (3) with $\det M_k = 1$ we arrive at

$$
\begin{aligned}
\text{tr}(M_{k+3}) &= \text{tr}((M_{k+1}\text{tr}(M_{k+1}) - I_2)M_k) \\
&= \text{tr}(M_{k+1})\text{tr}(M_{k+1}M_k) - \text{tr}(M_k) \\
&= \text{tr}(M_{k+1})\text{tr}(M_{k+2}) - \text{tr}(M_k)\,.
\end{aligned}
$$

Thus the recurrence relation follows.

Problem 18. The Lie group $SO(m, n)$ consists of all real matrices S that satisfy

$$
S^T g S = g
$$

where $\det S = 1$ and $g = \text{diag}(+1, +1, \ldots, +1, -1, \ldots, -1)$ with n $+1$'s and m -1's. Let V be a real symmetric positive definite matrix of dimension N. Show that for any choice of partition $N = m + n$, there exists an $S \in SO(m, n)$ such that

$$
S^T V S = D^2 = \text{diagonal (and } > 0)\,.
$$

Solution 18. Consider the matrix $V^{-1/2}gV^{-1/2}$ constructed from the given matrix V. Since $V^{-1/2}gV^{-1/2}$ is real symmetric, there exists a rotation matrix $R \in SO(N)$ which diagonalizes $V^{-1/2}gV^{-1/2}$

$$
R^T V^{-1/2}gV^{-1/2} R = \text{diagonal} \equiv \Lambda\,.
$$

This may be viewed also as a congruence of g using $V^{-1/2}R$, and signatures are preserved under congruence. As a consequence, the diagonal matrix Λ can be expressed as the product of a positive diagonal matrix and g

$$
R^T V^{-1/2}gV^{-1/2} R = D^{-2}g = D^{-1}gD^{-1}\,.
$$

Here D is diagonal and positive definite. Taking the inverse of the matrices on both sides of this equation we find that the diagonal entries of $gD^2 = D^2g$ are the eigenvalues of $V^{1/2}gV^{1/2}$ and that the columns of R are the eigenvectors of $V^{1/2}gV^{1/2}$. Since $V^{1/2}gV^{1/2}$, gV, and Vg are conjugate to one another, we conclude that D^2 is determined by the eigenvalues of $gV \sim Vg$. We define $S := V^{-1/2}RD$. Then S satisfies the two equations

$$
S^T g S = g, \qquad S^T V S = D^2 = \text{diagonal}\,.
$$

The first equation tells us that $S \in SO(m, n)$ and the second tells us that V is diagonalized through congruence by S.

Problem 19. The compact Lie group $SO(3, \mathbb{R})$ has a realization as the group of all 3×3 real matrices A such that $A^T A = I_3$ and $\det A = 1$. This is the natural realization of $SO(3, \mathbb{R})$ as the group of all rotations in the vector space $\mathbb{R}^3$ which leave the origin fixed. One convenient parameterization of $SO(3, \mathbb{R})$ is in terms of the *Euler angles*. A rotation through angle ϕ about the z-axis is given by

$$R_z(\phi) = \begin{pmatrix} \cos\phi & -\sin\phi & 0 \\ \sin\phi & \cos\phi & 0 \\ 0 & 0 & 1 \end{pmatrix} \in SO(3, \mathbb{R})$$

and rotations through angle ϕ about the x and y axis are given by

$$R_x(\phi) = \begin{pmatrix} 1 & 0 & 0 \\ 0 & \cos\phi & -\sin\phi \\ 0 & \sin\phi & \cos\phi \end{pmatrix} \in SO(3, \mathbb{R})$$

$$R_y(\phi) = \begin{pmatrix} \cos\phi & 0 & \sin\phi \\ 0 & 1 & 0 \\ -\sin\phi & 0 & \cos\phi \end{pmatrix} \in SO(3, \mathbb{R})$$

respectively. Differentiate each of these matrices with respect to ϕ and set $\phi = 0$ to find a basis L_x, L_y, L_z of the simple Lie algebra $so(3)$. Calculate the commutators. Find $\exp(\phi L_x)$, $\exp(\phi L_y)$, $\exp(\phi L_z)$.

Solution 19.   Since

$$\frac{d}{d\phi}\cos\phi = -\sin\phi \Rightarrow \left.\frac{d}{d\phi}\cos\phi\right|_{\phi=0} = 0$$

$$\frac{d}{d\phi}\sin\phi = \cos\phi \Rightarrow \left.\frac{d}{d\phi}\sin\phi\right|_{\phi=0} = 1$$

we obtain the skew-symmetric matrices

$$L_z = \begin{pmatrix} 0 & -1 & 0 \\ 1 & 0 & 0 \\ 0 & 0 & 0 \end{pmatrix}, \quad L_x = \begin{pmatrix} 0 & 0 & 0 \\ 0 & 0 & -1 \\ 0 & 1 & 0 \end{pmatrix}, \quad L_y = \begin{pmatrix} 0 & 0 & 1 \\ 0 & 0 & 0 \\ -1 & 0 & 0 \end{pmatrix}$$

with the commutators $[L_x, L_y] = L_z$, $[L_z, L_x] = L_y$, $[L_y, L_z] = L_x$ and

$$R_x(\phi) = \exp(\phi L_x), \qquad R_y(\phi) = \exp(\phi L_y), \qquad R_z(\phi) = \exp(\phi L_z).$$

Problem 20.   Consider the symplectic group, $Sp(2, \mathbb{R})$, consisting of 2×2 real matrices M satisfying

$$MNM^T = N, \qquad N = \begin{pmatrix} 0 & 1 \\ -1 & 0 \end{pmatrix}.$$

These matrices can also be characterized by the condition, $\det(M) = 1$, i.e. $Sp(2, \mathbb{R})$ is identical with the group $SL(2, \mathbb{R})$ of all 2×2 real matrices of determinant 1.

(i) Show that the matrix

$$A = \begin{pmatrix} 2 & 3 \\ 1 & 2 \end{pmatrix}$$

is an element of $Sp(2, \mathbb{R})$. Find $\det(A)$.

(ii) An element $M \in Sp(2, \mathbb{R})$ has the decomposition

$$M = \begin{pmatrix} 1 & 0 \\ -v & 1 \end{pmatrix} \begin{pmatrix} u^{-1/2} & 0 \\ 0 & u^{1/2} \end{pmatrix} \begin{pmatrix} \cos\theta & -\sin\theta \\ -\sin\theta & \cos\theta \end{pmatrix}$$

where $v \in \mathbb{R}$, $u > 0$, $0 < \theta \leq 2\pi$. Find the conditions on v, u, θ.

Solution 20. (i) Straightforward calculation yields

$$\begin{pmatrix} 2 & 3 \\ 1 & 2 \end{pmatrix} \begin{pmatrix} 0 & 1 \\ -1 & 0 \end{pmatrix} \begin{pmatrix} 2 & 1 \\ 3 & 2 \end{pmatrix} = \begin{pmatrix} 0 & 1 \\ -1 & 0 \end{pmatrix}.$$

We have $\det(A) = 1$.

(ii) Since

$$M = \begin{pmatrix} u^{-1/2}\cos\theta - u^{1/2}\sin\theta & -u^{-1/2}\sin\theta + u^{1/2}\cos\theta \\ -vu^{-1/2}\cos\theta - u^{1/2}\sin\theta & vu^{-1/2}\sin\theta + u^{1/2}\cos\theta \end{pmatrix}$$

we obtain the four conditions

$$u^{-1/2}\cos\theta - u^{1/2}\sin\theta = 2$$
$$-u^{-1/2}\sin\theta + u^{1/2}\cos\theta = 3$$
$$-vu^{-1/2}\cos\theta - u^{1/2}\sin\theta = 1$$
$$vu^{-1/2}\sin\theta + u^{1/2}\cos\theta = 2.$$

Problem 21. We know that all $n \times n$ unitary matrices form the Lie group $U(n)$ under matrix multiplication. Let U be an $n \times n$ unitary matrix. Is $\exp(U)$ unitary? Is $\exp(i(U + U^*))$ unitary?

Solution 21. Obviously $\exp(U)$ is not unitary. Consider for example the 2×2 identity matrix. Then

$$\exp(I_2) = \begin{pmatrix} e^1 & 0 \\ 0 & e^1 \end{pmatrix}$$

which is not unitary. Yes, $\exp(i(U + U^*))$ is unitary since $U + U^*$ is hermitian.

Problem 22.   Consider the Lie group $U(n)$. The eigenvalues of a unitary matrix take the form $e^{i\alpha}$, where $\alpha \in \mathbb{R}$. Consider the unitary matrix

$$U = \begin{pmatrix} 0 & i \\ i & 0 \end{pmatrix}.$$

Find the eigenvalues of U and the group generated by the eigenvalues under multiplication.

Solution 22.   The eigenvalues of U are $\pm i$. Thus the group generated by $\pm i$ is the set $\{\, 1, -1, i, -i \,\}$.

Problem 23.   (i) Let K be a skew-hermitian $n \times n$ matrix, i.e. $K^* = -K$, and $\epsilon \in \mathbb{R}$. Then

$$V(\epsilon) := \exp(\epsilon K)$$

is a unitary matrix. The unitary matrices form a Lie group under matrix multiplication. Given $V(\epsilon)$ reconstruct the skew-hermitian matrix K.
(ii) Apply the reconstruction technique to the unitary matrix

$$V(\epsilon) = \begin{pmatrix} \cos(\epsilon) & i\sin(\epsilon) \\ i\sin(\epsilon) & \cos(\epsilon) \end{pmatrix}.$$

(iii) Apply the reconstruction technique to the unitary matrix

$$V(\epsilon) = \begin{pmatrix} 0 & e^{i\epsilon} \\ e^{i\epsilon} & 0 \end{pmatrix}.$$

Solution 23.   (i) Differentiating $V(\epsilon)$ with respect to ϵ yields

$$\frac{dV(\epsilon)}{d\epsilon} = e^{\epsilon K} K = V(\epsilon) K.$$

Since the inverse of $V(\epsilon)$ exists we have

$$K = V^{-1}(\epsilon) \frac{dV(\epsilon)}{d\epsilon}.$$

(ii) Since

$$\frac{dV(\epsilon)}{d\epsilon} = \begin{pmatrix} -\sin(\epsilon) & i\cos(\epsilon) \\ i\cos(\epsilon) & -\sin(\epsilon) \end{pmatrix}, \quad V^{-1}(\epsilon) = \begin{pmatrix} \cos(\epsilon) & -i\sin(\epsilon) \\ -i\sin(\epsilon) & \cos(\epsilon) \end{pmatrix}$$

we obtain

$$K = V^{-1}(\epsilon) \frac{dV(\epsilon)}{d\epsilon} = \begin{pmatrix} 0 & i \\ i & 0 \end{pmatrix}.$$

(iii) Since

$$\frac{dV(\epsilon)}{d\epsilon} = \begin{pmatrix} 0 & ie^{i\epsilon} \\ ie^{i\epsilon} & 0 \end{pmatrix}$$

and

$$V^{-1}(\epsilon) = \begin{pmatrix} 0 & e^{-i\epsilon} \\ e^{-i\epsilon} & 0 \end{pmatrix}$$

we obtain

$$K = V^{-1}(\epsilon)\frac{dV(\epsilon)}{d\epsilon} = \begin{pmatrix} i & 0 \\ 0 & i \end{pmatrix}.$$

Thus

$$K = \begin{pmatrix} i & 0 \\ 0 & i \end{pmatrix}.$$

This matrix K of course does not generate $V(\epsilon) = e^{\epsilon K}$. What is the flaw in the derivation? In this case the K can be found as follows. We calculate the eigenvalues and normalized eigenvectors of $V(\epsilon)$. We obtain $e^{i\epsilon}$ and $-e^{i\epsilon} = e^{i(\epsilon-\pi)}$ for the eigenvalues with the corresponding normalized eigenvectors

$$\frac{1}{\sqrt{2}}\begin{pmatrix} 1 \\ 1 \end{pmatrix}, \quad \frac{1}{\sqrt{2}}\begin{pmatrix} 1 \\ -1 \end{pmatrix}.$$

Note that $V(\epsilon)$ is a normal matrix. Then applying the *spectral theorem* we find K as

$$K(\epsilon) = \frac{1}{2}\ln(e^{i\epsilon})\begin{pmatrix} 1 & 1 \\ 1 & 1 \end{pmatrix} + \ln(e^{i(\epsilon-\pi)})\frac{1}{2}\begin{pmatrix} 1 & -1 \\ -1 & 1 \end{pmatrix}$$
$$= \begin{pmatrix} i(\epsilon - \pi/2) & i\pi \\ i\pi & i(\epsilon - \pi/2) \end{pmatrix}.$$

Problem 24. (i) Consider the semisimple Lie group $SL(2,\mathbb{R})$. Let $A \in SL(2,\mathbb{R})$. Then $A^{-1} \in SL(2,\mathbb{R})$. Show that A and A^{-1} have the same eigenvalues.
(ii) Is this still true for $A \in SL(3,\mathbb{R})$?

Solution 24. (i) For $A \in SL(2,\mathbb{R})$ we have

$$\text{tr}(A) = \text{tr}(A^{-1}) = \lambda_1 + \lambda_2$$

and

$$\det(A) = \det(A^{-1}) = \lambda_1\lambda_2 = 1$$

where λ_1, λ_2 are the eigenvalues of A. Thus A and A^{-1} have the same eigenvalues.

(ii) This is not true in general if $A \in SL(3, \mathbb{R})$. Consider for example the diagonal matrix

$$A = \begin{pmatrix} 6 & 0 & 0 \\ 0 & 1/2 & 0 \\ 0 & 0 & 1/3 \end{pmatrix}$$

with eigenvalues 6, 1/2, 1/3. Now

$$A^{-1} = \begin{pmatrix} 1/6 & 0 & 0 \\ 0 & 2 & 0 \\ 0 & 0 & 3 \end{pmatrix}$$

with eigenvalues 1/6, 2 and 3.

Problem 25.   Consider the 2×2 matrices

$$A_j = \begin{pmatrix} 1 & 1/j \\ 0 & 1 \end{pmatrix}, \qquad j = 1, 2, \dots, n$$

and the 2×2 matrices

$$B_j = \begin{pmatrix} 1 & 0 \\ 1/j & 1 \end{pmatrix}, \qquad j = 1, 2, \dots, n.$$

Obviously, these matrices are elements of the Lie group $SL(2, \mathbb{R})$.
(i) Calculate the inverse matrices A_j^{-1}, B_j^{-1}.
(ii) Calculate the matrix products

$$A_1 B_1 A_1^{-1} B_1^{-1}, \qquad A_1 B_1 A_1^{-1} B_1^{-1} A_2 B_2 A_2^{-1} B_2^{-1}.$$

Solution 25.   (i) The inverse of A_j is given by

$$A_j^{-1} = \begin{pmatrix} 1 & -1/j \\ 0 & 1 \end{pmatrix}, \qquad j = 1, 2, \dots, n.$$

The inverse of B_j is given by

$$B_j^{-1} = \begin{pmatrix} 1 & 0 \\ -1/j & 1 \end{pmatrix}, \qquad j = 1, 2, \dots, n.$$

Thus

$$A_j B_j A_j^{-1} B_j^{-1} = \begin{pmatrix} \frac{j^4 + j^2 + 1}{j^4} & -\frac{1}{j^3} \\ \frac{1}{j^3} & 1 - \frac{1}{j^2} \end{pmatrix}.$$

(ii) We find

$$\Pi_1 = A_1 B_1 A_1^{-1} B_1^{-1} = \begin{pmatrix} 3 & -1 \\ 1 & 0 \end{pmatrix}$$

$$\Pi_2 = \prod_{j=1}^{j=2} A_j B_j A_j^{-1} B_j^{-1} = \begin{pmatrix} 61/16 & -9/8 \\ 21/16 & -1/8 \end{pmatrix}.$$

Problem 26. Let X, Y be real Banach spaces. Let $L(X,Y)$ be the space of bounded linear operators. If the operator f is differentiable at $u \in X$ then for every $v \in X$

$$Df(u)(v) = \frac{d}{d\epsilon} f(u + \epsilon v) \bigg|_{\epsilon=0}.$$

This is called the *directional derivative (Gateaux derivative)* of f at u in the direction of v. In the following we have $X = Y$. Consider the Banach space of all real $n \times n$ matrices with the norm induced by the *scalar product*

$$\langle A, B \rangle := \mathrm{tr}(AB^*).$$

Let f be the map $A \mapsto A^{-1}$ in the Lie group $GL(n, \mathbb{R})$. Show that

$$Df(A)(B) = -A^{-1}BA^{-1}.$$

Solution 26. Differentiation of

$$(A + \epsilon B)^{-1}(A + \epsilon B) = I_n$$

with respect to ϵ yields

$$\left(\frac{d}{d\epsilon}(A + \epsilon B)^{-1} \right)(A + \epsilon B) + (A + \epsilon B)^{-1}B = 0_n.$$

Thus

$$\frac{d}{d\epsilon}(A + \epsilon B)^{-1} = -(A + \epsilon B)^{-1}B(A + \epsilon B)^{-1}.$$

Thus with $\epsilon \to 0$ we obtain the result.

Problem 27. Show that $SU(1, 1, \mathbb{C})$ is *isomorphic* to $SL(2, \mathbb{R})$.

Solution 27. Denote elements of $SU(1, 1, \mathbb{C})$ as

$$g = \begin{pmatrix} a & b \\ c & d \end{pmatrix}.$$

Since $g \in SU(1, 1, \mathbb{C})$ we have $\det(g) = ad - bc = 1$. Let $\mathbf{z} \in \mathbb{C}^2$. Then we have

$$(az_1 + bz_2)(\bar{a}\bar{z}_1 + \bar{b}\bar{z}_2) - (cz_1 + dz_2)(\bar{c}\bar{z}_1 + \bar{d}\bar{z}_2) = z_1\bar{z}_1 - z_2\bar{z}_2.$$

It follows that

$$a\bar{a} - c\bar{c} = 1, \quad b\bar{b} - d\bar{d} = -1, \quad a\bar{b} - c\bar{d} = 0, \quad b\bar{a} - d\bar{c} = 0.$$

The third and the fourth equations are conjugated. We can solve them by substitution

$$b = \mu\bar{c}, \quad d = \mu\bar{a}.$$

We obtain from the second equation $\mu(a\bar{a} - c\bar{c}) = \mu = 1$. Therefore $b = \bar{c}$, $d = \bar{a}$. Consider the real and imaginary parts of a and c separately

$$a = p + iq, \quad c = r + is, \quad p, q, r, s \in \mathbb{R}.$$

Then the condition $ad - bc = 1$ gives $p^2 + q^2 - r^2 - s^2 = 1$. It follows that

$$(p + r)(p - r) - (s + q)(s - q) = 1.$$

Therefore the matrix

$$\begin{pmatrix} p + r & s + q \\ s - q & p - r \end{pmatrix}$$

belongs to the Lie group $SL(2, \mathbb{R})$. Thus we find a one-to-one mapping of $SU(1, 1, \mathbb{C})$ to $SL(2, \mathbb{R})$. It preserves matrix multiplication. In $SU(1, 1, \mathbb{C})$ we have

$$g_3 = g_1 g_2 = \begin{pmatrix} a_1 & \bar{c}_1 \\ c_1 & \bar{a}_1 \end{pmatrix} \begin{pmatrix} a_2 & \bar{c}_2 \\ c_2 & \bar{a}_2 \end{pmatrix} = \begin{pmatrix} a_1 a_2 + \bar{c}_1 c_2 & a_1 \bar{c}_2 + \bar{c}_1 \bar{a}_2 \\ c_1 a_2 + \bar{a}_1 c_2 & c_1 \bar{c}_2 + \bar{a}_1 \bar{a}_2 \end{pmatrix}.$$

Consider the first column of the multiplication result. Separating out the real and imaginary parts yields

$$p_3 = p_1 p_2 - q_1 q_2 + r_1 r_2 + s_1 s_2, \qquad q_3 = p_1 q_2 + q_1 p_2 + r_1 s_2 - s_1 r_2$$

$$r_3 = r_1 p_2 - s_1 q_2 + p_1 r_2 + q_1 s_2, \qquad s_3 = r_1 q_2 + s_1 p_2 + p_1 s_2 - q_1 r_2.$$

We have

$$p_3 + r_3 = (p_1 + r_1)(p_2 + r_2) + (s_1 + q_1)(s_2 - q_2)$$
$$p_3 - r_3 = (s_1 - q_1)(s_2 + q_2) + (p_1 - r_1)(p_2 - r_2)$$
$$s_3 + q_3 = (p_1 + r_1)(s_2 + q_2) + (s_1 + q_1)(p_2 - r_2)$$
$$s_3 - q_3 = (s_1 - q_1)(p_2 + r_2) + (p_1 - r_1)(s_2 - q_2).$$

We see from these equations that the one-to-one map preserves the group multiplication law. Thus $SU(1, 1, \mathbb{C})$ is isomorphic to $SL(2, \mathbb{R})$.

Problem 28.   Show that $SU(2)$ is the double covering group of $SO(3)$.

Solution 28. We introduce the polynomial representation of arbitrary element of $SO(3)$

$$Q = \begin{pmatrix} 1 - 2y^2 - 2z^2 & 2xy - 2zw & 2xz + 2yw \\ 2xy + 2zw & 1 - 2x^2 - 2z^2 & 2yz - 2xw \\ 2xz - 2yw & 2yz + 2xw & 1 - 2x^2 - 2y^2 \end{pmatrix} \tag{1}$$

where w, x, y, z satisfy the equation

$$w^2 + x^2 + y^2 + z^2 = 1. \tag{2}$$

To show that every element of $SO(3)$ has the form (1) we start from Euler's first parameterization by rotation angle ω and axis direction

$$(\cos(\alpha), \cos(\beta), \cos(\gamma))$$

and show that the substitution

$$p = \cos(\omega/2), \qquad x = \sin(\omega/2)\cos(\alpha),$$

$$y = \sin(\omega/2)\cos(\beta), \qquad z = \sin(\omega/2)\cos(\gamma)$$

transforms Euler's matrix to (1). Now w, x, y, z give following representation of arbitrary element of $SU(2)$

$$\begin{pmatrix} w + iz & -x - iy \\ x - iy & w - iz \end{pmatrix}.$$

Multiplying two $SU(2)$ matrices yields

$$\begin{pmatrix} w_1 + iz_1 & -x_1 - iy_1 \\ x_1 - iy_1 & w_1 - iz_1 \end{pmatrix} \begin{pmatrix} w_2 + iz_2 & -x_2 - iy_2 \\ x_2 - iy_2 & w_2 - iz_2 \end{pmatrix} = \begin{pmatrix} w_3 + iz_3 & -x_3 - iy_3 \\ x_3 - iy_3 & w_3 - iz_3 \end{pmatrix}$$

where

$$w_3 = w_1 w_2 - x_1 x_2 - y_1 y_2 - z_1 z_2,$$

$$x_3 = w_1 x_2 + x_1 w_2 + y_1 z_2 - z_1 y_2,$$

$$y_3 = w_1 y_2 + y_1 w_2 + z_1 x_2 - x_1 z_2,$$

$$z_3 = w_1 z_2 + z_1 w_2 + x_1 y_2 - y_1 x_2.$$

Thus we find the following group law for w, x, y, z. Multiplying two $SO(3)$ matrices we find the same group law for w, x, y, z. We obtain

$$Q_{11} = 1 - 2y_3^2 - 2z_3^2, \quad Q_{12} = 2x_3 y_3 - 2z_3 w_3, \quad Q_{13} = 2x_3 z_3 + 2y_3 w_3,$$

$$Q_{21} = 2x_3 y_3 + 2z_3 w_3, \quad Q_{22} = 1 - 2x_3^2 - 2z_3^2, \quad Q_{23} = 2y_3 z_3 - 2x_3 w_3,$$

$$Q_{31} = 2x_3 z_3 - 2y_3 w_3, \quad Q_{32} = 2y_3 z_3 + 2x_3 w_3, \quad Q_{33} = 1 - 2x_3^2 - 2y_3^2.$$

We see that correspondence between elements of $SU(2)$ and $SO(3)$ preserves the group multiplication law - it is a homomorphism. We also see that two different elements of $SU(2)$ $(w, x, y, z$ and $-w, -x, -y, -z)$ correspond to one element of $SO(3)$. Thus $SU(2)$ is the double covering group of $SO(3)$.

Problem 29. Let $t \in \mathbb{R}$. Consider the matrices

$$A_1(t) = \begin{pmatrix} e^{it} & 0 \\ 0 & 1 \end{pmatrix}, \qquad A_2(t) = \begin{pmatrix} 1 & t \\ 0 & 1 \end{pmatrix}, \qquad A_3(t) = \begin{pmatrix} 1 & it \\ 0 & 1 \end{pmatrix}$$

which are elements in the Lie groups $GL(2, \mathbb{R})$. Find matrices X_1, X_2, X_3 (*infinitesimal generators*) such that

$$A_1(t) = e^{tX_1}, \qquad A_2(t) = e^{tX_2}, \qquad A_3(t) = e^{tX_3} .$$

Solution 29. (i) Since

$$\left. \frac{dA_1}{dt} \right|_{t=0} = \begin{pmatrix} i & 0 \\ 0 & 0 \end{pmatrix}, \qquad \left. \frac{dA_2}{dt} \right|_{t=0} = \begin{pmatrix} 0 & 1 \\ 0 & 0 \end{pmatrix}, \qquad \left. \frac{dA_3}{dt} \right|_{t=0} = \begin{pmatrix} 0 & i \\ 0 & 0 \end{pmatrix}$$

we obtain the infinitesimal generators

$$X_1 = \begin{pmatrix} i & 0 \\ 0 & 0 \end{pmatrix}, \qquad X_2 = \begin{pmatrix} 0 & 1 \\ 0 & 0 \end{pmatrix}, \qquad X_3 = \begin{pmatrix} 0 & i \\ 0 & 0 \end{pmatrix} .$$

Problem 30. The following three matrices are elements of the Lie group $SL(2, \mathbb{R})$ with $t \in \mathbb{R}$

$$A_1(t) = \begin{pmatrix} 1 & t \\ 0 & 1 \end{pmatrix}, \qquad A_2(t) = \begin{pmatrix} 1 & 0 \\ t & 1 \end{pmatrix}, \qquad A_3(t) = \begin{pmatrix} e^t & 0 \\ 0 & e^{-t} \end{pmatrix} .$$

(i) Find 2×2 matrices X_1, X_2, X_3 such that

$$A_1(t) = e^{tX_1}, \qquad A_2(t) = e^{tX_2}, \qquad A_3(t) = e^{tX_3} .$$

(ii) Calculate the commutators $[X_1, X_2]$, $[X_3, X_2]$, $[X_3, X_1]$.

Solution 30. (i) We have

$$\left. \frac{dA_1}{dt} \right|_{t=0} = \begin{pmatrix} 0 & 1 \\ 0 & 0 \end{pmatrix}, \qquad \left. \frac{dA_2}{dt} \right|_{t=0} = \begin{pmatrix} 0 & 0 \\ 1 & 0 \end{pmatrix}, \qquad \left. \frac{dA_3}{dt} \right|_{t=0} = \begin{pmatrix} 1 & 0 \\ 0 & -1 \end{pmatrix} .$$

Thus we find

$$X_1 = \begin{pmatrix} 0 & 1 \\ 0 & 0 \end{pmatrix}, \qquad X_2 = \begin{pmatrix} 0 & 0 \\ 1 & 0 \end{pmatrix}, \qquad X_3 = \begin{pmatrix} 1 & 0 \\ 0 & -1 \end{pmatrix} .$$

(ii) For the commutators we obtain

$$[X_1, X_2] = X_3, \quad [X_3, X_2] = -2X_2, \quad [X_3, X_1] = 2X_1.$$

Problem 31. One-parameter subgroups of the Lie group $SL(2, \mathbb{R})$ are given by

$$A_1(t) = \begin{pmatrix} e^{t/2} & 0 \\ 0 & e^{-t/2} \end{pmatrix}$$

$$A_2(\tau) = \begin{pmatrix} \cosh(\tau/2) & \sinh(\tau/2) \\ \sinh(\tau/2) & \cosh(\tau/2) \end{pmatrix}$$

$$A_3(\phi) = \begin{pmatrix} \cos(\phi/2) & \sin(\phi/2) \\ -\sin(\phi/2) & \cos(\phi/2) \end{pmatrix}.$$

Find the infinitesimal generators.

Solution 31. We find the infinitesimal generators by

$$X_1 = \frac{d}{dt} A_1(t)|_{t=0}, \quad X_2 = \frac{d}{d\tau} A_2(\tau)|_{\tau=0}, \quad X_3 = \frac{d}{d\phi} A_3(\phi)|_{\phi=0}.$$

Thus we obtain the infinitesimal generators as

$$X_1 = \frac{1}{2} \begin{pmatrix} 1 & 0 \\ 0 & -1 \end{pmatrix}, \quad X_2 = \frac{1}{2} \begin{pmatrix} 0 & 1 \\ 1 & 0 \end{pmatrix}, \quad X_3 = \frac{1}{2} \begin{pmatrix} 0 & 1 \\ -1 & 0 \end{pmatrix}.$$

They satisfy the commutation relations

$$[X_1, X_2] = X_3, \quad [X_1, X_2] = X_2, \quad [X_2, X_3] = -X_1.$$

Problem 32. Consider the subgroups of the Lie group $SU(1, 1)$

$$\begin{pmatrix} e^{i\theta/2} & 0 \\ 0 & e^{-i\theta/2} \end{pmatrix}, \quad \begin{pmatrix} \cosh(t/2) & i\sinh(t/2) \\ -i\sinh(t/2) & \cosh(t/2) \end{pmatrix}, \quad \begin{pmatrix} 1 + \zeta i/2 & \zeta/2 \\ \zeta/2 & 1 - \zeta i/2 \end{pmatrix}$$

where $\theta, t, \zeta \in \mathbb{R}$. Find the infinitesimal generators.

Solution 32. For the first matrix we find

$$\begin{pmatrix} i/2 & 0 \\ 0 & -i/2 \end{pmatrix}.$$

For the second matrix we find

$$\begin{pmatrix} 0 & i/2 \\ -i/2 & 0 \end{pmatrix}$$

For the third matrix we find

$$\begin{pmatrix} i/2 & 1/2 \\ 1/2 & -i/2 \end{pmatrix}.$$

Problem 33.   Consider the subgroups of the Lie group $SU(2)$

$$\begin{pmatrix} e^{i\theta/2} & 0 \\ 0 & e^{-i\theta/2} \end{pmatrix}, \quad \begin{pmatrix} \cos(t/2) & -\sin(t/2) \\ \sin(t/2) & \cos(t/2) \end{pmatrix}, \quad \begin{pmatrix} \cos(t/2) & i\sin(t/2) \\ i\sin(t/2) & 1\cos(t/2) \end{pmatrix}$$

where $t \in \mathbb{R}$. Find the infinitesimal generators.

Solution 33.   The infinitesimal generators are

$$X_1 = \frac{i}{2} \begin{pmatrix} 1 & 0 \\ 0 & -1 \end{pmatrix}, \quad X_2 = \frac{1}{2} \begin{pmatrix} 0 & -1 \\ 1 & 0 \end{pmatrix}, \quad X_3 = \frac{i}{2} \begin{pmatrix} 0 & 1 \\ 1 & 0 \end{pmatrix}.$$

Problem 34.   Consider 2×2 real matrices A satisfying the following two conditions. First A is a rotation matrix, i.e. its rows (and therefore its columns) are pairwise orthonormal. Second we impose $A^2 = -I_2$. Find all these matrices.

Solution 34.   Since A is a rotation matrix we can write

$$A = \begin{pmatrix} \cos\alpha & \sin\alpha \\ -\sin\alpha & \cos\alpha \end{pmatrix}$$

which is an element of the compact Lie group $SO(2)$. From the condition $A^2 = -I_2$ we obtain

$$A^2 = \begin{pmatrix} \cos(2\alpha) & \sin(2\alpha) \\ -\sin(2\alpha) & \cos(2\alpha) \end{pmatrix} = \begin{pmatrix} -1 & 0 \\ 0 & -1 \end{pmatrix}$$

using

$$\cos^2\alpha - \sin^2\alpha \equiv \cos(2\alpha), \qquad 2\sin\alpha\cos\alpha \equiv \sin(2\alpha).$$

Thus we have the two equations $\cos(2\alpha) = -1$, $\sin(2\alpha) = 0$ with the solutions $\alpha = \pi/2$ and $\alpha = 3\pi/2$. This leads to the two matrices

$$\begin{pmatrix} 0 & 1 \\ -1 & 0 \end{pmatrix}, \qquad \begin{pmatrix} 0 & -1 \\ 1 & 0 \end{pmatrix}.$$

These two matrices together with the 2×2 identity matrix form a group under matrix multiplication.

Problem 35. Let $\alpha \in \mathbb{R}$. The 4×4 matrices

$$A(\alpha) = \begin{pmatrix} \cosh \alpha & 0 & 0 & -\sinh \alpha \\ 0 & 1 & 0 & 0 \\ 0 & 0 & 1 & 0 \\ -\sinh \alpha & 0 & 0 & \cosh \alpha \end{pmatrix}$$

form a Lie group. Find the infinitesimal generator.

Solution 35. We have

$$B = \left.\frac{dA(\alpha)}{d\alpha}\right|_{\alpha=0} = \begin{pmatrix} 0 & 0 & 0 & -1 \\ 0 & 0 & 0 & 0 \\ 0 & 0 & 0 & 0 \\ -1 & 0 & 0 & 0 \end{pmatrix}$$

with $A(\alpha) = e^{\alpha B}$.

Problem 36. Let $\alpha, \beta \in \mathbb{R}$. Let T be the 2×2 unitary diagonal matrix

$$T(\alpha, \beta) = \begin{pmatrix} e^{i\alpha} & 0 \\ 0 & e^{i\beta} \end{pmatrix}.$$

Find the condition on a unitary 2×2 matrix U such that

$$UT(\alpha, \beta)U^{-1} = T(\alpha', \beta') \equiv \begin{pmatrix} e^{i\alpha'} & 0 \\ 0 & e^{i\beta'} \end{pmatrix}.$$

Solution 36. Denote an arbitrary element of the Lie group $U(2)$ by

$$U = \begin{pmatrix} a & b \\ c & d \end{pmatrix}.$$

The inverse matrix is

$$U^{-1} = U^* = \begin{pmatrix} \bar{a} & \bar{c} \\ \bar{b} & \bar{d} \end{pmatrix}.$$

From $UU^* = I_2$ we obtain the conditions

$$a\bar{a} + b\bar{b} = 1, \quad c\bar{c} + d\bar{d} = 1, \quad a\bar{c} + b\bar{d} = 0, \quad c\bar{a} + d\bar{b} = 0.$$

If U is a diagonal matrix, then $b = \bar{b} = c = \bar{c} = 0$. It follows that $a\bar{a} = b\bar{b} = 1$. Thus U has the form T. However

$$T(\alpha_1, \beta_1)\, T(\alpha_2, \beta_2) = \begin{pmatrix} e^{i(\alpha_1 + \alpha_2)} & 0 \\ 0 & e^{i(\beta_1 + \beta_2)} \end{pmatrix} = T(\alpha_1 + \alpha_2, \beta_1 + \beta_2).$$

This is the first case, when

$$UT(\alpha, \beta)U^{-1} = T(\alpha', \beta') = \begin{pmatrix} e^{i\alpha'} & 0 \\ 0 & e^{i\beta'} \end{pmatrix}.$$

Now

$$UTU^{-1} = \begin{pmatrix} a & b \\ c & d \end{pmatrix} \begin{pmatrix} e^{i\alpha}\bar{a} & e^{i\alpha}\bar{c} \\ e^{i\beta}\bar{b} & e^{i\beta}\bar{d} \end{pmatrix} = \begin{pmatrix} e^{i\alpha}a\bar{a} + e^{i\beta}b\bar{b} & e^{i\alpha}a\bar{c} + e^{i\beta}b\bar{d} \\ e^{i\alpha}c\bar{a} + e^{i\beta}d\bar{b} & e^{i\alpha}c\bar{c} + e^{i\beta}d\bar{d} \end{pmatrix}.$$

Equating the nondiagonal elements to zero yields

$$e^{i\alpha}a\bar{c} + e^{i\beta}b\bar{d} = 0, \quad e^{i\alpha}c\bar{a} + e^{i\beta}d\bar{b} = 0.$$

If $a\bar{c} \neq 0$ then $e^{i\alpha} - e^{i\beta} = 0$ or $\alpha = \beta + 2k\pi$ ($k \in \mathbb{Z}$). This is the second case. The third case is $a = d = 0$. This case is similar to the first case

$$U = \begin{pmatrix} 0 & e^{i\gamma} \\ e^{i\delta} & 0 \end{pmatrix}.$$

There are three cases
1) U is a diagonal matrix and T is arbitrary. In this case $UTU^{-1} = T$.
2) U is an arbitrary unitary matrix, and $\alpha = \beta$. In this case $UTU^{-1} = T$.
3) U is a nondiagonal matrix and T is arbitrary. In this case

$$UTU^{-1} = \begin{pmatrix} e^{i\beta} & 0 \\ 0 & e^{i\alpha} \end{pmatrix}.$$

Problem 37.   Consider the compact Lie group $SO(2, \mathbb{R})$. Let

$$A(\phi) = \begin{pmatrix} \cos\phi & \sin\phi \\ -\sin\phi & \cos\phi \end{pmatrix}$$

be an element of $SO(2, \mathbb{R})$. Find a 2×2 matrix B such that $B^{-1}A(\phi)B$ is diagonal. To find B calculate the eigenvalues and normalized eigenvectors of $A(\phi)$. Is B an element of the Lie group $SO(2, \mathbb{R})$?

Solution 37.   The eigenvalues are

$$\lambda_+ = \cos\phi + i\sin\phi \equiv e^{i\phi}, \qquad \lambda_- = \cos\phi - i\sin\phi \equiv e^{-i\phi}$$

with the corresponding normalized eigenvectors

$$\frac{1}{\sqrt{2}}\begin{pmatrix} -i \\ 1 \end{pmatrix}, \qquad \frac{1}{\sqrt{2}}\begin{pmatrix} i \\ 1 \end{pmatrix}.$$

This leads to the matrix

$$B = \frac{1}{\sqrt{2}} \begin{pmatrix} -i & i \\ 1 & 1 \end{pmatrix} \Rightarrow B^{-1} = B^* = \frac{1}{\sqrt{2}} \begin{pmatrix} i & 1 \\ -i & 1 \end{pmatrix}$$

with

$$B^{-1}A(\phi)B = \begin{pmatrix} e^{i\phi} & 0 \\ 0 & e^{-i\phi} \end{pmatrix}.$$

Obviously B is not an element of $SO(2, \mathbb{R})$, but of the Lie goup $U(2)$. Note that B does not depend on ϕ and $\det(B) = -i$.

Problem 38. Consider the Lie group $SO(1, 1, \mathbb{R})$. Let

$$A(\phi) = \begin{pmatrix} \cosh\phi & \sinh\phi \\ \sinh\phi & \cosh\phi \end{pmatrix}$$

be an element of $SO(1, 1, \mathbb{R})$. Find a 2×2 matrix B such that $B^{-1}A(\phi)B$ is diagonal. To find B calculate the eigenvalues and normalized eigenvectors of $A(\phi)$. Is B an element of the Lie group $SO(1, 1, \mathbb{R})$?

Solution 38. The eigenvalues are

$$\lambda_+ = \cosh\phi + \sinh\phi \equiv e^\phi, \qquad \lambda_- = \cosh\phi - \sinh\phi \equiv e^{-\phi}$$

with the corresponding normalized eigenvectors

$$\frac{1}{\sqrt{2}} \begin{pmatrix} 1 \\ 1 \end{pmatrix}, \qquad \frac{1}{\sqrt{2}} \begin{pmatrix} 1 \\ -1 \end{pmatrix}.$$

This leads to the matrix

$$B = \frac{1}{\sqrt{2}} \begin{pmatrix} 1 & 1 \\ 1 & -1 \end{pmatrix} = B^{-1}$$

with

$$B^{-1}A(\phi)B = \begin{pmatrix} e^\phi & 0 \\ 0 & e^{-\phi} \end{pmatrix}.$$

Note that the matrix B does not depend on ϕ and $\det(B) = -1$.

Problem 39. Consider the Lie group $GL(2, \mathbb{R})$. Show that the Lie group $SL(2, \mathbb{R})$ is a normal subgroup of $GL(2, \mathbb{R})$. The group elements of $SL(2, \mathbb{R})$ have determinant 1. Apply the theorem:

A subgroup H of the group G is *normal* in G if and only if $gHg^{-1} \subseteq H$ for all g in G.

Solution 39.   Let $g \in GL(2, \mathbb{R})$ and $h \in SL(2, \mathbb{R})$. Then since $\det(h) = 1$

$$\det(ghg^{-1}) = \det(g)\det(h)\det(g^{-1}) = \det(g)\det(g^{-1})$$
$$= \det(gg^{-1}) = \det(I_2) = 1.$$

Thus $ghg^{-1} \in SL(2, \mathbb{R})$. Consequently $gHg^{-1} \subseteq H$. Thus $SL(n, \mathbb{R})$ is a $(n^2 - 1)$-dimensional subgroup of $GL(n, \mathbb{R})$.

Problem 40.   Consider the Lie group $SL(2, \mathbb{R})$. Show that each element of $SL(2, \mathbb{R})$ can be generated by the two matrices

$$\begin{pmatrix} 1 & t \\ 0 & 1 \end{pmatrix}, \qquad \begin{pmatrix} 1 & 0 \\ t & 1 \end{pmatrix}$$

where $t \in \mathbb{R}$.

Solution 40.   Consider

$$\begin{pmatrix} a & b \\ c & d \end{pmatrix}, \qquad ad - cb = 1, \quad a, b, c, d \in \mathbb{R}.$$

If $c \neq 0$, then

$$\begin{pmatrix} a & b \\ c & d \end{pmatrix} = \begin{pmatrix} 1 & (a-1)c^{-1} \\ 0 & 1 \end{pmatrix} \begin{pmatrix} 1 & 0 \\ c & 1 \end{pmatrix} \begin{pmatrix} 1 & (d-1)c^{-1} \\ 0 & 1 \end{pmatrix}.$$

If $b \neq 0$, then

$$\begin{pmatrix} a & b \\ c & d \end{pmatrix} = \begin{pmatrix} 1 & 0 \\ (d-1)b^{-1} & 1 \end{pmatrix} \begin{pmatrix} 1 & b \\ 0 & 1 \end{pmatrix} \begin{pmatrix} 1 & 0 \\ (a-1)b^{-1} & 1 \end{pmatrix}.$$

If $b = c = 0$, then

$$\begin{pmatrix} a & 0 \\ 0 & a^{-1} \end{pmatrix} = \begin{pmatrix} 1 & 0 \\ a^{-1} - 1 & 1 \end{pmatrix} \begin{pmatrix} 1 & 1 \\ 0 & 1 \end{pmatrix} \begin{pmatrix} 1 & 0 \\ a - 1 & 1 \end{pmatrix} \begin{pmatrix} 1 & -a^{-1} \\ 0 & 1 \end{pmatrix}.$$

Problem 41.   (i) Find an invertible 2×2 matrix X, i.e. $X \in GL(2, \mathbb{C})$, such that

$$X \begin{pmatrix} a & b \\ c & d \end{pmatrix} X^{-1} = \begin{pmatrix} a & -b \\ -c & d \end{pmatrix}.$$

(ii) Find an invertible 2×2 matrix Y, i.e. $Y \in GL(2, \mathbb{C})$, such that

$$Y \begin{pmatrix} a & b \\ c & d \end{pmatrix} Y^{-1} = \begin{pmatrix} d & -c \\ -b & a \end{pmatrix}.$$

Solution 41. (i) We obtain $X = \sigma_z = \sigma_z^{-1}$.
(ii) We obtain

$$Y = \begin{pmatrix} 0 & 1 \\ -1 & 0 \end{pmatrix} = i\sigma_y$$

with $Y^{-1} = -Y$.

Problem 42. The Pauli spin matrices are elements of the Lie group $U(2)$. Additionally they are hermitian.
(i) Let $z \in \mathbb{C}$. Find

$$\exp(z\sigma_x), \quad \exp(z\sigma_y), \quad \exp(z\sigma_z).$$

(ii) Find

$$\exp(-z\sigma_x)\sigma_y\exp(z\sigma_x), \quad \exp(-z\sigma_y)\sigma_z\exp(z\sigma_y), \quad \exp(-z\sigma_z)\sigma_x\exp(z\sigma_z).$$

Solution 42. (i) Since $\sigma_x^2 = \sigma_y^2 = \sigma_z^2 = I_2$ we find for the exponential map of the Pauli spin matrices

$$\exp(z\sigma_x) = I_2 \cosh(z) + \sigma_x \sinh(z)$$
$$\exp(z\sigma_y) = I_2 \cosh(z) + \sigma_y \sinh(z)$$
$$\exp(z\sigma_z) = I_2 \cosh(z) + \sigma_z \sinh(z).$$

(ii) We have

$$\exp(-z\sigma_x)\sigma_y\exp(z\sigma_x) = \sigma_y \cosh(2z) - i\sigma_z \sinh(2z)$$
$$\exp(-z\sigma_y)\sigma_z\exp(z\sigma_y) = \sigma_z \cosh(2z) - i\sigma_x \sinh(2z)$$
$$\exp(-z\sigma_z)\sigma_x\exp(z\sigma_z) = \sigma_x \cosh(2z) - i\sigma_y \sinh(2z).$$

Problem 43. The Pauli spin matrices are elements of the Lie group $U(2)$. Additionally they are hermitian. Let $\otimes$ be the Kronecker product. Then $\sigma_x \otimes \sigma_x$, $\sigma_y \otimes \sigma_y$, $\sigma_z \otimes \sigma_z$ are elements of the Lie group $U(4)$. The matrices are also hermitian. Let $z \in \mathbb{C}$. Find

$$\exp(z\sigma_x \otimes \sigma_x), \quad \exp(z\sigma_y \otimes \sigma_y), \quad \exp(z\sigma_z \otimes \sigma_z).$$

Solution 43. Since $(\sigma_x \otimes \sigma_x)^2 = I_2 \otimes I_2$, $(\sigma_y \otimes \sigma_y)^2 = I_2 \otimes I_2$, $(\sigma_z \otimes \sigma_z)^2 = I_2 \otimes I_2$ we find

$$\exp(z\sigma_x \otimes \sigma_x) = (I_2 \otimes I_2)\cosh(z) + (\sigma_x \otimes \sigma_x)\sinh(z)$$
$$\exp(z\sigma_y \otimes \sigma_y) = (I_2 \otimes I_2)\cosh(z) + (\sigma_y \otimes \sigma_y)\sinh(z)$$
$$\exp(z\sigma_z \otimes \sigma_z) = (I_2 \otimes I_2)\cosh(z) + (\sigma_z \otimes \sigma_z)\sinh(z).$$

Consider also the exponential functions of the terms $(\sigma_x \oplus \sigma_x)$, $(\sigma_y \oplus \sigma_y)$, $(\sigma_z \oplus \sigma_z)$, where $\oplus$ is the direct sum. These terms are also elements of the Lie group $U(4)$.

Problem 44.   Consider the Lie group $U(n)$ of the unitary $n \times n$ matrices. Let $\mathbb{T}^n$ be the n-torus of diagonal unitary $n \times n$ matrices. Obviously $\mathbb{T}^n$ is a subgroup of $U(n)$. Describe the homogeneous space $U(n)/\mathbb{T}^n$.

Solution 44.   The homogeneous space $U(n)/\mathbb{T}^n$ is the subset of $U(n)$, which contains only matrices with real diagonal elements. This subset contains the group unity.

Problem 45.   (i) Let G be a semisimple Lie group and H a maximal compact Lie subgroup of G. Give an example for G and H.
(ii) Find the maximal compact subgroup K of $SL(2,\mathbb{C})$.

Solution 45.   (i) Let $G = SL(2,\mathbb{R})$. Then $H = SO(2,\mathbb{R})$ is a maximal compact Lie subgroup.
(ii) We obtain $K = SU(2)$. Obviously $SU(2)$ is a subset of $SL(2,\mathbb{C})$.

Problem 46.   Let A, B be $n \times n$ matrices over $\mathbb{C}$. If $AB = BA$, then $\exp(A+B) = \exp(A)\exp(B)$. Let $t \in \mathbb{R}$. Consider the 2×2 matrices

$$C = \begin{pmatrix} 0 & 1 \\ 0 & 0 \end{pmatrix}, \qquad D = \begin{pmatrix} 0 & 0 \\ 0 & 1 \end{pmatrix}.$$

Then $\exp(tC)$ is an element of $SL(2,\mathbb{R})$ and $\exp(tD)$ is a diagonal matrix with determinant e^t, i.e. $\exp(tD)$ is an element of $GL(2,\mathbb{R})$.
(i) Show that $CD \neq DC$.
(ii) Calculate

$$e^{tC}e^{tD} \qquad \text{and} \qquad e^{t(C+D)}.$$

Discuss.

Solution 46.   (i) We find $[C, D] = CD - DC = C$.
(ii) Since

$$C + D = \begin{pmatrix} 0 & 1 \\ 0 & 1 \end{pmatrix} \Rightarrow (C+D)^n = C + D, \quad n \in \mathbb{N}$$

we obtain

$$e^{t(C+D)} = \begin{pmatrix} 1 & e^t - 1 \\ 0 & e^t \end{pmatrix}.$$

On the other hand we have

$$e^{tC}e^{tD} = \begin{pmatrix} 1 & t \\ 0 & 1 \end{pmatrix} \begin{pmatrix} 1 & 0 \\ 0 & e^t \end{pmatrix} = \begin{pmatrix} 1 & te^t \\ 0 & e^t \end{pmatrix}.$$

Thus $\exp(t(C+D)) \neq \exp(tC)\exp(tD)$.

Problem 47. Given two 2×2 matrices

$$A = \begin{pmatrix} a_{11} & a_{12} \\ a_{21} & a_{22} \end{pmatrix}, \qquad B = \begin{pmatrix} b_{11} & b_{12} \\ b_{21} & b_{22} \end{pmatrix}$$

in the Lie group $GL(2,\mathbb{C})$, i.e. $\det(A) \neq 0$, $\det(B) \neq 0$. The *Kronecker product* of A and B is given by the 4×4 matrix

$$A \otimes B = \begin{pmatrix} a_{11}B & a_{12}B \\ a_{21}B & a_{22}B \end{pmatrix}.$$

The action of an $n \times n$ matrix $M = (m_{jk})$ on a polynomial $p(\mathbf{x}) \in \mathbf{R}_n = \mathbb{C}[x_1, x_2, \ldots, x_n]$ may be defined by setting

$$T_M p(\mathbf{x}) := p(\mathbf{x}M)$$

where $\mathbf{x}M$ is the multiplication of the row n-vector $\mathbf{x}$ by the $n \times n$ matrix M. Thus $\mathbf{x}M$ is a row vector again. We denote by

$$\mathbf{R}_4^{GL(2,\mathbb{C}) \otimes GL(2,\mathbb{C})}$$

the ring of polynomials in $\mathbf{R}_4$ that are invariant under the action of $A \otimes B$ for all pairs $A, B \in GL(2,\mathbb{C})$, i.e.

$$\mathbf{R}_4^{GL(2,\mathbb{C}) \otimes GL(2,\mathbb{C})} := \{\, p \in \mathbf{R}_4 \,:\, T_{A \otimes B} p(\mathbf{x}) = p(\mathbf{x}) \,\}.$$

The action preserves degree and homogeneity. Consider

$$p(x_1, x_2, x_3, x_4) = x_1^2 + x_2^2 + x_3^2 + x_4^2 + x_1 x_2 + x_2 x_3 + x_3 x_4 + x_4 x_1$$

and

$$A = B = \begin{pmatrix} 0 & 1 \\ 1 & 0 \end{pmatrix}.$$

Show that $T_{A \otimes B} p(\mathbf{x}) = p(\mathbf{x})$.

Solution 47. Since

$$(\, x_1 \;\; x_2 \;\; x_3 \;\; x_4 \,)(A \otimes B) = (\, x_1 \;\; x_2 \;\; x_3 \;\; x_4 \,) \begin{pmatrix} 0 & 0 & 0 & 1 \\ 0 & 0 & 1 & 0 \\ 0 & 1 & 0 & 0 \\ 1 & 0 & 0 & 0 \end{pmatrix}$$

$$= (\, x_4 \;\; x_3 \;\; x_2 \;\; x_1 \,)$$

we have

$$T_{A\otimes B}p(\mathbf{x}) = p(\mathbf{x}(A \otimes B)) = p(x_4, x_3, x_2, x_1) = p(x_1, x_2, x_3, x_4).$$

Problem 48. Let $D \geq 4$. Show that $SO(3) \otimes SO(D-3)$ is a subgroup of $SO(D)$.

Solution 48. Let $A, B \in SO(3)$ and $C, D \in SO(D-3)$. Then

$$(A \otimes C)(B \otimes D) = (AB) \otimes (CD)$$

where $\otimes$ denotes the Kronecker product. Now $AB \in SO(3)$, $CD \in SO(D-3)$ and $(AB) \otimes (CD) \in SO(3) \otimes SO(D-3)$. We have

$$(A \otimes C)^{-1} = A^{-1} \otimes C^{-1}$$

with $A^{-1} \in SO(3)$, $C^{-1} \in SO(D-3)$ and $A^{-1} \otimes C^{-1} \in SO(3) \otimes SO(D-3)$.

Problem 49. Consider the compact Lie groups $H = SU(2)$ and $G = SO(3)$. Show that we have a *homeomorphism* $\psi : SU(2) \to SO(3)$.

Solution 49. The Lie group H acts on the three-dimensional space V of hermitian matrices

$$A = \begin{pmatrix} x & y+iz \\ y-iz & -x \end{pmatrix}, \qquad x, y, z \in \mathbb{R}$$

of trace 0 by $S : A \to SAS^{-1} \equiv SAS^*$ $(S \in H)$ and

$$A \mapsto -\det(A) = x^2 + y^2 + z^2$$

is an invariant positive definite quadratic form on V invariant under this action. Consequently, the map $A \mapsto SAS^{-1}$ of V is orthogonal and we have a homeomorphism $\psi : SU(2) \to SO(3)$. Both groups are three-dimensional and ψ is a local homeomorphism at the identity.

Problem 50. The circle

$$S^1 := \{ (x_1, x_2) : x_1^2 + x_2^2 = 1 \}$$

is a C^∞-manifold which is a one-dimensional submanifold of the vector space $\mathbb{R}^2$. An atlas is given by

$$\begin{aligned}
U_1 &:= \{ (x_1, x_2) \in S^1 : x_1 > 0 \} & \phi_1(x_1, x_2) &= x_2 \\
U_2 &:= \{ (x_1, x_2) \in S^1 : x_1 < 0 \} & \phi_2(x_1, x_2) &= x_2 \\
U_3 &:= \{ (x_1, x_2) \in S^1 : x_2 > 0 \} & \phi_3(x_1, x_2) &= x_1 \\
U_4 &:= \{ (x_1, x_2) \in S^1 : x_2 < 0 \} & \phi_4(x_1, x_2) &= x_1.
\end{aligned}$$

Show that the overlap functions are differentiable.

Solution 50.   Note that $U_1 \cap U_2 = \emptyset$ and $U_3 \cap U_4 = \emptyset$. Consider as an example the overlap of U_1 and U_3. In $U_1 \cap U_3$ we have

$$x_2 = \sqrt{1 - x_1^2} \quad \text{with} \quad 0 < x_2 < 1 \quad \text{and} \quad 0 < x_1 < 1.$$

Therefore

$$\phi_3(x_1)^{-1} = \left(x_1, \sqrt{1 - x_1^2}\right).$$

It follows that

$$\phi_1 \circ \phi_3^{-1}(x_1) = \sqrt{1 - x_1^2}$$

which is infinitely differentiable on this range of values for x_1 and x_2.

Problem 51.   Consider the Lie group $GL(n, \mathbb{R})$. The connected component of $GL(n, \mathbb{R})$ is defined by

$$GL^+(n, \mathbb{R}) := \{\, A \in GL(n, \mathbb{R}) \, : \, \det(A) > 0 \,\}$$

and is itself a differentiable manifold of dimension n^2. Show that $GL^+(n, \mathbb{R})$ is a subgroup of $GL(n, \mathbb{R})$.

Solution 51.   Obviously $I_n \in GL^+(n, \mathbb{R})$ with $\det(I_n) = 1$. Since

$$\det(AB) = \det(A)\det(B)$$

we find that $A, B \in GL^+(n, \mathbb{R})$ implies that $AB \in GL^+(n, \mathbb{R})$. Since

$$\det(A^{-1}) = (\det(A))^{-1}$$

we find that $A \in GL^+(n, \mathbb{R})$ implies that $A^{-1} \in GL^+(n, \mathbb{R})$. These three conditions imply that $GL^+(n, \mathbb{R})$ is a subgroup of $GL(n, \mathbb{R})$. $GL^+(n, \mathbb{R})$ is also a submanifold of $GL(n, \mathbb{R})$.

Problem 52.   We define

$$O(n, \mathbb{R}) := \{\, A \in GL(n, \mathbb{R}) \, : \, AA^T = I_n \,\}$$

where ^{T} stands for transpose. This is the real orthogonal group. Show that $O(n, \mathbb{R})$ is a subgroup of $GL(n, \mathbb{R})$. Give the condition $AA^T = I_n$.

Solution 52.   Since $I_n I_n^T = I_n$ the identity element belongs to $O(n, \mathbb{R})$. If $AA^T = I_n$ and $BB^T = I_n$, then

$$(AB)(AB)^T = ABB^T A^T = AA^T = I_n.$$

Therefore $O(n, \mathbb{R})$ is closed under matrix multiplication. If $AA^T = I_n$ then $A^{-1} = A^T$. Consequently

$$A^{-1}(A^{-1})^T = A^T A = A^{-1} A = I_n \,.$$

Therefore $A \in O(n, \mathbb{R})$ implies that $A^{-1} \in O(n, \mathbb{R})$. The constraint $AA^T = I_n$ can be written as

$$\sum_{k=1}^{n} a_{ik} a_{jk} = \delta_{ij}, \quad i, j = 1, \ldots, n \,.$$

This imposes $(n^2 + n)/2$ polynomial constraints on the matrices regarded as elements of the vector space of all $n \times n$ matrices $M(n, \mathbb{R})$. We have $\dim(GL(n, \mathbb{R})) = n^2$. Then $O(n, \mathbb{R})$ is a submanifold of $GL(n, \mathbb{R})$ of dimension

$$n^2 - \frac{1}{2}(n^2 + n) \equiv \frac{1}{2} n(n - 1) \,.$$

The constraint also implies that

$$\sum_{i=1}^{n} \sum_{j=1}^{n} (a_{ij})^2 = n \,.$$

Problem 53. Consider the Lie group $O(n, \mathbb{R})$. Show that the n-sphere

$$S^n := \{ (x_1, \ldots, x_n, x_{n+1}) : x_1^2 + \cdots + x_{n+1}^2 = 1 \}$$

is diffeomorphic to $O(n + 1, \mathbb{R})/O(n, \mathbb{R})$.

Solution 53. Consider the real vector space $\mathbb{R}^{n+1}$. Let $\mathbf{v}, \mathbf{w} \in \mathbb{R}^{n+1}$ with the scalar product defined by

$$\langle \mathbf{v}, \mathbf{w} \rangle := \sum_{j=1}^{n+1} v_j w_j$$

where v_j $(j = 1, \ldots, n+1)$ are the components of the vector $\mathbf{v}$ with respect to the standard basis. The linear action of any element of $O(n + 1, \mathbb{R})$ on this vector space leaves the scalar product invariant. Consequently the vectors of length one which are elements of S^n are left invariant. Next we show that the group acts transitively. Let $\mathbf{w} := (1, 0, \ldots, 0) \in S^n \subset \mathbb{R}^{n+1}$. Let $\mathbf{v}$ be any other unit length vector. Choose a set of orthonormal vectors $\{ \mathbf{e}_1, \ldots, \mathbf{e}_n \}$ such that the complete set $\{ \mathbf{v}, \mathbf{e}_1, \ldots, \mathbf{e}_n \}$ is an orthonormal basis in the vector space $\mathbb{R}^{n+1}$. Then we have the matrix equation $\mathbf{v} = A\mathbf{w}$

$$
\begin{pmatrix} v_1 \\ v_2 \\ \vdots \\ v_{n+1} \end{pmatrix} = \begin{pmatrix} v_1 & e_{1,1} & e_{2,1} & \cdots & e_{n,1} \\ v_2 & e_{1,2} & e_{2,2} & \cdots & e_{n,2} \\ \vdots & \vdots & \vdots & & \vdots \\ v_{n+1} & e_{1,n+1} & e_{2,n+1} & \cdots & e_{n,n+1} \end{pmatrix} \begin{pmatrix} 1 \\ 0 \\ \vdots \\ 0 \end{pmatrix}.
$$

The choice of the unit vectors $\{\, e_1, \ldots, e_n \,\}$ is such that the $(n+1) \times (n+1)$ matrix A satisfies

$$
A^T A = A A^T = I_{n+1} .
$$

Therefore $A \in O(n+1, \mathbb{R})$. Since the vector $\mathbf{v}$ is any unit vector in the n-sphere we find that the action of $O(n+1, \mathbb{R})$ is transitive on the vector space $\mathbb{R}^{n+1}$. The isotropy group at the vector $\mathbf{w}$ are all $(n+1) \times (n+1)$ matrices of the form $(1) \oplus B$, where $\oplus$ denotes the direct sum. The $n \times n$ matrix B is any such matrix satisfying $BB^T = B^T B = I_n$. Thus the isotropy group is this particular $O(n, \mathbb{R})$ subgroup.

Problem 54. Consider the linear action of $SU(n+1)$ on the vector space $\mathbb{C}^{n+1}$. This maps complex lines into complex lines and therefore passes to an action on the complex projective space $\mathbb{C}P^n$ which is defined to be the vector space of all such lines in $\mathbb{C}^{n+1}$. Let $n = 1$, i.e. we consider $SU(2)$. Find an element A in $SU(2)$ such that

$$
A \frac{1}{\sqrt{2}} \begin{pmatrix} i \\ -i \end{pmatrix} = \frac{1}{\sqrt{2}} \begin{pmatrix} 1 \\ 1 \end{pmatrix} .
$$

Solution 54. An element is

$$
A = \begin{pmatrix} -i & 0 \\ 0 & i \end{pmatrix} .
$$

Are there other elements? Note that

$$
A = \begin{pmatrix} 0 & i \\ -i & 0 \end{pmatrix}
$$

also satisfies the equation, but is not an element of $SU(2)$ ($\det(A) = -1$). It is an element of the Lie group $U(2)$.

Problem 55. Consider the 2×2 matrix

$$
S = \begin{pmatrix} 1 & 0 \\ 0 & -1 \end{pmatrix}
$$

(parity reversal) where $S \in O(2)$ but $S \notin SO(2)$. Show that every $A \in O(2)$, $A \notin SO(2)$ can be written as $A = \widetilde{A}S$ with $\widetilde{A} \in SO(2)$.

Solution 55. All elements Q of the Lie group $O(2)$ satisfy $QQ^T = I_2$. It follows that

$$(\det(Q))^2 = 1.$$

This gives for $\det(Q)$ two possibilities: $\det(Q) = 1$ and $\det(Q) = -1$. Elements of $SO(2)$ satisfy the first possibility, all another elements $A \in O(2)$, $A \notin SO(2)$ satisfy the second. Consider the determinant of the expression $A = \widetilde{A}S$ with $\widetilde{A} \in SO(2)$

$$\det(A) = \det(\widetilde{A})\det(S) = -1.$$

Thus $A \in O(2), A \notin SO(2)$. For every A we can get $\widetilde{A} = AS^{-1} = AS$. In the same way we calculate the determinants and obtain $\widetilde{A} \in SO(2)$. Thus every element $A \in O(2)$, $A \notin SO(2)$ can be represented in the form $A = \widetilde{A}S$.

Problem 56. Let

$$A = \begin{pmatrix} a_{11} & a_{12} \\ a_{21} & a_{22} \end{pmatrix} \in GL(2, \mathbb{R})$$

i.e. $\det(A) = a_{11}a_{22} - a_{12}a_{21} \neq 0$. Find the conditions on the entries a_{jk} such that

$$\begin{pmatrix} 0 & 1 \\ 1 & 0 \end{pmatrix} A \begin{pmatrix} 0 & 1 \\ 1 & 0 \end{pmatrix} = A.$$

Do these matrices form a subgroup of $GL(2, \mathbb{R})$ under matrix multiplication?

Solution 56. The condition yields $a_{11} = a_{22}$, $a_{12} = a_{21}$. Thus we have the matrices

$$\widetilde{A} = \begin{pmatrix} a_{11} & a_{12} \\ a_{12} & a_{11} \end{pmatrix}.$$

Since $a_{11}a_{22} - a_{12}a_{21} \neq 0$ we also have $a_{11}^2 - a_{12}^2 \neq 0$. The neutral element is $a_{11} = 1$, $a_{12} = 0$. Now

$$\begin{pmatrix} a_{11} & a_{12} \\ a_{12} & a_{22} \end{pmatrix} \begin{pmatrix} b_{11} & b_{12} \\ b_{12} & b_{22} \end{pmatrix} = \begin{pmatrix} a_{11}b_{11} + a_{12}b_{12} & a_{11}b_{12} + a_{12}b_{11} \\ a_{11}b_{12} + a_{12}b_{11} & a_{11}b_{11} + a_{12}b_{12} \end{pmatrix}.$$

Thus the set is closed under matrix multiplication. The inverse of $\widetilde{A}$ is

$$\widetilde{A}^{-1} = \frac{1}{a_{11}^2 - a_{12}^2} \begin{pmatrix} a_{11} & -a_{12} \\ -a_{12} & a_{11} \end{pmatrix}.$$

Thus these matrices form a subgroup under matrix multiplication. Is the subgroup commutative?

Problem 57. The Lie groups $SO(3)$ and $SU(2)$ have *isomorphic* Lie algebras $so(3)$ and $su(2)$, respectively. A basis for the Lie algebra $su(2)$ is given by

$$\tau_1 = \frac{1}{2}\begin{pmatrix} 0 & -i \\ -i & 0 \end{pmatrix}, \quad \tau_2 = \frac{1}{2}\begin{pmatrix} 0 & -1 \\ 1 & 0 \end{pmatrix}, \quad \tau_3 = \frac{1}{2}\begin{pmatrix} -i & 0 \\ 0 & i \end{pmatrix}$$

with $[\tau_1, \tau_2] = \tau_3$, $[\tau_2, \tau_3] = \tau_1$, $[\tau_3, \tau_1] = \tau_2$. Construct a homomorphism

$$\varphi : SU(2) \to SO(3).$$

Solution 57. We start with an auxiliary function

$$f : \mathbb{R}^3 \to su(2)$$

$$\mathbf{x} = \begin{pmatrix} x_1 \\ x_2 \\ x_3 \end{pmatrix} \mapsto \sum_{j=1}^{3} x_j \tau_j = -\frac{1}{2}i\begin{pmatrix} x_3 & x_1 - ix_2 \\ x_1 + ix_2 & -x_3 \end{pmatrix}.$$

Let $g \in SU(2)$. We define $\varphi(g)$ by

$$\varphi(g)\mathbf{x} := f^{-1}(gf(\mathbf{x})g^{-1}).$$

For all $\mathbf{x} \in \mathbb{R}^3$ the 2×2 matrix $gf(\mathbf{x})g^{-1}$ is in the Lie algebra $su(2)$. Thus $\varphi(g)$ is a well-defined linear map from the vector space $\mathbb{R}^3$ into itself. To show that this is a rotation we utilize

$$\det(f(\mathbf{x})) = \frac{1}{4}(x_1^2 + x_2^2 + x_3^2) = \frac{1}{4}|\mathbf{x}|^2.$$

Now $\varphi(g)$ preserves the length

$$|\varphi(g)\mathbf{x}|^2 = 4\det(gf(\mathbf{x})g^{-1}) = 4\det(f(\mathbf{x})) = |\mathbf{x}|^2.$$

Now $\varphi(g)$ is orientation-preserving and hence an element of the Lie group $SO(3)$. The linear map φ is a surjective group homomorphism. The map φ is not injective. To find the kernel we have to find g such that $\varphi(g)\mathbf{x} = \mathbf{x}$ for all vectors $x \in \mathbb{R}^3$. It follows that

$$gf(\mathbf{x})g^{-1} = f(\mathbf{x})$$

or $g\tau = \tau g$ for all $\tau \in su(2)$. Thus $g \in SU(2)$ must be a multiple of the 2×2 identity matrix and $\ker\varphi = \{I_2, -I_2\}$. Hence φ is two-to-one. We have

$$\varphi(g) = \varphi(-g).$$

Note that the manifold of Lie group $SU(2)$ is S^3 which is simply connected. The Lie group $SO(3)$ is connected, but not simply connected. Its manifold is the three-dimensional projective space.

Problem 58. Let $S \in SL(2, \mathbb{C})$ and

$$E = \begin{pmatrix} 0 & 1 \\ -1 & 0 \end{pmatrix}.$$

Obviously, $E \in SL(2, \mathbb{C})$.
(i) Find SES^T, where ^{T} denotes the transpose.
(ii) Let $A, B \in SL(2, \mathbb{C})$. Find $(A \otimes B)(E \otimes E)(A \otimes B)^T$, where $\otimes$ denotes the Kronecker product.

Solution 58. (i) Let

$$S = \begin{pmatrix} a & b \\ c & d \end{pmatrix}$$

with $a, b, c, d \in \mathbb{C}$ and $ad - bc = 1$. Then

$$SES^T = \begin{pmatrix} a & b \\ c & d \end{pmatrix} \begin{pmatrix} 0 & 1 \\ -1 & 0 \end{pmatrix} \begin{pmatrix} a & c \\ b & d \end{pmatrix} = \begin{pmatrix} 0 & 1 \\ -1 & 0 \end{pmatrix} = E.$$

(ii) We have

$$(A \otimes B)(E \otimes E)(A \otimes B)^T = (A \otimes B)(E \otimes E)(A^T \otimes B^T)$$
$$= (AEA^T) \otimes (BEB^T)$$
$$= E \otimes E$$

where we used the result from (i).

Problem 59. A *quandle*, X, is a set with a binary operation $(a, b) \mapsto a * b$ such that
(I) For any $a \in X$, $a * a = a$.
(II) For any $a, b \in X$, there is a unique $c \in X$ such that $a = c * b$.
(III) For any $a, b, c \in X$, we have $(a * b) * c = (a * c) * (b * c)$.

Examples of quandles are: (1) A group $X = G$ with conjugation as the quandle operation $a * b := bab^{-1}$. We denote by $\text{Conj}(G)$ the quandle defined for a group G by $a * b = bab^{-1}$. Any subset that is closed under such conjugation is also a quandle. (2) Any $\Lambda(= \mathbb{Z}[t, t^{-1}])$-module M is a quandle with $a * b := ta + (1-t)b$, where $a, b \in M$, that is called an *Alexander quandle*. (3) Let n be a positive integer, and for elements $i, j \in \{0, 1, \ldots, n-1\}$, define

$$i * j := 2j - i \pmod{n}.$$

Then $*$ defines a quandle structure called the *dihedral quandle*.
Let $A, B \in SL(2, \mathbb{R})$ with

$$A = \begin{pmatrix} 1 & 0 \\ 1 & 1 \end{pmatrix}, \qquad B = \begin{pmatrix} 1 & 1 \\ 0 & 1 \end{pmatrix}.$$

Calculate $A * B$ and $B * A$. Is $A * B = B * A$?

Solution 59. Since

$$A^{-1} = \begin{pmatrix} 1 & 0 \\ -1 & 1 \end{pmatrix}, \qquad B^{-1} = \begin{pmatrix} 1 & -1 \\ 0 & 1 \end{pmatrix}$$

we obtain

$$A * B = BAB^{-1} = \begin{pmatrix} 2 & -1 \\ 1 & 0 \end{pmatrix}, \qquad B * A = ABA^{-1} = \begin{pmatrix} 0 & 1 \\ -1 & 2 \end{pmatrix}.$$

Thus we see that $A * B \neq B * A$. However we can find a permutation matrix such that $P(A * B)P^{-1} = B * A$, namely

$$\begin{pmatrix} 0 & 1 \\ 1 & 0 \end{pmatrix} \begin{pmatrix} 2 & -1 \\ 1 & 0 \end{pmatrix} \begin{pmatrix} 0 & 1 \\ 1 & 0 \end{pmatrix} = \begin{pmatrix} 0 & 1 \\ -1 & 2 \end{pmatrix}.$$

Problem 60. Let $GL(1, \mathbb{C})$ be the multiplicative group of nonzero complex numbers also denoted by $\mathbb{C}^*$. Show that $GL(1, \mathbb{C})$ is a Lie group.

Solution 60. Topologically $\mathbb{C}^*$ is identical to $\mathbb{R}^2 \setminus \{ (0, 0) \}$. The coordinates of its elements $x + iy$ are the (x, y). Thus $\mathbb{C}^*$ is a two-dimensional smooth manifold. Suppose that $z_k = x_k + iy_k$, $k = 1, 2$, $z_k \neq 0$. Then the product of the z_j in terms of coordinates is

$$(x_1, y_1) \cdot (x_2, y_2) = (x_1 x_2 - y_1 y_2, x_1 y_2 + x_2 y_1)$$

and the inverse of $z = x + iy$ is

$$(x, y)^{-1} = \left(\frac{x}{x^2 + y^2}, -\frac{y}{x^2 + y^2} \right).$$

Now the multiplication and inverse maps are both smooth. Thus $\mathbb{C}^*$ is a two-dimensional Lie group.

Problem 61. $GL(n, \mathbb{R})$ is the set of nondegenerate (determinant nonzero) $n \times n$ matrices with matrix multiplication for its group operation. Show that $GL(n, \mathbb{R})$ is a Lie group.

Solution 61.   Since $GL(n, \mathbb{R})$ is an open subset of the vector space $\mathbb{R}^{n^2}$ it has the differentiable structure induced from $\mathbb{R}^{n^2}$. Now let

$$A, B \in GL(n, \mathbb{R})$$

with $A = (a_{ij})$, $B = (b_{ij})$ and $i, j = 1, 2, \ldots, n$. Then the product of A and B is given by

$$(AB)_{ij} = \sum_{k=1}^{n} a_{ik} b_{kj} .$$

The right-hand side is a polynomial of the elements of the $n \times n$ matrices A and B. Therefore the map

$$\varphi(A, B) = AB$$

is smooth. Moreover since the elements of the inverse matrix A^{-1} of A are rational functions of the elements a_{ij} the inverse must also be smooth. Thus $GL(n, \mathbb{R})$ is a Lie group.

Problem 62.   Let $V \in M(2n, \mathbb{R})$, $V^T = V$ and $V > 0$. Show that there exists $S \in Sp(2n, \mathbb{R})$ and $D \in M(n, \mathbb{R})$ diagonal and positive definite such that

$$V = S^T \begin{pmatrix} D & 0_n \\ 0_n & D \end{pmatrix} S \tag{1}$$

where 0_n is the $n \times n$ zero matrix. The matrices S and D are unique up to a permutation of the elements of D.

Solution 62.   Let $\oplus$ be the direct sum. Equation (1) implies that

$$S = (D \oplus D)^{-1/2} O V^{-1/2}$$

where O is an $2n \times 2n$ orthogonal matrix. Requiring symplecticity of the matrix S means that

$$O V^{-1/2} J V^{-1/2} O^T = \begin{pmatrix} 0_n & D^{-1} \\ -D^{-1} & 0_n \end{pmatrix} \tag{2}$$

where

$$J = J_{2n} = \begin{pmatrix} 0_n & I_n \\ -I_n & 0_n \end{pmatrix}$$

with I_n the $n \times n$ identity matrix. Since V is symmetric and J is anti-symmetric it follows that the matrix $V^{-1/2} J V^{-1/2}$ are antisymmetric. Thus there exist a unique orthogonal matrix O such that equation (2) holds.

Problem 63. The n-sphere defined by

$$S^n := \{\, \mathbf{x} = (x_1, \ldots, x_{n+1}) \in \mathbb{R}^{n+1} \; : \; x_1^2 + \cdots + x_{n+1}^2 = 1 \,\}$$

is a manifold of dimension n. Find two different atlases.

Solution 63. One possible atlas is

$$\mathcal{A}_1 = \{\, (S^n \setminus (0, \ldots, 0, 1), \varphi) \,,\, (S^n \setminus \{(0, \ldots, 0, -1)\}, \psi) \,\}$$

where the map φ is given by

$$\varphi(x_1, \ldots, x_{n+1}) = \left(\frac{2x_1}{1 - x_{n+1}}, \ldots, \frac{2x_n}{1 - x_{n+1}} \right)$$

and the map ψ is given by

$$\psi(x_1, \ldots, x_{n+1}) = \left(\frac{2x_1}{1 + x_{n+1}}, \ldots, \frac{2x_n}{1 + x_{n+1}} \right).$$

The maps are the *stereographic projections* from the north and south pole, respectively. Another atlas is

$$\mathcal{A}_2 = \{\, (U_j, \varphi_j), (V_j, \psi_j) \; : \; 1 \le j \le n+1 \,\}$$

where, for each $1 \le j \le n+1$

$$U_j := \{\, (x_1, \ldots, x_{n+1}) \in S^n \; : \; x_j > 0 \,\}$$

with

$$\varphi_j(x_1, \ldots, x_{n+1}) = (x_1, \ldots, x_{j-1}, x_{j+1}, \ldots, x_{n+1})$$

and

$$V_j := \{\, (x_1, \ldots, x_{n+1}) \in S^n \; : \; x_j < 0 \,\}$$

with

$$\psi_j(x_1, \ldots, x_{n+1}) = (x_1, \ldots, x_{j-1}, x_{j+1}, \ldots, x_{n+1}).$$

Thus both φ_j and ψ_j project onto $\mathbb{R}^n$, which can be considered as the hyperplane $x_j = 0$ in $\mathbb{R}^{n+1}$.

Problem 64. Consider the manifold $\mathbb{R}^2$. Consider the analytic function $\mathbf{f} : \mathbb{R}^2 \to \mathbb{R}^2$

$$\mathbf{f}(\mathbf{x}) = \begin{pmatrix} e^{x_1} \cos x_2 \\ e^{x_1} \sin x_2 \end{pmatrix}.$$

Show that $\mathbf{f}$ satisfies the conditions of the *inverse function theorem* on $\mathbb{R}^2$, but the function is not $1-1$ on $\mathbb{R}^2$.

Solution 64.   The *Jacobian matrix* is given by

$$J(\mathbf{f})(\mathbf{x}) = \begin{pmatrix} \partial f_1/\partial x_1 & \partial f_1/\partial x_2 \\ \partial f_2/\partial x_1 & \partial f_2/\partial x_2 \end{pmatrix} = \begin{pmatrix} e^{x_1}\cos x_2 & -e^{x_1}\sin x_2 \\ e^{x_1}\sin x_2 & e^{x_1}\cos x_2 \end{pmatrix}.$$

Thus the Jacobian determinant is given by

$$\det(J(\mathbf{f})(\mathbf{x})) = e^{2x_1} \neq 0$$

for all $\mathbf{x}$. However, since $\mathbf{f}(x_1, x_2 + 2\pi) = \mathbf{f}(x_1, x_2)$ the function $\mathbf{f}$ is not $1-1$ on $\mathbb{R}^2$.

Problem 65.   Let $\alpha, \beta, \gamma \in \mathbb{R}$. Consider the invertible 3×3 matrices

$$S_1(\alpha) = \begin{pmatrix} \cosh\alpha & 0 & \sinh\alpha \\ 0 & 1 & 0 \\ \sinh\alpha & 0 & \cosh\alpha \end{pmatrix}, \quad S_2(\beta) = \begin{pmatrix} 1 & 0 & 0 \\ 0 & \cosh\beta & \sinh\beta \\ 0 & \sinh\beta & \cosh\beta \end{pmatrix}$$

$$R(\gamma) = \begin{pmatrix} \cos\gamma & -\sin\gamma & 0 \\ \sin\gamma & \cos\gamma & 0 \\ 0 & 0 & 1 \end{pmatrix}.$$

Find matrices K_1, K_2, L such that

$$S_1 = \exp(-i\alpha K_1), \quad S_2 = \exp(-i\beta K_2), \quad R = \exp(-i\gamma L).$$

Solution 65.   We obtain

$$K_1 = \begin{pmatrix} 0 & 0 & i \\ 0 & 0 & 0 \\ i & 0 & 0 \end{pmatrix}, \quad K_2 = \begin{pmatrix} 0 & 0 & 0 \\ 0 & 0 & i \\ 0 & i & 0 \end{pmatrix}$$

and

$$L = \begin{pmatrix} 0 & -i & 0 \\ i & 0 & 0 \\ 0 & 0 & 0 \end{pmatrix}.$$

The commutators are $[K_1, K_2] = -iL$, $[K_1, L] = -iK_2$, $[K_2, L] = iK_1$.

Problem 66.   Consider the rotation matrix around the z-axis

$$A(\phi) = \begin{pmatrix} \cos\phi & -\sin\phi & 0 \\ \sin\phi & \cos\phi & 0 \\ 0 & 0 & 1 \end{pmatrix}.$$

Let

$$G = \begin{pmatrix} 1 & 0 & 0 \\ 0 & 0 & -1 \\ 0 & 1 & 0 \end{pmatrix}.$$

Find $GA(\phi)G^{-1}$. Discuss.

Solution 66. Since

$$G^{-1} = \begin{pmatrix} 1 & 0 & 0 \\ 0 & 0 & 1 \\ 0 & -1 & 0 \end{pmatrix}$$

we have

$$GA(\phi)G^{-1} = \begin{pmatrix} \cos\phi & 0 & -\sin\phi \\ 0 & 1 & 0 \\ \sin\phi & 0 & \cos\phi \end{pmatrix}.$$

This is a matrix for the rotation around the y-axis.

Problem 67. The Lie group $U(2,2)$ consists of all linear transformations of $\mathbb{C}^4$ which preserves a non-degenerate hermitian form with two positive and two negative eigenvalues. For a suitable choice of basis in $\mathbb{C}^4$ we can characterize a $g \in U(2,2)$ by

$$g^* Jg = J, \qquad J = \begin{pmatrix} 0 & 0 & -1 & 0 \\ 0 & 0 & 0 & -1 \\ -1 & 0 & 0 & 0 \\ 0 & -1 & 0 & 0 \end{pmatrix} \equiv \begin{pmatrix} 0 & 1 \\ 1 & 0 \end{pmatrix} \otimes \begin{pmatrix} -1 & 0 \\ 0 & -1 \end{pmatrix}.$$

If we write the 4×4 matrix $g \in U(2,2)$ in a 2×2 block matrix form we have

$$g = \begin{pmatrix} A & B \\ C & D \end{pmatrix}$$

where

$$B^*D + D^*B = 0_2, \quad A^*C + C^*A = 0_2, \quad B^*D + D^*B = 0_2, \quad A^*D + C^*B = I_2.$$

(i) Is

$$U = \begin{pmatrix} 0 & 1 \\ 1 & 0 \end{pmatrix} \otimes \begin{pmatrix} 0 & 1 \\ 1 & 0 \end{pmatrix}$$

an element of $U(2,2)$?
(ii) Is

$$\begin{pmatrix} \cos(\alpha/2) & i\sin(\alpha/2) \\ i\sin(\alpha/2) & \cos(\alpha/2) \end{pmatrix} \otimes \begin{pmatrix} \cos(\beta/2) & i\sin(\beta/2) \\ i\sin(\beta/2) & \cos(\beta/2) \end{pmatrix}$$

an element of $U(2,2)$?

Solution 67. (i) We start from relation $g^* Jg = J$ with

$$g = \begin{pmatrix} A & B \\ C & D \end{pmatrix}.$$

Then we find

$$g^* J g = \begin{pmatrix} -A^*C - C^*A & -A^*D - C^*B \\ -B^*C - D^*A & -B^*D - D^*B \end{pmatrix}.$$

From

$$g^* J g = J = \begin{pmatrix} 0_2 & -I_2 \\ -I_2 & 0_2 \end{pmatrix}$$

we obtain the conditions

$$A^*C + C^*A = 0_2, \quad A^*D + C^*B = I_2, \quad B^*C + D^*A = I_2, \quad B^*D + D^*B = 0_2.$$

For the matrix U we have

$$A = 0_2, \quad B = \begin{pmatrix} 0 & 1 \\ 1 & 0 \end{pmatrix}, \quad C = \begin{pmatrix} 0 & 1 \\ 1 & 0 \end{pmatrix}, \quad D = 0_2.$$

Therefore $A^*C + C^*A = 0_2$, $B^*D + D^*B = 0_2$, $BC = CB = B^*C = C^*B = I_2$. Thus the matrix U is an element of the Lie group $U(2,2)$.

(ii) For the matrix

$$\begin{pmatrix} \cos(\alpha/2) & i\sin(\alpha/2) \\ i\sin(\alpha/2) & \cos(\alpha/2) \end{pmatrix} \otimes \begin{pmatrix} \cos(\beta/2) & i\sin(\beta/2) \\ i\sin(\beta/2) & \cos(\beta/2) \end{pmatrix}$$

we have

$$A = \cos(\alpha/2) \begin{pmatrix} \cos(\beta/2) & i\sin(\beta/2) \\ i\sin(\beta/2) & \cos(\beta/2) \end{pmatrix}$$

$$B = i\sin(\alpha/2) \begin{pmatrix} \cos(\beta/2) & i\sin(\beta/2) \\ i\sin(\beta/2) & \cos(\beta/2) \end{pmatrix}$$

$$C = i\sin(\alpha/2) \begin{pmatrix} \cos(\beta/2) & i\sin(\beta/2) \\ i\sin(\beta/2) & \cos(\beta/2) \end{pmatrix}$$

$$D = \cos(\alpha/2) \begin{pmatrix} \cos(\beta/2) & i\sin(\beta/2) \\ i\sin(\beta/2) & \cos(\beta/2) \end{pmatrix}.$$

For the following calculations we utilize that

$$\begin{pmatrix} \cos(\beta/2) & -i\sin(\beta/2) \\ -i\sin(\beta/2) & \cos(\beta/2) \end{pmatrix} \begin{pmatrix} \cos(\beta/2) & i\sin(\beta/2) \\ i\sin(\beta/2) & \cos(\beta/2) \end{pmatrix} = I_2.$$

Then

$$A^*C + C^*A = 0_2, \quad A^*D + C^*B = I_2, \quad B^*C + D^*A = I_2, \quad B^*D + D^*B = 0_2.$$

Thus the matrix is an element of the Lie group $U(2,2)$.

Problem 68. Let $0 \leq \tau < \infty$ and S be positive real number. Consider the vector

$$\mathbf{S}_\tau := (\tau/(1+\tau), i\tau/(1+\tau), 1/(1+\tau))S.$$

For $\tau = 0$ we have $\mathbf{S}_0 = (0, 0, S)$ and for $\tau \to \infty$ we have $\mathbf{S}_\infty = (S, iS, 0)$. Let σ_1, σ_2, σ_3 be the Pauli spin matrices. We define the 2×2 matrix

$$\hat{S}_\tau := S_{1\tau}\sigma_1 + S_{2\tau}\sigma_2 + S_{3\tau}\sigma_3.$$

(i) Calculate the matrix $(S_{1\tau}\sigma_1)^2 + (S_{2\tau}\sigma_2)^2 + (S_{3\tau}\sigma_3)^2$.
(ii) Let $\sigma_+ := \sigma_1 + i\sigma_2$. Let $\phi := \phi_1 + i\phi_2$, where ϕ_1, ϕ_2 are real. Calculate

$$H_\tau(\phi) := \exp\left(-i\frac{\phi}{2}(\sigma_3 + \tau\sigma_+)\frac{1}{1+\tau}\right).$$

(iii) Calculate $H_\tau(\phi)\hat{S}_\tau H_\tau^{-1}(\phi)$. Discuss.

Solution 68. (i) We obtain

$$(S_{\tau 1}^2 + S_{\tau 2}^2 + S_{\tau 3}^2)^{1/2} = \frac{S}{1+\tau}I_2.$$

(ii) We obtain the triangular matrix

$$H_\tau(\phi) = \begin{pmatrix} \exp\left(-i\frac{1}{1+\tau}\frac{\phi}{2}\right) & -2i\tau\sin\left(\frac{1}{1+\tau}\frac{\phi}{2}\right) \\ 0 & \exp\left(i\frac{1}{1+\tau}\frac{\phi}{2}\right) \end{pmatrix}.$$

(iii) Straightforward calculation yields $H_\tau(\phi)\hat{S}_\tau H_\tau^{-1}(\phi) = \hat{S}_\tau$. The range of ϕ_1 and ϕ_2 is given by

$$-2\pi(1+\tau) \leq \phi_1 < 2\pi(1+\tau), \qquad -\infty < \phi_2 < \infty.$$

If $\tau = 0$ we have

$$H_0(\phi) = \begin{pmatrix} e^{-i\phi/2} & 0 \\ 0 & e^{i\phi/2} \end{pmatrix}.$$

For $\tau \to \infty$ we have

$$H_\infty(\phi) = \begin{pmatrix} 1 & -i\phi \\ 0 & 1 \end{pmatrix}.$$

Problem 69. Consider the Lie group $SL(2, \mathbb{R})$. Consider the *Jacobi elliptic functions* $\text{sn}(x, k)$, $\text{cn}(x, k)$, $\text{dn}(x, k)$, where $x \in \mathbb{R}$ and $k \in [0, 1]$.
(i) Show that the matrix ($\epsilon \in \mathbb{R}$)

$$\begin{pmatrix} \text{cn}(\epsilon, k) & \text{sn}(\epsilon, k) \\ -\text{sn}(\epsilon, k) & \text{cn}(\epsilon, k) \end{pmatrix}$$

is an element of $SL(2, \mathbb{R})$. Find the inverse of the matrix.
(ii) Study the limiting cases with $k = 0$ and $k = 1$. Are the matrix elements bounded?

Solution 69.   (i) We have the identity $\mathrm{cn}^2(\epsilon, k) + \mathrm{sn}^2(\epsilon, k) = 1$. Thus the matrix is an element of $SL(2, \mathbb{R})$. The inverse matrix is given by

$$\begin{pmatrix} \mathrm{cn}(\epsilon, k) & -\mathrm{sn}(\epsilon, k) \\ \mathrm{sn}(\epsilon, k) & \mathrm{cn}(\epsilon, k) \end{pmatrix}.$$

(ii) For $k = 0$ we obtain the rotation matrix

$$\begin{pmatrix} \cos(\epsilon) & \sin(\epsilon) \\ -\sin(\epsilon) & \cos(\epsilon) \end{pmatrix}.$$

The matrix elements are bounded. We have $-1 \leq \cos(\epsilon) \leq +1$ and $-1 \leq \sin(\epsilon) \leq +1$. For $k = 1$ we find the matrix

$$\begin{pmatrix} \mathrm{sech}(\epsilon) & \tanh(\epsilon) \\ -\tanh(\epsilon) & \mathrm{sech}(\epsilon) \end{pmatrix}$$

where $\mathrm{sech}(\epsilon) := 1/\cosh(\epsilon)$. The matrix elements are bounded. We have $-1 \leq \tanh(\epsilon) \leq +1$ and $0 \leq \mathrm{sech}(\epsilon) \leq +1$.

Problem 70.   Consider the Lie group $SL(2, \mathbb{R})$. Consider the Jacobi elliptic functions $\mathrm{sn}(x, k)$, $\mathrm{cn}(x, k)$, $\mathrm{dn}(x, k)$, where $x \in \mathbb{R}$ and $k \in [0, 1]$.
(i) Show that the matrix ($\epsilon \in \mathbb{R}$)

$$\begin{pmatrix} \mathrm{dn}(\epsilon, k) & k\mathrm{sn}(\epsilon, k) \\ -k\mathrm{sn}(\epsilon, k) & \mathrm{dn}(\epsilon, k) \end{pmatrix}$$

is an element of $SL(2, \mathbb{R})$. Find the inverse of the matrix.
(ii) Study the limiting cases with $k = 0$ and $k = 1$. Are the matrix elements bounded?

Solution 70.   (i) We have the identity $\mathrm{dn}^2(\epsilon, k) + k^2\mathrm{sn}^2(\epsilon, k) = 1$. Thus the matrix is an element of $SL(2, \mathbb{R})$. The inverse matrix is given by

$$\begin{pmatrix} \mathrm{dn}(\epsilon, k) & -k\mathrm{sn}(\epsilon, k) \\ k\mathrm{sn}(\epsilon, k) & \mathrm{dn}(\epsilon, k) \end{pmatrix}.$$

(ii) For $k = 0$ we obtain the identity matrix I_2. Obviously the matrix elements are bounded. For $k = 1$ we find the matrix

$$\begin{pmatrix} \mathrm{sech}(\epsilon) & \tanh(\epsilon) \\ -\tanh(\epsilon) & \mathrm{sech}(\epsilon) \end{pmatrix}.$$

The matrix elements are bounded. We have $-1 \leq \tanh(\epsilon) \leq +1$ and $0 \leq \text{sech}(\epsilon) \leq +1$.

Problem 71. Consider the Lie group $SL(2, \mathbb{R})$ and the matrix $(x \in \mathbb{R})$

$$M(x) = \begin{pmatrix} \sqrt{\pi}\text{Ai}(x) & \sqrt{\pi}\text{Bi}(x) \\ \sqrt{\pi}\text{Ai}'(x) & \sqrt{\pi}\text{Bi}'(x) \end{pmatrix}$$

where Ai and Bi are *Airy functions* which are solutions of the linear second order differential equation $d^2u/dx^2 = xu$ and $'$ denotes the derivative. The Airy functions are defined by the improper integrals

$$\text{Ai}(x) = \frac{1}{\pi} \int_0^\infty \cos\left(\frac{1}{3}t^3 + xt\right) dt$$

$$\text{Bi}(x) = \frac{1}{\pi} \int_0^\infty \left(\exp\left(-\frac{1}{3}t^3 + xt\right) + \sin\left(\frac{1}{3}t^3 + xt\right)\right) dt .$$

The matrix $M(x)$ is an element of $SL(2, \mathbb{R})$ since

$$\text{Ai}(x)\text{Bi}'(x) - \text{Ai}'(x)\text{Bi}(x) = \frac{1}{\pi} .$$

(i) Find the inverse of $M(x)$.
(ii) Find $M(0)$.

Solution 71. (i) The inverse is given by

$$M^{-1}(x) = \begin{pmatrix} \sqrt{\pi}\text{Bi}'(x) & -\sqrt{\pi}\text{Bi}(x) \\ -\sqrt{\pi}\text{Ai}'(x) & \sqrt{\pi}\text{Ai}(x) \end{pmatrix} .$$

(ii) We have

$$\text{Ai}(0) = \frac{1}{3^{2/3}\Gamma(2/3)}, \qquad \text{Ai}'(0) = -\frac{1}{3^{1/3}\Gamma(1/3)}$$

$$\text{Bi}(0) = \frac{1}{3^{1/6}\Gamma(2/3)}, \qquad \text{Bi}'(0) = \frac{3^{1/6}}{\Gamma(1/3)} .$$

For the calculation of $\Gamma(1/3)\Gamma(2/3)$ one uses the identity

$$\Gamma(p)\Gamma(1-p) \equiv \frac{\pi}{\sin(p\pi)} .$$

Find $M(x)$ for $x \to \infty$ using the asymptotic expansion

$$\text{Ai}(x) \sim \frac{e^{-2x^{3/2}/3}}{2\sqrt{\pi}x^{1/4}}, \qquad \text{Bi}(x) \sim \frac{e^{2x^{3/2}/3}}{\sqrt{\pi}x^{1/4}} .$$

Problem 72. Consider the group G of $n \times n$ real matrices under matrix addition. Show that this is an n^2-dimensional Lie group.

Solution 72. The group identity element is the zero matrix. The inverse matrix is $-g$. Thus the inverse element exists for all $n \times n$ matrices. Associative and commutative laws follow from the same laws of matrix addition. Thus the set G of $n \times n$ real matrices forms a commutative n^2-dimensional Lie group under matrix addition. The chart $G \to \mathbb{R}^{n^2}$ is trivial, namely the coordinates are matrix components. One chart is sufficient. Therefore the group manifold is isomorphic to the whole of $\mathbb{R}^{n^2}$.

Problem 73. Let $M(n, \mathbb{R})$ be the vector space of all $n \times n$ matrices over the real number $\mathbb{R}$. Show that the real general linear group

$$GL(n, \mathbb{R}) := \{\, g \in M(n, \mathbb{R}) \,:\, \det(g) \neq 0 \,\}$$

which is the group of all non-singular real $n \times n$ matrices is a Lie group.

Solution 73. The set $GL(n, \mathbb{R})$ can be viewed as a subset of the vector space $\mathbb{R}^{n^2}$. Since the function $\det : \mathbb{R}^{n^2} \to \mathbb{R}$ is continuous and $\mathbb{R} \setminus \{0\}$ is an open set in $\mathbb{R}$, then $\det^{-1}(\mathbb{R} \setminus \{0\})$ is open in $\mathbb{R}^{n^2}$ and therefore can be given a C^∞ structure. Thus it is an open submanifold of the topological space $\mathbb{R}^{n^2}$. We have a manifold. The maps

$$\phi : GL(n, \mathbb{R}) \times GL(n, \mathbb{R}) \to GL(n, \mathbb{R}), \quad (A, B) \to \phi(A, B) \to AB$$

and

$$\psi : GL(n, \mathbb{R}) \to GL(n, \mathbb{R}), \quad AMA^{-1}$$

are both differentiable. Let $A \in GL(n, \mathbb{R})$ $(A = (a_{ij}))$ and set $x_{ij}(A) = a_{ij} \in \mathbb{R}$ for $i, j = 1, 2, \ldots, n$. Consequently (x_{ij}) is a coordinate system on $GL(n, \mathbb{R})$. We define

$$x_{ij} \circ \phi = \phi_{ij}, \qquad x_{ij} \circ \psi = \psi_{ij}.$$

It follows that $\phi_{ij}(A, B)$ and $\psi_{ij}(A)$ are the (i, j) entry of the matrix AB and A^{-1}, respectively. Therefore

$$\phi_{ij}(A, B) = \sum_{k=1}^{n} x_{ik}(A) x_{kj}(B)$$

shows that the function $\phi_{ij}(A, B)$ is a polynomial in the coordinates of the matrices A and B. It follows that ϕ_{ij} is of class C^∞ and $\phi \in C^\infty$. Let

$$\psi_{ij}(A) = f_{ij}(A)/\det(A)$$

where $f_{ij}(A)$ is the matrix obtained from the matrix A be deleting row i and column j. From $\psi_{ij}(A)$ we find by Cramer's rule the (i, j)th entry of the inverse matrix A^{-1}. The denominator and numerator are polynomials in the coordinates $x_{ij}(A)$ of A. The denominator is not zero at any point of $GL(n, \mathbb{R})$. Therefore ψ_{ij} is of class C^∞ and finally $\psi \in C^\infty$.

Problem 74. Consider the group of $n \times n$ upper triangular real matrices under matrix multiplication with nonzero determinants. Show that this is an $n(n+1)/2$-dimensional Lie group. Describe its group manifold.

Solution 74. The product of two upper triangular matrices is again an upper triangular matrix. The group unity element is the identity matrix I_n. The invertibility condition is $\det(A) \neq 0$. If the inverse matrix exists, then it is an upper triangular matrix. The associative law follows from the same law of matrix multiplication. The set of $n \times n$ upper triangular matrices, which satisfy invertibility condition, forms a noncommutative $n(n+1)/2$-dimensional Lie group under matrix multiplication G. The chart $G \to \mathbb{R}^{n(n+1)/2}$ is trivial: coordinates are the matrix components. One chart is sufficient. Therefore the group manifold is isomorphic to

$$\{ A \in \mathbb{R}^{n(n+1)/2} : a_{jj} \neq 0 \}$$

(a_{jj} - any diagonal matrix element). This is an open set, because matrices A, for which $a_{ii} = 0$ for some i, are limit points of G, but they do not belong to G. This set is unbounded, because for each matrix $A \in G$ there exist matrices $\{\lambda A : \lambda \neq 0\}$, which also belong to G because their diagonal elements are products of λ and A diagonal elements.

Problem 75. Let

$$S^n := \{ (x_1, x_2, \ldots, x_{n+1}) \in \mathbb{R}^{n+1} : x_1^2 + x_2^2 + \cdots + x_{n+1}^2 = 1 \}.$$

(i) Show that S^3 can be considered as a subset of $\mathbb{C}^2$ ($\mathbb{C}^2 \cong \mathbb{R}^4$)

$$S^3 = \{ (z_1, z_2) \in \mathbb{C}^2 : |z_1|^2 + |z_2|^2 = 1 \}.$$

(ii) The *Hopf map* $\pi : S^3 \to S^2$ is defined by

$$\pi(z_1, z_2) := (\bar{z}_1 z_2 + \bar{z}_2 z_1, -i\bar{z}_1 z_2 + i\bar{z}_2 z_1, |z_1|^2 - |z_2|^2).$$

Find the parameterization of S^3, i.e. find $z_1(\theta, \phi)$, $z_2(\theta, \phi)$ and thus show that indeed π maps S^3 onto S^2.
(iii) Show that $\pi(z_1, z_2) = \pi(z_1', z_2')$ if and only if $z_j' = e^{i\alpha} z_j$ ($j = 1, 2$) and $\alpha \in \mathbb{R}$.

Solution 75.   (i) Let $z_1 = x_1 + iy_1$ and $z_2 = x_2 + iy_2$, where $x_1, x_2, y_1, y_2 \in \mathbb{R}$. Then from $|z_1|^2 + |z_2|^2 = 1$ it follows that

$$x_1^2 + y_1^2 + x_2^2 + y_2^2 = 1.$$

(ii) Since $|z_1|^2 + |z_2|^2 = 1$ we have the parameterization

$$z_1(\theta, \phi) = \cos(\theta/2)e^{i\phi_1}, \qquad z_2(\theta, \phi) = \sin(\theta/2)e^{i\phi_2}$$

where $0 \leq \theta \leq \pi$ and $\phi_1, \phi_2 \in \mathbb{R}$. Thus

$$\pi(\cos(\theta/2)e^{i\phi_1}, \sin(\theta/2)e^{i\phi_2}) = (\sin\theta\cos(\phi_2 - \phi_1), \sin\theta\sin(\phi_2 - \phi_1), \cos\theta).$$

(iii) From $\pi(z_1, z_2) = \pi(z_1', z_2')$ we obtain the three equations

$$\bar{z}_1 z_2 + \bar{z}_2 z_1 = \bar{z}_1' z_2' + \bar{z}_2' z_1'$$
$$-\bar{z}_1 z_2 + \bar{z}_2 z_1 = -\bar{z}_1' z_2' + \bar{z}_2' z_1'$$
$$|z_1|^2 - |z_2|^2 = |z_1'|^2 - |z_2'|^2.$$

Inserting $|z_1|^2 + |z_2|^2 = |z_1'|^2 + |z_2'|^2 = 1$ into the last equation yields $|z_1| = |z_1'|$ and $|z_2| = |z_2'|$. Adding the first two equations provides $z_1\bar{z}_2 = z_1'\bar{z}_2'$. Thus we obtain the solution $z_1' = e^{i\alpha}z_1$ and $z_2' = e^{i\alpha}z_2$, where $\alpha \in \mathbb{R}$.

Problem 76.   Consider the compact differentiable manifold

$$S^2 := \{ (x_1, x_2, x_3) : x_1^2 + x_2^2 + x_3^2 = 1 \}.$$

An element $\eta \in S^2$ can be written as

$$\eta = (\cos\phi\sin\theta, \sin\phi\sin\theta, \cos\theta)$$

where $\phi \in [0, 2\pi)$ and $\theta \in [0, \pi]$. The *stereographic projection* is a map

$$\Pi : S^2 \setminus \{ (0, 0, -1) \} \to \mathbb{R}^2$$

given by

$$x_1(\theta, \phi) = \frac{2\sin(\theta)\cos(\phi)}{1 + \cos(\theta)}, \qquad x_2(\theta, \phi) = \frac{2\sin(\theta)\sin(\phi)}{1 + \cos(\theta)}.$$

(i) Let $\theta = 0$ and ϕ arbitrary. Find x_1, x_2. Give a geometric interpretation.
(ii) Find the inverse of the map, i.e., find

$$\Pi^{-1} : \mathbb{R}^2 \to S^2 \setminus \{ (0, 0, -1) \}.$$

Solution 76.   (i) Since $\sin(0) = 0$ we find $x_1 = x_2 = 0$, i.e. the point $(0, 0, 1)$ is mapped to the origin $(0, 0)$.

(ii) Using division we find

$$\phi(x_1, x_2) = \arctan\left(\frac{x_2}{x_1}\right).$$

Since

$$x_1^2 + x_2^2 = \frac{4\sin^2\theta}{(1+\cos\theta)^2}$$

and

$$\tan\left(\frac{\theta}{2}\right) = \frac{\sin\theta}{1+\cos\theta}$$

we obtain

$$\theta(x_1, x_2) = 2\arctan\left(\frac{\sqrt{x_1^2 + x_2^2}}{2}\right).$$

Problem 77. One natural parameterization of the Lie algebra $su(1,1)$ as a sub Lie algebra of $sp(2,2)$ is given by the 4×4 matrices

$$R_1 = \begin{pmatrix} 0 & 0 & 0 & 1 \\ 0 & 0 & 1 & 0 \\ 0 & 1 & 0 & 0 \\ 1 & 0 & 0 & 0 \end{pmatrix}, \quad R_2 = \begin{pmatrix} i & 0 & 0 & 0 \\ 0 & -i & 0 & 0 \\ 0 & 0 & i & 0 \\ 0 & 0 & 0 & -i \end{pmatrix}, \quad R_3 = \begin{pmatrix} 0 & 0 & 0 & i \\ 0 & 0 & -i & 0 \\ 0 & i & 0 & 0 \\ -i & 0 & 0 & 0 \end{pmatrix}$$

with $[R_1, R_2] = -2R_3$, $[R_3, R_1] = 2R_2$, $[R_2, R_3] = -2R_1$. Find $g_2 g_1 g_2$, where $g_1 = \exp(\alpha R_1)$ and $g_2 = \exp(\beta R_2/2)$.

Solution 77. We obtain

$$g_2 g_1 g_2 = \begin{pmatrix} e^{i\beta}\cosh(\alpha) & 0 & 0 & \sinh(\alpha) \\ 0 & e^{-i\beta}\cosh(\alpha) & \sinh(\alpha) & 0 \\ 0 & \sinh(\alpha) & e^{i\beta}\cosh(\alpha) & 0 \\ \sinh(\alpha) & 0 & 0 & e^{-i\beta}\cosh(\alpha) \end{pmatrix}.$$

Problem 78. Find the factor group $GL(n, \mathbb{R})/SL(n, \mathbb{R})$.

Solution 78. Let g be given element of $GL(n, \mathbb{R})$ and h be arbitrary element of $SL(n, \mathbb{R})$. Then

$$\det(gh) = \det(g)\det(h) = \det(g).$$

When h runs over all elements of $SL(n, \mathbb{R})$, the product gh runs over all elements of the $(n^2 - 1)$-dimensional submanifold of $n \times n$ matrices with constant determinant $\{A : \det(A) = \det(g)\}$. This submanifold is called the left coset of $SL(n, \mathbb{R})$, generated by the fixed element g. We can identify left cosets by the value of matrix determinant. Specifically we identify the left coset by its element $f = cI_n$. This is a bijection and $\det(f) = c^n$. Then the set of all left cosets is the group $F = GL(n, \mathbb{R})/SL(n, \mathbb{R})$. We can take its group law directly from the $GL(n, \mathbb{R})$ group law

$$f_1 f_2 = (c_1 I_n)(c_2 I_n) = (c_1 c_2) I_n.$$

Thus F is a 1-dimensional Lie group isomorphic to multiplication of the group $\mathbb{R} \setminus \{0\}$ group. $GL(n, \mathbb{R})$ is the direct product of $SL(n, \mathbb{R})$ and $GL(n, \mathbb{R})/SL(n, \mathbb{R})$, since each element of $GL(n, \mathbb{R})$ can be represented as a product $hf = fh$ of the normal subgroup $SL(n, \mathbb{R})$ and the factor group $GL(n, \mathbb{R})/SL(n, \mathbb{R})$ elements. The elements f and h are the same in both parts of this expression, because the unit matrix I_n commutate with any $n \times n$ matrix.

Problem 79. The $n \times n$ unitary matrices form a Lie group. Do the $n \times n$ unitary matrices satisfying the additional condition $U = U^T$ form a sub Lie group?

Solution 79. The answer is no. For example consider

$$U_1 = \begin{pmatrix} 0 & 1 \\ 1 & 0 \end{pmatrix}, \quad U_2 = \begin{pmatrix} 1 & 0 \\ 0 & -1 \end{pmatrix}.$$

Then

$$U_1 U_2 = \begin{pmatrix} 0 & -1 \\ 1 & 0 \end{pmatrix}.$$

Therefore $(U_1 U_2)^T \neq U_1 U_2$.

Problem 80. The $n \times n$ unitary matrices form a Lie group $U(n)$ under matrix multiplication. Let 0_n be the $n \times n$ zero matrix. Do the $2n \times 2n$ matrices

$$\begin{pmatrix} 0_n & U \\ U & 0_n \end{pmatrix}$$

form a group under matrix multiplication?

Solution 80. No. The product of such two matrices is the diagonal matrix

$$\begin{pmatrix} U_1 U_2 & 0_n \\ 0_n & U_1 U_2 \end{pmatrix}.$$

Thus the set is not closed under matrix multiplication.

Problem 81. Let x be a fixed element of a group G. The *centralizer* of x in G, $C(x)$, is the set of all elements in G that commute with x, i.e.

$$C(x) := \{\, g \in G \ : \ gx = xg \,\}.$$

Consider the Lie group $SL(2, \mathbb{R})$ and

$$A = \begin{pmatrix} 0 & 1 \\ -1 & 0 \end{pmatrix}.$$

Then $A \in SL(2, \mathbb{R})$ since $\det(A) = 1$. Find the centralizer $C(A)$ of A.

Solution 81. Let

$$g = \begin{pmatrix} a & b \\ c & d \end{pmatrix}, \qquad ad - bc = 1.$$

From the condition $gA = Ag$ we obtain the four equations

$$-b - c = 0, \quad a - d = 0, \quad -d + a = 0, \quad c + b = 0.$$

Only two are linearly independent, say $a - d = 0$, $c + b = 0$. Thus $d = a$ and $c = -b$. We obtain the matrix

$$g = \begin{pmatrix} a & b \\ -b & a \end{pmatrix}$$

with $a^2 + b^2 = 1$. Thus we find the parameterization for g

$$g = \begin{pmatrix} \cos\phi & \sin\phi \\ -\sin\phi & \cos\phi \end{pmatrix}.$$

Thus the centralizer $C(A)$ consists of the rotation matrices which form the compact Lie group $SO(2, \mathbb{R})$.

Problem 82. The *kernel* of a homomorphism ϕ from a group G with identity e is the set

$$\{\, g \in G \ : \ \phi(g) = e \,\}.$$

The kernel of ϕ is denoted by Kerϕ. Let $\mathbb{R}^*$ be the abelian group of nonzero real numbers under multiplication. Let $g \in GL(2, \mathbb{R})$. The mapping (determinant) $g \to \det(g)$ is a homomorphism from $GL(2, \mathbb{R})$ to $\mathbb{R}^*$. Find the kernel.

Solution 82. The kernel is the group $SL(2, \mathbb{R})$, since for all elements in $SL(2, \mathbb{R})$ the determinant is 1. The unity element of the group $\mathbb{R}^*$ is 1.

Therefore the kernel of homomorphism $\psi : g \to \det(g)$ is the set of matrices with unity determinant $SL(2, \mathbb{R})$.

Problem 83.   (i) Consider the compact Lie group $SO(3, \mathbb{R})$. Let $A \in SO(3, \mathbb{R})$. Find the *Frobenius norm*

$$\|A\| = \operatorname{tr}(AA^T).$$

(ii) Find the norm

$$\|A\| = \sup_{\|\mathbf{x}\|=1} \|A\mathbf{x}\|.$$

Solution 83.   (i) Since $AA^T = I_3$ we obtain $\operatorname{tr}(AA^T) = 3$.
(ii) We obtain $\|A\| = 1$ since $AA^T = I_3$ and the largest eigenvalue of I_3 is 1.

Problem 84.   The vector space of all $n \times n$ matrices over $\mathbb{C}$ form a *Hilbert space* with the scalar product

$$\langle A, B \rangle := \operatorname{tr}(AB^*).$$

This implies a norm $\|A\|^2 = \operatorname{tr}(AA^*)$.
(i) Let U be a unitary matrix. Find $\langle U, U \rangle$.
(ii) Consider the 2×2 unitary matrices

$$U_1 = \begin{pmatrix} 0 & 1 \\ 1 & 0 \end{pmatrix}, \qquad U_2 = \frac{1}{\sqrt{2}} \begin{pmatrix} 1 & 1 \\ 1 & -1 \end{pmatrix}.$$

Find the norm $\|U_1 - U_2\|$.
(iii) Let V be a 2×2 unitary matrix. Let $\phi \in \mathbb{R}$. Find the norm $\|V - e^{i\phi}V\|$.

Solution 84.   (i) We obtain

$$\langle U, U \rangle = \operatorname{tr}(UU^*) = \operatorname{tr}(I_n) = n.$$

(ii) Using the the trace is linear we obtain

$$\begin{aligned} \langle U_1 - U_2, U_1 - U_2 \rangle &= \operatorname{tr}((U_1 - U_2)(U_1^* - U_2^*)) \\ &= \operatorname{tr}(U_1 U_1^*) + \operatorname{tr}(U_2 U_2^*) - \operatorname{tr}(U_1 U_2^*) - \operatorname{tr}(U_2 U_1^*) \\ &= 2 + 2 - \sqrt{2} - \sqrt{2} \\ &= 4 - 2\sqrt{2}. \end{aligned}$$

Thus $\|U_1 - U_2\|^2 = 4 - 2\sqrt{2}$.

(iii) We find
$$\|V - e^{i\phi}V\|^2 = 4 - 4\cos\phi.$$
Obviously if $\phi = 0$ we find that the norm is 0.

Problem 85. Consider the Lie group $U(n)$ of all $n \times n$ unitary matrices. Let K be an $n \times n$ skew-hermitian matrix with eigenvalues $\mu_1, \ldots, \mu_n$ (counted according to multiplicity) and the corresponding normalized eigenvectors $\mathbf{u}_1, \ldots, \mathbf{u}_n$, where $\mathbf{u}_j^*\mathbf{u}_k = 0$ for $k \neq j$. Then K can be written as

$$K = \sum_{j=1}^{n} \mu_j \mathbf{u}_j \mathbf{u}_j^*$$

and $\mathbf{u}_j \mathbf{u}_j^* \mathbf{u}_k \mathbf{u}_k^* = 0$ for $k \neq j$ and $j, k = 1, 2, \ldots, n$. Note that the matrices $\mathbf{u}_j \mathbf{u}_j^*$ are projection matrices and

$$\sum_{j=1}^{n} \mathbf{u}_j \mathbf{u}_j^* = I_n.$$

(i) Calculate $\exp(K)$.
(ii) Every $n \times n$ unitary matrix can be written as $U = \exp(K)$, where K is a skew-hermitian matrix. Find U from a given K.
(iii) Use the result from (ii) to find for a given U a possible K.
(iv) Apply the result from (ii) and (iii) to the unitary 2×2 matrix

$$U(\theta) = \begin{pmatrix} \cos\theta & \sin\theta \\ -\sin\theta & \cos\theta \end{pmatrix}.$$

(v) Apply the result from (ii) and (iii) to the 2×2 unitary matrix

$$V(\theta, \phi) = \begin{pmatrix} \cos\theta & -e^{i\phi}\sin\theta \\ e^{-i\phi}\sin\theta & \cos\theta \end{pmatrix}.$$

(vi) Every hermitian matrix H can be written as $H = iK$, where K is a skew-hermitian matrix. Find H for the examples given above.

Solution 85. (i) Using the properties of $\mathbf{u}_j \mathbf{u}_j^*$ we find

$$\exp(K) = \exp\left(\sum_{j=1}^{n} \mu_j \mathbf{u}_j \mathbf{u}_j^*\right) = \sum_{j=1}^{n} e^{\mu_j} \mathbf{u}_j \mathbf{u}_j^*.$$

(ii) From $U = \exp(K)$ we find

$$U = \sum_{j=1}^{n} e^{\mu_j} \mathbf{u}_j \mathbf{u}_j^*$$

where $\mathbf{u}_j$ $(j = 1, 2, \ldots, n)$ are the normalized eigenvectors of U.
(iii) The matrix K is given by

$$K = \sum_{j=1}^{n} \ln(\lambda_j)\mathbf{u}_j\mathbf{u}_j^*$$

where λ_j $(j = 1, 2, \ldots, n)$ are the eigenvalues if U and $\mathbf{u}_j$ are the normalized eigenvectors of U. Note that the eigenvalues of U are of the form $\exp(i\alpha)$ with $\alpha \in \mathbb{R}$. Thus we have $\ln(e^{i\alpha}) = i\alpha$.
(iv) The eigenvalues of the matrix $U(\theta)$ are $e^{i\theta}$ and $e^{-i\theta}$ with the corresponding normalized eigenvectors

$$\mathbf{u}_1 = \frac{1}{\sqrt{2}} \begin{pmatrix} 1 \\ i \end{pmatrix}, \qquad \mathbf{u}_2 = \frac{1}{\sqrt{2}} \begin{pmatrix} 1 \\ -i \end{pmatrix}.$$

Thus

$$K(\theta) = \ln(e^{i\theta})\mathbf{u}_1\mathbf{u}_1^* + \ln(e^{i\theta})\mathbf{u}_2\mathbf{u}_2 = \theta \begin{pmatrix} 0 & 1 \\ -1 & 0 \end{pmatrix}.$$

(v) For the matrix $V(\theta, \phi)$ the eigenvalues are $e^{-i\theta}$ and $e^{i\theta}$ with the corresponding normalized eigenvectors

$$\frac{1}{\sqrt{2}} \begin{pmatrix} 1 \\ ie^{-i\phi} \end{pmatrix}, \qquad \frac{1}{\sqrt{2}} \begin{pmatrix} 1 \\ -ie^{-i\phi} \end{pmatrix}.$$

Thus

$$K(\theta, \phi) = \ln(e^{-i\theta})\mathbf{u}_1\mathbf{u}_1^* + \ln(e^{i\theta})\mathbf{u}_2\mathbf{u}_2^* = \begin{pmatrix} 0 & -\theta e^{i\phi} \\ \theta e^{-i\phi} & 0 \end{pmatrix}.$$

(vi) For $U(\theta)$ we find

$$i\theta \begin{pmatrix} 0 & 1 \\ -1 & 0 \end{pmatrix}.$$

For $V(\theta, \phi)$ we find

$$\begin{pmatrix} 0 & -i\theta e^{i\phi} \\ i\theta e^{-i\phi} & 0 \end{pmatrix}.$$

Problem 86. An *action* or realization of a group G on a set M is defined as map φ_g

$$\varphi_g : \mathbf{x} \mapsto \mathbf{x}' = \varphi_g(\mathbf{x})$$

where $\mathbf{x}, \mathbf{x}' \in M$ and $g \in G$. An element $\mathbf{x}$ of the manifold M is a *fixed point* if

$$\varphi_g(\mathbf{x}) = \mathbf{x} \quad \text{for all} \quad g \in G.$$

(i) Find the fixed points for the Lie group $SO(3)$ on $M = \mathbb{R}^3$.

(ii) Find the fixed points for $SO(3)$ on $M = S^2$.

Solution 86. (i) The only fixed point is the zero vector $\mathbf{0} \in \mathbb{R}^3$.

(ii) There are no fixed points.

Problem 87. Consider the subgroups of the Lie group $SU(1,1)$

$$\begin{pmatrix} e^{i\theta/2} & 0 \\ 0 & e^{-i\theta/2} \end{pmatrix}, \quad \begin{pmatrix} \cosh(t/2) & i\sinh(t/2) \\ -i\sinh(t/2) & \cosh(t/2) \end{pmatrix}, \quad \begin{pmatrix} 1+\zeta i/2 & \zeta/2 \\ \zeta/2 & 1-\zeta i/2 \end{pmatrix}$$

where $\theta, t, \zeta \in \mathbb{R}$. The Lie group $SU(1,1)$ acts as a linear group of transformations of linear vector space $\mathbb{R}^2$. Find the infinitesimal generators of the group as vector fields.

Solution 87. The first matrix generates the transformation in $\mathbb{R}^2$

$$x_1' = e^{i\theta/2}x_1, \qquad x_2' = e^{-i\theta/2}x_2.$$

Thus the corresponding infinitesimal generator (vector field) is

$$V_1 = \frac{i}{2}x_1\frac{\partial}{\partial x_1} - \frac{i}{2}x_2\frac{\partial}{\partial x_2}.$$

The second matrix generates the transformation in $\mathbb{R}^2$

$$x_1' = \cosh(t/2)x_1 + i\sinh(t/2)x_2, \quad x_2' = -i\sinh(t/2)x_1 + \cosh(t/2)x_2.$$

Thus the corresponding infinitesimal generator (vector field) is

$$V_2 = \frac{i}{2}x_2\frac{\partial}{\partial x_1} - \frac{i}{2}x_1\frac{\partial}{\partial x_2}.$$

The third matrix generates the transformation in $\mathbb{R}^2$

$$x_1' = (1+\frac{i}{2}\zeta)x_1 + \frac{\zeta}{2}x_2, \quad x_2' = \frac{\zeta}{2}x_1 + (1-\frac{i}{2}\zeta)x_2.$$

Thus the corresponding infinitesimal generator (vector field) is

$$V_3 = \left(\frac{i}{2}x_1 + \frac{1}{2}x_2\right)\frac{\partial}{\partial x_1} + \left(\frac{1}{2}x_1 - \frac{i}{2}x_2\right)\frac{\partial}{\partial x_2}.$$

Problem 88. (i) The representation of the Lie group $SO(2)$ by 2×2 matrices is

$$\begin{pmatrix} \cos(\phi) & -\sin(\phi) \\ \sin(\phi) & \cos(\phi) \end{pmatrix}, \quad \phi \in \mathbb{R}.$$

Show that the quadratic form (Euclid's metric) $s^2 = x^2 + y^2$ is an invariant of the group action of $SO(2)$ on vectors (x, y) on the plane.

(ii) The representation of the Lie group $SH(2)$ group by 2×2 matrices is

$$\begin{pmatrix} \cosh(t) & \sinh(t) \\ \sinh(t) & \cosh(t) \end{pmatrix}, \quad t \in \mathbb{R}.$$

Show that the quadratic form (Minkowski's metric) $s^2 = x^2 - y^2$ is an invariant of the group action of $SH(2)$ on vectors (x, y) on the plane.

Solution 88. (i) By direct calculation we have

$$\begin{pmatrix} \cos(\phi) & -\sin(\phi) \\ \sin(\phi) & \cos(\phi) \end{pmatrix} \begin{pmatrix} x \\ y \end{pmatrix} = \begin{pmatrix} \cos(\phi)x - \sin(\phi)y \\ \sin(\phi)x + \cos(\phi)y \end{pmatrix}.$$

Thus the transformed quadratic form value is

$$(s^2)' = (\cos(\phi)x - \sin(\phi)y)^2 + (\sin(\phi)x + \cos(\phi)y)^2 = x^2 + y^2 = s^2.$$

(ii) By direct calculation we have

$$\begin{pmatrix} \cosh(t) & \sinh(t) \\ \sinh(t) & \cosh(t) \end{pmatrix} \begin{pmatrix} x \\ y \end{pmatrix} = \begin{pmatrix} \cosh(t)x + \sinh(t)y \\ \sinh(t)x + \cosh(t)y \end{pmatrix}.$$

Thus the transformed quadratic form value is

$$(s^2)' = (\cosh(t)x + \sinh(t)y)^2 - (\sinh(t)x + \cosh(t)y)^2 = x^2 - y^2 = s^2.$$

Problem 89. Let $z \in \mathbb{C}$. Consider the complex form $SO(2, \mathbb{C})$ of the $SO(2)$ group

$$\begin{pmatrix} \cos(z) & -\sin(z) \\ \sin(z) & \cos(z) \end{pmatrix}.$$

Show that $SO(2, \mathbb{R})$ is an abelian subgroup of $SO(2, \mathbb{C})$. Show that $SH(2)$ is isomorphic to a subgroup of $SO(2, \mathbb{C})$.

Solution 89. Consider product of two elements of the Lie group $SO(2, \mathbb{C})$

$$\begin{pmatrix} \cos(z_1) & -\sin(z_1) \\ \sin(z_1) & \cos(z_1) \end{pmatrix} \begin{pmatrix} \cos(z_2) & -\sin(z_2) \\ \sin(z_2) & \cos(z_2) \end{pmatrix} = \begin{pmatrix} \cos(z_1 + z_2) & -\sin(z_1 + z_2) \\ \sin(z_1 + z_2) & \cos(z_1 + z_2) \end{pmatrix}$$

where we used the identities

$$\sin(z_1)\cos(z_2) + \cos(z_1)\sin(z_2) \equiv \sin(z_1 + z_2)$$
$$\cos(z_1)\cos(z_2) - \sin(z_1)\sin(z_2) \equiv \cos(z_1 + z_2).$$

Thus the group multiplication law results in the addition of the complex parameter z. The parameter z can be written in polar form

$$z = e^{i\alpha}t, \quad t, \alpha \in \mathbb{R}.$$

We see that $t = 0$ provides the group unity. The inverse element is obtained from $t' = -t$. For $\alpha = 0$ we find the subgroup described by

$$\begin{pmatrix} \cos(t) & -\sin(t) \\ \sin(t) & \cos(t) \end{pmatrix}, \quad t \in \mathbb{R}$$

i.e. we have the subgroup $SO(2, \mathbb{R})$. For $\alpha = \pi/2$ we find the subgroup described by

$$g(t) = \begin{pmatrix} \cos(it) & -\sin(it) \\ \sin(it) & \cos(it) \end{pmatrix}, \quad t \in \mathbb{R}.$$

Now $\cos(it) \equiv \cosh(t)$, $\sin(it) \equiv i\sinh(t)$. Thus

$$g(t) = \begin{pmatrix} \cosh(t) & -i\sinh(t) \\ i\sinh(t) & \cosh(t) \end{pmatrix}.$$

We introduce the map $\phi(g(t)) := Ag(t)A^{-1}$, where $A \in U(2)$ is the matrix

$$A = \begin{pmatrix} 1 & 0 \\ 0 & -i \end{pmatrix}.$$

It follows that

$$\phi(g(t)) = \begin{pmatrix} \cosh(t) & -i\sinh(t) \\ \sinh(t) & -i\cosh(t) \end{pmatrix} \begin{pmatrix} 1 & 0 \\ 0 & i \end{pmatrix} = \begin{pmatrix} \cosh(t) & \sinh(t) \\ \sinh(t) & \cosh(t) \end{pmatrix}.$$

Thus the map ϕ is an isomorphism between a subgroup (described by $g(t)$) of $SO(2, \mathbb{C})$ and $SH(2)$. The invariant quadratic form of $SO(2)$ $x^2 + y^2$ transforms to $x^2 + y^2 = (x')^2 + (iy')^2 = (x')^2 - (y')^2$.

Problem 90. Let $R = \mathbb{C}[x, y]$ be the *polynomial ring* in two variables x, y over the field $\mathbb{C}$. Let q be a positive integer and R_q be the sub vector space of polynomials which are homogeneous of degree q. Then R_q has the basis

$$x^q, \ x^{q-1}y, \ x^{q-2}y^2, \ \dots, y^q$$

and dimension $q + 1$. An element of $SL(2, \mathbb{C})$

$$\begin{pmatrix} a & b \\ c & d \end{pmatrix}, \quad \cdot \quad ad - bc = 1$$

operates on R by the transformation $x \to ax + cy$, $y \to bx + dy$ and R_q is the representation space for $SL(2, \mathbb{C})$ for each q under this operation. We could also write in matrix form

$$(x \quad y) \begin{pmatrix} a & b \\ c & d \end{pmatrix} = (ax + cy \quad bx + dy) \,.$$

Let $p_j := x^j y^{q-j}$, $j = 0, 1, \ldots, q$. Show that the images of p_j under the transformation by

$$\begin{pmatrix} 1 & t \\ 0 & 1 \end{pmatrix} \quad \text{and} \quad \begin{pmatrix} 1 & 0 \\ t & 1 \end{pmatrix}$$

are given by

$$\begin{pmatrix} 1 & t \\ 0 & 1 \end{pmatrix} \cdot p_j = \sum_{k=0}^{q-j} \binom{q-j}{k} t^k p_{j+k}$$

$$\begin{pmatrix} 1 & 0 \\ t & 1 \end{pmatrix} \cdot p_j = \sum_{k=0}^{j} \binom{j}{k} t^k p_{j-k} \,.$$

Solution 90.   Since

$$\begin{pmatrix} 1 & t \\ 0 & 1 \end{pmatrix} \cdot x = x, \qquad \begin{pmatrix} 1 & t \\ 0 & 1 \end{pmatrix} \cdot y = tx + y$$

and

$$\begin{pmatrix} 1 & 0 \\ t & 1 \end{pmatrix} \cdot x = x + ty, \qquad \begin{pmatrix} 1 & t \\ 0 & 1 \end{pmatrix} \cdot y = y$$

we find the desired result.

Problem 91.   (i) Let $t \in \mathbb{R}$. Show that the matrix

$$g_t = \begin{pmatrix} \cosh t & i \sinh t \\ -i \sinh t & \cosh t \end{pmatrix}$$

has an inverse and find g_t^{-1}. g_t is an element of $SO(1, 1)$.
(ii) The symmetric space $SL(2, \mathbb{C})/SU(2)$ can be identified with the space H^3 of positive 2×2 matrices with determinant 1

$$A = \begin{pmatrix} a+b & x+iy \\ x-iy & a-b \end{pmatrix}, \qquad a, b, c, d \in \mathbb{R}$$

i.e. $\det A = a^2 - b^2 - x^2 - y^2 = 1$. The group action is defined by

$$g_t \cdot A := g_t A g_t^* \,.$$

Find $g_t \cdot A$.

(iii) Show that on the hyperboloid $a^2 = 1 + b^2 + x^2 + y^2$, the matrix g_t only changes the coordinates y and a.

Solution 91. (i) We have $\det g_t = \cosh^2 t - \sinh^2 t = 1$. Thus the inverse exists and is given by

$$g_t^{-1} = \begin{pmatrix} \cosh t & -i \sinh t \\ i \sinh t & \cosh t \end{pmatrix} = g_t^T.$$

(ii) We obtain

$$g_t \cdot A = \begin{pmatrix} a \cosh(2t) + y \sinh(2t) + b & x + i(y \cosh(2t) + a \sinh(2t)) \\ x - i(y \cosh(2t) + a \sinh(2t)) & a \cosh(2t) + y \sinh(2t) - b \end{pmatrix}.$$

(iii) It follows that

$$a' = \sinh(2t)y + \cosh(2t)a, \quad b' = b$$

$$x' = x, \quad y' = \cosh(2t)y + \sinh(2t)a.$$

Thus x and b are invariants, the matrix g_t only changes the coordinates y and a. The hyperboloid $a^2 = 1 + b^2 + x^2 + y^2$ is invariant, because $\det g_t = 1$. We have

$$a'^2 - y'^2 = (\sinh(2t)y + \cosh(2t)a)^2 - (\cosh(2t)y + \sinh(2t)a)^2 = a^2 - y^2.$$

Therefore

$$b^2 - a^2 + x^2 + y^2 = b'^2 - a'^2 + x'^2 + y'^2.$$

Problem 92. Find an abelian Lie subgroup of the Lie group $U(n)$.

Solution 92. $G = \mathbb{T}^n$ is an abelian Lie subgroup of $U(n)$ consisting of the diagonal matrices of the form

$$Z = \mathrm{diag}(e^{i\theta_1}, \ldots, e^{i\theta}) $$

where $\theta_j \in \mathbb{R}$. This group is topologically $S^1 \times \cdots \times S^1$, the topological product of n copies of the unit circle. Thus it is an n-torus.

Problem 93. We consider the Lie groups $U(2,2)$, $SU(2,2)$ and the corresponding Lie algebras $u(2,2)$, $su(2,2)$, respectively. We define the matrix

$$L := \begin{pmatrix} -1 & 0 & 0 & 0 \\ 0 & -1 & 0 & 0 \\ 0 & 0 & 1 & 0 \\ 0 & 0 & 0 & 1 \end{pmatrix}.$$

Then

$$U(2,2) := \{\, G \in GL(4, \mathbb{C}) \ : \ G^* L G = L \,\}$$
$$SU(2,2) := \{\, G \in U(2,2) \ : \ \det G = 1 \,\}$$
$$u(2,2) := \{\, X \in M_4(\mathbb{C}) \ : \ XL + LX = 0_4 \,\}$$
$$su(2,2) := \{\, X \in u(2,2) \ : \ \mathrm{tr}(X) = 0 \,\}$$

where $M_4(\mathbb{C})$ is the vector space of the 4×4 matrices over the complex numbers.
(i) Show that

$$X = \begin{pmatrix} 0 & 0 & 0 & 1 \\ 0 & 0 & 1 & 0 \\ 0 & 1 & 0 & 0 \\ 1 & 0 & 0 & 0 \end{pmatrix}$$

is an element of $u(2,2)$ and $su(2,2)$.
(ii) Let $t \in \mathbb{R}$. Calculate $\exp(tX)$ to find an element of $SU(2,2)$.

Solution 93.   (i) We have $X^* = X$ and $X^* L + LX = 0_4$. Moreover $\mathrm{tr}(X) = 0$. Thus X is an element of $su(2,2)$.
(ii) Since $X^2 = I_4$ we obtain the matrix

$$G = \begin{pmatrix} \cosh(t) & 0 & 0 & \sinh(t) \\ 0 & \cosh(t) & \sinh(t) & 0 \\ 0 & \sinh(t) & \cosh(t) & 0 \\ \sinh(t) & 0 & 0 & \cosh(t) \end{pmatrix}$$

with $G^* = G$, $\det G = 1$ and $G^* L G = L$.

Problem 94.   Show that the Lie group $U(n)$ is connected.

Solution 94.   Obviously $\mathbb{T}^n$ is a Lie subgroup (and thus a subset) of $U(n)$. Note that a unitary matrix is a normal matrix. Any $g \in U(n)$ can be diagonalized by an $n \times n$ unitary matrix. Thus given $g \in U(n)$, there exists an $h \in U(n)$ such that

$$h^{-1} g h = \mathrm{diag}(e^{i\theta_1}, \dots, e^{i\theta}).$$

It follows that

$$g = h \mathrm{diag}(e^{i\theta_1}, \dots, e^{i\theta}) h^{-1}.$$

This means that $g \in h\mathbb{T}^n h^{-1}$, i.e., the group element g lies on the diffeomorphic copy of $\mathbb{T}^n$ that results from left translating $\mathbb{T}^n$ by h and then right translating by h^{-1}. Therefore the group element g can be joined by the identity by a curve setting $\theta_j(\epsilon) = (1 - \epsilon)\theta_j$. Consequently the Lie group $U(n)$ is connected.

Problem 95. Consider the Lie group T_3 consisting of all 4×4 matrices

$$
g(z_1, z_2, z_3) = \begin{pmatrix} 1 & 0 & 0 & z_3 \\ 0 & e^{-z_3} & 0 & z_2 \\ 0 & 0 & e^{z_3} & z_1 \\ 0 & 0 & 0 & 1 \end{pmatrix}, \qquad z_1, z_2, z_3 \in \mathbb{C}.
$$

(i) Find the inverse $g^{-1}(z_1, z_2, z_3)$.
(ii) Find the product $g(z_1, z_2, z_3)g(w_1, w_2, w_3)$.
(iii) Find a basis for the Lie algebra t_3 of the Lie group T_3 and the commutators.

Solution 95. (i) Obviously the inverse exists since $\det(g(z_1, z_2, z_3)) = 1$. We find

$$
g^{-1}(z_1, z_2, z_3) = \begin{pmatrix} 1 & 0 & 0 & -z_3 \\ 0 & e^{z_3} & 0 & -e^{z_3}z_2 \\ 0 & 0 & e^{-z_3} & -e^{-z_3}z_1 \\ 0 & 0 & 0 & 1 \end{pmatrix}.
$$

(ii) We find

$$
g(z_1, z_2, z_3)g(w_1, w_2, w_3) = \begin{pmatrix} 1 & 0 & 0 & z_3 + w_3 \\ 0 & e^{-z_3 - w_3} & 0 & z_2 + e^{-z_3}w_2 \\ 0 & 0 & e^{z_3 + w_3} & z_1 + e^{z_3}w_1 \\ 0 & 0 & 0 & 1 \end{pmatrix}.
$$

(iii) We have

$$
\frac{\partial}{\partial z_1} g(z_1, z_2, z_3)\bigg|_{z_1 = z_2 = z_3 = 0} = J_+ = \begin{pmatrix} 0 & 0 & 0 & 0 \\ 0 & 0 & 0 & 0 \\ 0 & 0 & 0 & 1 \\ 0 & 0 & 0 & 0 \end{pmatrix}
$$

$$
\frac{\partial}{\partial z_2} g(z_1, z_2, z_3)\bigg|_{z_1 = z_2 = z_3 = 0} = J_- = \begin{pmatrix} 0 & 0 & 0 & 0 \\ 0 & 0 & 0 & 1 \\ 0 & 0 & 0 & 0 \\ 0 & 0 & 0 & 0 \end{pmatrix}
$$

$$
\frac{\partial}{\partial z_3} g(z_1, z_2, z_3)\bigg|_{z_1 = z_2 = z_3 = 0} = J_3 = \begin{pmatrix} 0 & 0 & 0 & 1 \\ 0 & -1 & 0 & 0 \\ 0 & 0 & 1 & 0 \\ 0 & 0 & 0 & 0 \end{pmatrix}
$$

with the commutation relations

$$
[J_3, J_+] = J_+, \qquad [J_3, J_-] = -J_-, \qquad [J_+, J_-] = 0.
$$

Problem 96. Consider the vector space V_2^n of homogeneous polynomials of two complex variables. An element of this vector space can be written as

$$p(z_1, z_2) = c_0 z_1^n + c_1 z_1^{n-1} z_2 + \cdots + c_n z_2^n.$$

The compact Lie group $SU(2)$ acts on this vector space through the action of $U \in SU(2)$ as a linear transform on the vector $\mathbf{z} = (z_1, z_2)^T$ by

$$\pi(U)f(\mathbf{z}) = f(U^{-1}\mathbf{z}).$$

Show that this a group homomorphism.

Solution 96. Let $U_1, U_2 \in SU(2)$. Note that $(U_1 U_2)^{-1} = U_2^{-1} U_1^{-1}$. We have

$$\pi(U_1)(\pi(U_2)f)(\mathbf{z}) = (\pi(U_2)f)(U_1^{-1}\mathbf{z}) = f(U_2^{-1} U_1^{-1}\mathbf{z}) = \pi(U_1 U_2)f(\mathbf{z}).$$

Problem 97. Consider all 2×2 matrices with $UU^* = I_2$, $\det U = 1$ i.e., $U \in SU(2)$. Then U can be written as

$$U = \begin{pmatrix} a & b \\ -b^* & a^* \end{pmatrix}, \qquad a, b \in \mathbb{C}$$

with the constraint $aa^* + bb^* = 1$. Let

$$\begin{pmatrix} z_1' \\ z_2' \end{pmatrix} = \begin{pmatrix} a & b \\ -b^* & a^* \end{pmatrix} \begin{pmatrix} z_1 \\ z_2 \end{pmatrix}.$$

Show that

$$(z_1')(z_1')^* + (z_2')(z_2')^* = z_1 z_1^* + z_2 z_2^*.$$

(ii) Consider

$$\begin{pmatrix} z_1' \\ z_2' \end{pmatrix} = \begin{pmatrix} a & b \\ -b^* & a^* \end{pmatrix} \begin{pmatrix} z_1 \\ z_2 \end{pmatrix}.$$

Show that $dz_1' \wedge dz_2' = dz_1 \wedge dz_2$, where $\wedge$ is the *exterior product* and d the *exterior derivative*. Note that d is linear.

Solution 97. (i) We have

$$z_1' = az_1 + bz_2, \qquad z_2' = -b^* z_1 + a^* z_2.$$

Thus

$$(z_1')(z_1')^* + (z_2')(z_2')^* = (aa^* + bb^*)z_1 z_1^* + (bb^* + aa^*)z_2 z_2^* = z_1 z_1^* + z_2 z_2^*.$$

(ii) We have

$$dz_1' = d(az_1 + bz_2) = adz_1 + bdz_2$$
$$dz_2' = d(-b^*z_1 + a^*z_2) = -b^*dz_1 + a^*dz_2.$$

Since $dz_1 \wedge dz_1 = 0$, $dz_2 \wedge dz_2 = 0$ and $dz_1 \wedge dz_2 = -dz_2 \wedge dz_1$ we obtain

$$dz_1' \wedge dz_2' = (aa^* + bb^*)(dz_1 \wedge dz_2) = dz_1 \wedge dz_2.$$

Problem 98. (i) Consider the 2×2 matrix

$$A = \begin{pmatrix} 0 & 1 \\ -1 & 0 \end{pmatrix}.$$

Let $t \in \mathbb{R}$. Find $\exp(tA)$.
(ii) Consider the *vector field*

$$V = x_2 \frac{\partial}{\partial x_1} - x_1 \frac{\partial}{\partial x_2}$$

defined on $\mathbb{R}^2$. Let $t \in \mathbb{R}$. Calculate

$$\exp(tV) \begin{pmatrix} x_1 \\ x_2 \end{pmatrix} \equiv \begin{pmatrix} \exp(tV)x_1 \\ \exp(tV)x_2 \end{pmatrix}, \qquad \begin{pmatrix} x_1 \\ x_2 \end{pmatrix} \in \mathbb{R}^2.$$

(iii) Use the results from either (i) or (ii) to solve the initial value problem of the system of linear first order differential equations

$$\frac{dx_1}{dt} = x_2, \qquad \frac{dx_2}{dt} = -x_1.$$

Solution 98. (i) Since $A^2 = -I_2$, $A^3 = -A$ we obtain

$$\exp(tA) = I_2 \cos(t) + A \sin(t) = \begin{pmatrix} \cos(t) & \sin(t) \\ -\sin(t) & \cos(t) \end{pmatrix}.$$

(ii) We obtain

$$\begin{pmatrix} x_1(t) \\ x_2(t) \end{pmatrix} = \begin{pmatrix} e^{tV}x_1 \\ e^{tV}x_2 \end{pmatrix} = \begin{pmatrix} x_1 \cos(t) + x_2 \sin(t) \\ -x_1 \sin(t) + x_2 \cos(t) \end{pmatrix}.$$

(iii) From (ii) we obtain the solution of the initial value problem

$$x_1(t) = x_1(0)\cos(t) + x_2(0)\sin(t), \qquad x_2(t) = x_2(0)\cos(t) - x_1(0)\sin(t).$$

Problem 99.   Consider the vector fields

$$x\frac{d}{dx}, \qquad \frac{d}{dx}.$$

The manifold is $M = \mathbb{R}$, i.e. $x \in \mathbb{R}$. Calculate

$$\exp\left(tx\frac{d}{dx}\right)\exp\left(t\frac{d}{dx}\right)x, \qquad \exp\left(t\frac{d}{dx}\right)\exp\left(tx\frac{d}{dx}\right)x.$$

Solution 99.   For the first case we have

$$\exp\left(tx\frac{d}{dx}\right)\exp\left(t\frac{d}{dx}\right)x = \exp\left(tx\frac{d}{dx}\right)\left(1 + \frac{t}{1!}\frac{d}{dx} + \frac{t^2}{2!}\frac{d^2}{dx^2} + \cdots\right)x$$

$$= \exp\left(tx\frac{d}{dx}\right)(x+t)$$

$$= \exp\left(tx\frac{d}{dx}\right)x + \exp\left(tx\frac{d}{dx}\right)t$$

$$= xe^t + t.$$

Analogously for the second case we find

$$\exp\left(t\frac{d}{dx}\right)\exp\left(tx\frac{d}{dx}\right)x = e^t(x+t).$$

Problem 100.   Consider the nonlinear ordinary differential equation

$$\frac{dx}{dt} = \cos^2(x)$$

with the initial condition $x(t = 0) = x_0 = 0$. Find the solution of the initial value problem using the exponential map (*Lie series*)

$$x(t) = \exp(tV)x|_{x \to x_0}$$

where V is the corresponding vector field

$$V = \cos^2(x)\frac{d}{dx}.$$

Solution 100.   We find $x(t) = \arctan(t)$ with $x(0) = 0$. For $t \to \infty$ we obtain $x(t \to \infty) = \pi/2$.

Programming Problems

Problem 101. Consider the 2×2 matrix

$$A(\alpha_0, \alpha_1, \alpha_2) =$$

$$\begin{pmatrix} (\cos\alpha_0 \cos\alpha_1 + i\sin\alpha_0 \sin\alpha_1)e^{i\alpha_2} & -\cos\alpha_0 \sin\alpha_1 + i\sin\alpha_0 \cos\alpha_1 \\ \cos\alpha_0 \sin\alpha_1 + i\sin\alpha_0 \cos\alpha_1 & (\cos\alpha_0 \cos\alpha_1 - i\sin\alpha_0 \sin\alpha_1)e^{-i\alpha_2} \end{pmatrix}$$

where $-\pi \leq \alpha_0 < \pi$, $-\pi \leq \alpha_1 < \pi$, $0 \leq \alpha_2 \leq \pi$. Write a Computer algebra program in SymbolicC++ that calculates

$$\left. \frac{\partial}{\partial \alpha_j} A(\alpha_0, \alpha_1, \alpha_2) \right|_{\alpha_0=0, \alpha_1=0, \alpha_2=0}$$

and thus finds the generators

$$X_0 = \begin{pmatrix} 0 & i \\ i & 0 \end{pmatrix}, \quad X_1 = \begin{pmatrix} 0 & -1 \\ 1 & 0 \end{pmatrix}, \quad X_2 = \begin{pmatrix} i & 0 \\ 0 & -i \end{pmatrix}.$$

Thus

$$A(\alpha_0, \alpha_1, \alpha_2) = \exp(\alpha_0 X_0 + \alpha_1 X_1 + \alpha_2 X_2).$$

Solution 101. df indicates differentiation.

```
// generator.cpp

#include <iostream>
#include "symbolicc++.h"
using namespace std;

int main(void)
{
 using SymbolicConstant::i;
 Symbolic alpha("alpha",3);
 Symbolic A("A",2,2);

 A(0,0) = (cos(alpha(0))*cos(alpha(1))+i*sin(alpha(0))
          *sin(alpha(1)))*exp(i*alpha(2));
 A(0,1) = -cos(alpha(0))*sin(alpha(1))
          +i*sin(alpha(0))*cos(alpha(1));
 A(1,0) = cos(alpha(0))*sin(alpha(1))
          +i*sin(alpha(0))*cos(alpha(1));
 A(1,1) = (cos(alpha(0))*cos(alpha(1))-i*sin(alpha(0))
          *sin(alpha(1)))*exp(-i*alpha(2));

 cout << df(A,alpha(0))[alpha(0)==0,alpha(1)==0,alpha(2)==0]
```

```
       << endl;
cout << df(A,alpha(1))[alpha(0)==0,alpha(1)==0,alpha(2)==0]
       << endl;
cout << df(A,alpha(2))[alpha(0)==0,alpha(1)==0,alpha(2)==0]
       << endl;
return 0;
}
```

Problem 102.   Let $\alpha \in \mathbb{R}$. Given the matrix

$$A(\alpha) = \begin{pmatrix} \cosh\alpha & 0 & 0 & \sinh\alpha \\ 0 & \cos\alpha & \sin\alpha & 0 \\ 0 & -\sin\alpha & \cos\alpha & 0 \\ \sinh\alpha & 0 & 0 & \cosh\alpha \end{pmatrix}.$$

Write a SymbolicC++ program that calculates the vector

$$A(\alpha) \begin{pmatrix} x_0 \\ x_1 \\ x_2 \\ x_3 \end{pmatrix}$$

and shows that $x_0^2 + x_1^2 + x_2^2 - x_3^2$ is invariant.

Solution 102.   Note that $\sin^2\alpha + \cos^2\alpha = 1$ and $\cosh^2\alpha - \sinh^2\alpha = 1$.

```
// invariant.cpp

#include <iostream>
#include "symbolicc++.h"
using namespace std;

int main(void)
{
 Symbolic alpha("alpha");
 Symbolic A = ((cosh(alpha),0,0, sinh(alpha)),
               (Symbolic(0),cos(alpha),sin(alpha),0),
               (Symbolic(0),-sin(alpha),cos(alpha),0),
               (sinh(alpha),0,0, cosh(alpha)));
 Symbolic x("x",4);
 Symbolic y = A*x;
 Symbolic r = (y(0)^2) + (y(1)^2) + (y(2)^2) - (y(3)^2);
 r = r[(cos(alpha)^2)==1-(sin(alpha)^2),
       (cosh(alpha)^2)==1+(sinh(alpha)^2)];
 cout << r << endl;
 return 0;
}
```

Supplementary Problems

Problem 103. Show that the compact Lie groups $SO(3)$ and $SU(2)$ have isomorphic Lie algebras. A basis for $so(3)$ is given by the skew-symmetric matrices

$$T_1 = \begin{pmatrix} 0 & 0 & 0 \\ 0 & 0 & -1 \\ 0 & 1 & 0 \end{pmatrix}, \quad T_2 = \begin{pmatrix} 0 & 0 & 1 \\ 0 & 0 & 0 \\ -1 & 0 & 0 \end{pmatrix}, \quad T_3 = \begin{pmatrix} 0 & -1 & 0 \\ 1 & 0 & 0 \\ 0 & 0 & 0 \end{pmatrix}.$$

A basis for $su(2)$ is given by skew-hermitian matrices

$$\tau_1 = \frac{1}{2} \begin{pmatrix} 0 & -i \\ -i & 0 \end{pmatrix}, \quad \tau_2 = \frac{1}{2} \begin{pmatrix} 0 & -1 \\ 1 & 0 \end{pmatrix}, \quad \tau_3 = \frac{1}{2} \begin{pmatrix} -i & 0 \\ 0 & i \end{pmatrix}.$$

Thus $\tau_1 = -i\sigma_1/2$, $\tau_2 = -i\sigma_2/2$, $\tau_3 = -i\sigma_3/2$, where σ_1, σ_2, σ_3 denote the Pauli spin matrices.

Problem 104. Let $n \geq 1$. We define the $(2n+1)$-sphere

$$S^{2n+1} := \{\, \mathbf{z} \in \mathbb{C}^{n+1} : \sum_{j=1}^{n+1} |z_j|^2 = 1 \,\}.$$

Show that

$$S^{2n+1} \cong U(n+1)/U(n) \cong SU(n+1)/SU(n).$$

For $n = 1$ one has $S^3 \cong U(2)/U(1) \cong SU(2)$.

Problem 105. (i) Consider the Pauli spin matrices

$$\sigma_1 = \begin{pmatrix} 0 & 1 \\ 1 & 0 \end{pmatrix}, \quad \sigma_2 = \begin{pmatrix} 0 & -i \\ i & 0 \end{pmatrix}, \quad \sigma_3 = \begin{pmatrix} 1 & 0 \\ 0 & -1 \end{pmatrix}.$$

Let $\alpha_1, \alpha_2, \alpha_3 \in \mathbb{R}$. Calculate

$$e^{i\alpha_1\sigma_1 + i\alpha_2\sigma_2 + i\alpha_3\sigma_3} \quad \text{and} \quad e^{i\alpha_1\sigma_1} e^{i\alpha_2\sigma_2} e^{i\alpha_3\sigma_3}.$$

Discuss.

(ii) A basis for $so(3)$ is given by the skew-symmetric matrices

$$T_1 = \begin{pmatrix} 0 & 0 & 0 \\ 0 & 0 & -1 \\ 0 & 1 & 0 \end{pmatrix}, \quad T_2 = \begin{pmatrix} 0 & 0 & 1 \\ 0 & 0 & 0 \\ -1 & 0 & 0 \end{pmatrix}, \quad T_3 = \begin{pmatrix} 0 & -1 & 0 \\ 1 & 0 & 0 \\ 0 & 0 & 0 \end{pmatrix}.$$

Let $\alpha_1, \alpha_2, \alpha_3 \in \mathbb{R}$. Calculate

$$e^{\alpha_1 T_1 + \alpha_2 T_2 + \alpha_3 T_3} \qquad \text{and} \qquad e^{i\alpha_1 T_1} e^{i\alpha_2 T_2} e^{i\alpha_3 T_3} .$$

Discuss.

Problem 106. Show that any element U of $SU(n)$ can be diagonalized by an $n \times n$ unitary matrix and the diagonal elements are given by $\exp(i\phi_{kk})$.

Problem 107. The Pauli matrices σ_1, σ_2, σ_3 and the 2×2 identity matrix $I_2 = \sigma_0$ are elements of the Lie group $U(2)$. What is the condition on $c_j \in \mathbb{C}$ $(j = 0, 1, 2, 3)$ such that

$$\sum_{j=0}^{3} c_j \sigma_j \otimes \sigma_j$$

is an invertible matrix, i.e. an element of $GL(2, \mathbb{C})$?

Problem 108. The matrix

$$U = \begin{pmatrix} 0 & 0 & i \\ 0 & i & 0 \\ i & 0 & 0 \end{pmatrix}$$

is an element of the Lie group $U(3)$. Find a 3×3 unitary matrix V such that $V^{-1}UV$ is a diagonal matrix.

Problem 109. (i) Show that the group $SO^+(3, 1)$ is isomorphic to $SL(2, \mathbb{C})/\{\pm \mathrm{id}\}$ the group of 2×2 complex matrices with determinant 1. The quotient means that two matrices differing by an overall sign are identified.
(ii) Show that $SL(2, \mathbb{C})/\{\mathrm{id}\}$ is isomorphic to the group of complex Möbius transformations in the extended complex plane $\mathbb{C} \cup \{\infty\}$.

Problem 110. Let G denote the five-dimensional manifold $\mathbb{C} \times \mathbb{C} \times \mathbb{R}$ with multiplication defined as follows

$$(c_1, c_2, r)(c_1', c_2', r') := (c_1 + e^{2\pi i r} c_1', c_2 + e^{2\pi i h r} c_2', r + r')$$

where h is a fixed irrational number and $c_1, c_1', c_2, c_2' \in \mathbb{C}$, $r, r' \in \mathbb{R}$. Show that G is a Lie group.

Chapter 3

Lie Algebras

A Lie algebra L is a vector space over a field $\mathbb{F}$ together with a bilinear map, $[\,,\,] : L \times L \to L$, called the Lie bracket, such that the following identities hold for all $a, b, c \in L$

$$[a, a] = 0$$

and the so-called *Jacobi identity*

$$[a, [b, c]] + [c, [a, b]] + [b, [c, a]] = 0\,.$$

If $\mathrm{char}\mathbb{F} \neq 2$ it follows that $[b, a] = -[a, b]$. Unless otherwise stated we assume that $\mathrm{char}\mathbb{F} \neq 2$. A Lie algebra is said to be abelian (or commutative) if $[a, b] = 0$ for all $x, y \in L$.

If $\mathcal{A}$ is an associative algebra over a field $\mathbb{F}$ (for example, the $n \times n$ matrices over $\mathbb{C}$ and matrix multiplication) with the definition

$$[a, b] := ab - ba, \quad a, b \in \mathcal{A}$$

then $\mathcal{A}$ acquires the structure of a Lie algebra.

If $X \subseteq L$ then $\langle X \rangle$ denotes the *Lie subalgebra* generated by X, that is, the smallest Lie subalgebra of L containing X. It consists of all elements obtainable from X by a finite sequence of vector space operations and Lie bracket operations. A set of generators for L is a subset $X \subseteq L$ such that $L = \langle X \rangle$. If L has a finite set of generators we say that it is finitely generated.

A subspace I of a Lie algebra L is called an *ideal* of the Lie algebra L if $a \in L$, $b \in I$ together imply $[a, b] \in I$. Now 0 (the sub vector space consisting only of the zero element) and L itself are ideals of L. The *center* $Z(L)$ of a Lie algebra L is defined by

$$Z(L) := \{ c \in L : [c, a] = 0 \text{ for all } a \in L \}.$$

Obviously the center is an ideal.

If a non-abelian Lie algebra L has no ideals except itself and 0 and if $[L, L] \neq 0$ one calls L *simple*. Thus a Lie algebra L is called simple if L is non-abelian and has no proper ideals.

A Lie algebra is called *semisimple* if it is a direct sum of simple Lie algebras, i.e. non-abelian Lie algebras whose only ideals are $\{ 0 \}$ and the Lie algebra L itself. A Lie algebra is called *reductive* if it is the sum of a semisimple and an abelian Lie algebra.

The *radical* of a Lie algebra is the largest solvable ideal of the Lie algebra. If the radical is not zero, then there is a nonzero solvable ideal. The subalgebra is solvable if its derived series terminates in the zero subalgebra. Consider the next to last element of the derived series. The commutators of all its elements are zero. Therefore it is an abelian subalgebra. Thus an algebra, which contains a solvable ideal, also contains an abelian ideal. If L has an abelian ideal, it is not semisimple.

Let L be a Lie algebra. The set of commutators $[L, L]$ is an ideal of L, called $L^{(1)}$. Similarly $L^{(2)} = [L^{(1)}, L^{(1)}]$ is an ideal of $L^{(1)}$. One defines

$$L^{(n+1)} := [L^{(n)}, L^{(n)}].$$

If this sequence terminates in the zero element, one says that L is *solvable*. Let L be a Lie algebra. Consider a sequence of ideals given by

$$L_{(1)} = [L, L], \quad L_{(2)} = [L, L_{(1)}], \ldots, L_{(n+1)} = [L, L_{(n)}].$$

We say that L is *nilpotent* if this sequence terminates in zero.

The direct sum of a semisimple and a commutative Lie algebra is called a reductive Lie algebra.

The *Levi decomposition theorem* states that every finite dimensional Lie algebra can be written as the semidirect product of a solvable ideal (its radicals) and a semisimple Lie algebra.

For semisimple Lie algebras one can introduce new basis elements H_j $(j = 1, \ldots, r)$, $E_{\boldsymbol{\alpha}}$, $E_{-\boldsymbol{\alpha}}$ such that

$$[H_j, H_k] = 0$$
$$[H_j, E_{\boldsymbol{\alpha}}] = \alpha_j E_{\boldsymbol{\alpha}}$$
$$[E_{\boldsymbol{\alpha}}, E_{\boldsymbol{\beta}}] = C_{\boldsymbol{\alpha}\boldsymbol{\beta}}^{\boldsymbol{\alpha}+\boldsymbol{\beta}} E_{\boldsymbol{\alpha}+\boldsymbol{\beta}}$$
$$[E_{\boldsymbol{\alpha}}, E_{-\boldsymbol{\alpha}}] = \sum_{j=1}^{r} \alpha_j H_j$$

where

$$\boldsymbol{\alpha} = (\alpha_1, \alpha_2, \ldots, \alpha_r)^T \in \mathbb{F}^n \,.$$

This is called a *Cartan-Weyl basis*. We can write

$$\begin{pmatrix} [H_1, E_{\boldsymbol{\alpha}}] \\ [H_2, E_{\boldsymbol{\alpha}}] \\ \vdots \\ [H_r, E_{\boldsymbol{\alpha}}] \end{pmatrix} = \begin{pmatrix} \alpha_1 \\ \alpha_2 \\ \vdots \\ \alpha_r \end{pmatrix} E_{\boldsymbol{\alpha}} = \boldsymbol{\alpha} E_{\boldsymbol{\alpha}} \,.$$

One calls $\boldsymbol{\alpha}$ a root and $E_{\boldsymbol{\alpha}}$ a *ladder operator*. The rank of a semisimple Lie algebra is r, i.e. the number of the elements H_j.

Given two Lie algebras L_1 and L_2. A *homomorphism* of Lie algebras is a function, $f : L_1 \to L_2$, that is a linear map between vector spaces L_1 and L_2 and that preserves Lie brackets, i.e.,

$$f([a, b]) = [f(a), f(b)]$$

for all $a, b \in L_1$.

A representation of a Lie algebra L on a vector space V is a map of Lie algebras

$$\rho : L \to g\ell(V) = \mathrm{End}(V)$$

i.e., a linear map that preserves the commutators, or an action of L on V such that

$$[X, Y](v) = X(Y(v)) - Y(X(v)), \quad v \in V \,.$$

Representations of a connected and simple connected Lie group are in one-to-one correspondence with representations of its Lie algebra.

If L is a Lie algebra and $x \in L$, the linear operator $\mathrm{ad}x$ that maps $y \in L$ to $[x, y]$ is a linear transformation of L onto itself, i.e.

$$(\mathrm{ad}\,x)y := [x, y] \,.$$

Then $x \mapsto \text{ad } x$ is a representation of L with L itself considered as the vector space of the representation. This is called the *adjoint representation*.

The *Killing form* of a Lie algebra L is the map $\kappa : L \times L \to \mathbb{F}$ such that $(a, b \in L)$
$$\kappa(a, b) = \text{tr}((\text{ad}(a))(\text{ad}(b)))$$
where tr denotes the trace of a square matrix.

The *Cartan matrix* of a rank r root system is an $r \times r$ matrix whose entries are derived from the simple roots. The entries of the Cartan matrix are given by
$$A_{jk} = 2\frac{(\alpha_j, \alpha_k)}{(\alpha_j, \alpha_j)}$$
where $(\cdot, \cdot)$ is the Euclidean inner product and α_j are the simple roots. A Cartan matrix is a square matrix $A = (a_{ij})$ with integer entries such that
1) For diagonal entries, $a_{jj} = 2$.
2) For non-diagonal entries $a_{jk} \leq 0$.
3) $a_{jk} = 0$ if and only if $a_{kj} = 0$.
4) A can be written as DS, where D is a diagonal matrix, and S is a symmetric matrix.
The third condition is not independent but is a consequence of the first and fourth conditions.

The finite dimensional simple Lie algebras have been classified around 1900 by Killing and Cartan. In this classification four infinite series and five exceptional Lie algebras occur. The Lie algebras of the four infinite series are called classical Lie algebras. They are

$$
\begin{array}{lll}
A_n & \text{or} & s\ell(n+1) \\
B_n & \text{or} & so(2n+1) \\
C_n & \text{or} & sp(2n) \\
D_n & \text{or} & so(2n) \,.
\end{array}
$$

The five exceptional simple Lie algebras are e_6 (dimension 78), e_7 (dimension 133), e_8 (dimension 248), f_4 (dimension 52) and g_2 (dimension 14).

Lie group theory reduces locally to the theory of Lie algebras.

Problem 1. Let L be a Lie algebra and $a, b, h \in L$. Assume that

$$[a, h] = 0, \qquad [b, h] = 0.$$

Calculate $[[a, b], h]$.

Solution 1. Using the Jacobi identity we obtain $[[a, b], h] = 0$.

Problem 2. Let e_1, e_2, e_3 be the generators of a three dimensional Lie algebra with the commutation relation

$$[e_1, e_3] = e_1, \quad [e_2, e_3] = e_1 + e_2, \quad [e_1, e_2] = 0.$$

Find

$$[e_1, [e_2, e_3]] + [e_3, [e_1, e_2]].$$

Solution 2. From the Jacobi identity we have

$$[e_1, [e_2, e_3]] + [e_2, [e_3, e_1]] + [e_3, [e_1, e_2]] = 0.$$

This means

$$[e_1, [e_2, e_3]] + [e_3, [e_1, e_2]] = -[e_2, [e_3, e_1]] = -[e_2, -e_1] = -[e_1, e_2] = 0.$$

Thus we have a two dimensional abelian sub Lie algebra consisting of the elements e_1 and e_2.

Problem 3. Let L be a three dimensional Lie algebra with basis a, b, c and commutation relations (solvable Lie algebra)

$$[a, b] = 0, \quad [b, c] = a + kb, \quad [c, a] = -ka + b, \qquad k > 0.$$

(i) Find the commutator $[[b, c], [c, a]]$.
(ii) Check the Jacobi identity

$$[[a, b], c] + [[c, a], b] + [[b, c], a] = 0.$$

Solution 3. (i) We find

$$[[b, c], [c, a]] = [a + kb, -ka + b] = -k[a, a] + [a, b] - k^2[b, a] + k[b, b] = 0.$$

(ii) We have

$$\begin{aligned}
[[a, b], c] + [[c, a], b] + [[b, c], a] &= [0, c] + [-ka + b, b] + [a + kb, a] \\
&= [-ka, b] + [b, b] + [a, a] + [kb, a] \\
&= 0.
\end{aligned}$$

Problem 4.   Let L be a Lie algebra and let I be an ideal of L. We define the bracket $[\,,\,] : L/I \to L/I$ by

$$[a + I, b + I] := [a, b] + I$$

where $a, b \in L$. Show that the quotient vector space becomes a Lie algebra under this bracket.

Solution 4.   Let

$$a + I = a' + I \in L/I, \qquad b + I = b' + I \in L/I.$$

Then

$$[a, b] + I = [a + I, b + I] = [a' + I, b' + I] = [a', b'] + I.$$

Thus $[a, b] - [a', b'] \in I$ and $[a, b] = [a', b'] + i$ for some $i \in I$. Thus $[a', b'] \in [a, b] + I$. Now

$$[a + I, a + I] = [a, a] + I = I$$

and (Jacobi identity)

$$[a + I, [b + I, c + I]] + [c + I, [a + I, b + I]] + [b + I, [c + I, a + I]] =$$

$$[a, [b, c]] + [c, [a, b]] + [b, [c, a]] + I = I.$$

Problem 5.   Let L be a Lie algebra. A linear map $D : L \to L$ is called a *derivation* if

$$D[a, b] = [Da, b] + [a, Db]$$

for all $a, b \in L$. The collection of all derivatives of L is denoted by $\mathrm{Der}(L)$. Let $n \in \mathbb{N}$. Show that (*Leibniz rule*)

$$D^n[a, b] = \sum_{j=0}^{n} {}^n C_j [D^j a, D^{n-j} b]$$

where the *binomial coefficients* ${}^n C_j$ are defined for an arbitrary field $\mathbb{F}$

$${}^n C_0 = {}^n C_n = 1, \qquad {}^n C_j = {}^{n-1} C_{j-1} + {}^{n-1} C_j \quad \text{for} \quad 0 < j < n.$$

Solution 5.   The proof is by induction. The basis of induction is the definition of the derivation, where the binomial coefficients for $n = 1$ are ${}^1 C_0 = {}^1 C_1 = 1$. The induction step uses the special form of the definition

$$D\left([D^j a, D^{n-j} b]\right) = [D^{j+1} a, D^{n-j} b] + [D^j a, D^{n-j+1} b].$$

It follows that

$$D^{n+1}[a,b] = D\left(D^n[a,b]\right) = D\left(\sum_{j=0}^{n} {}^nC_j\,[D^j a, D^{n-j}b]\right)$$

$$= \sum_{j=0}^{n} {}^nC_j\left([D^{j+1}a, D^{n-j}b] + [D^j a, D^{n-j+1}b]\right)$$

$$= \sum_{j=0}^{n+1}[D^j a, D^{n-j+1}b]\,({}^nC_{j-1} + {}^nC_j)$$

$$= \sum_{j=0}^{n+1} {}^{n+1}C_j\,[D^j a, D^{n-j+1}b].$$

Problem 6. A Lie algebra L is called simple if L is non-abelian and has no proper ideals. Show that for a simple Lie algebra the derived Lie algebra $L^{(1)} = [L, L]$ is equal to L, i.e. $L^{(1)} = L$.

Solution 6. The derived Lie algebra is an ideal. Since L is simple, $L^{(1)}$ has to be a trivial ideal. One has only two alternatives: either $L^{(1)} = 0$ or $L^{(1)} = L$. The first option is not possible since L is non-abelian. Thus $L^{(1)} = L$ follows.

Problem 7. A Lie algebra L is called *semisimple* if $L \neq 0$ and L has no abelian ideals $\neq 0$. Show that a finite dimensional Lie algebra L is semisimple if and only if $\mathrm{rad}L = 0$.

Solution 7. We first show that $\mathrm{rad}L = 0$ implies that L is semisimple. Assume that $I \neq 0$ is an abelian ideal in L. Then I is a solvable ideal and it follows that

$$\mathrm{rad}L \supset I \neq 0\,.$$

However this contradicts $\mathrm{rad}L = 0$. Thus $I = 0$ and L is semisimple. Now assume that L is semisimple. Suppose that $R \equiv \mathrm{rad}L \neq 0$. Since R is solvable there exists in its derived sequence an ideal $R_{n-1} \neq 0$ such that

$$R_n = [R_{n-1}, R_{n-1}] = 0\,.$$

From this equation we see that $R_{n-1} \neq 0$ is abelian. This contradicts the semisimplicity of the Lie algebra L.

Problem 8. Show that the adjoint representation of a semisimple Lie algebra is faithful. A faithful representation of a Lie algebra is a linear representation in which different elements of L are represented by distinct linear mappings.

Solution 8.   There is a known theorem according which the adjoint representation of a Lie algebra L is isomorphic to $L/Z(L)$, where $Z(L)$ is the center of L. A semisimple algebra has not any proper abelian ideal, therefore its center $Z(L)$ is zero. Therefore the adjoint representation of a semisimple Lie algebra is isomorphic to L.

Problem 9.   Consider the Lie algebra with basis $\{e_1, e_2, e_3\}$ and the commutators

$$[e_1, e_2] = e_3, \qquad [e_2, e_3] = e_1, \qquad [e_3, e_1] = e_2.$$

Find the adjoint representation.

Solution 9.   From

$$(e_1 \; e_2 \; e_3)\text{ade}_1 = (0 \; -e_3 \; e_2)$$
$$(e_1 \; e_2 \; e_3)\text{ade}_2 = (e_3 \; 0 \; -e_1)$$
$$(e_1 \; e_2 \; e_3)\text{ade}_3 = (-e_2 \; e_1 \; 0)$$

we obtain

$$\text{ade}_1 = \begin{pmatrix} 0 & 0 & 0 \\ 0 & 0 & 1 \\ 0 & -1 & 0 \end{pmatrix}, \quad \text{ade}_2 = \begin{pmatrix} 0 & 0 & -1 \\ 0 & 0 & 0 \\ 1 & 0 & 0 \end{pmatrix}, \quad \text{ade}_3 = \begin{pmatrix} 0 & 1 & 0 \\ -1 & 0 & 0 \\ 0 & 0 & 0 \end{pmatrix}.$$

These 3×3 matrices are skew-symmetric over $\mathbb{R}$.

Problem 10.   Apply the *Levi decomposition theorem* to the finite dimensional Lie algebra $g\ell(2, \mathbb{R})$. Consider as basis

$$X = \begin{pmatrix} 0 & 1 \\ 0 & 0 \end{pmatrix}, \quad Y = \begin{pmatrix} 0 & 0 \\ 1 & 0 \end{pmatrix}, \quad H = \begin{pmatrix} 1 & 0 \\ 0 & -1 \end{pmatrix}, \quad I_2 = \begin{pmatrix} 1 & 0 \\ 0 & 1 \end{pmatrix}.$$

Solution 10.   The algebra is not semisimple. To show this we find the *adjoint representation*, the *Killing form* and then check that the Killing form is degenerate. The commutators are

$$[X, X] = 0, \quad [X, Y] = H, \quad [X, H] = -2X, \quad [X, I_2] = 0$$

$$[Y, X] = -H, \quad [Y, Y] = 0, \quad [Y, H] = 2Y, \quad [Y, I_2] = 0$$

$$[H, X] = 2X, \quad [H, Y] = -2Y, \quad [H, H] = 0, \quad [H, I_2] = 0$$
$$[I_2, X] = 0, \quad [I_2, Y] = 0, \quad [I_2, H] = 0, \quad [I_2, I_2] = 0.$$

Thus the adjoint representation is

$$\mathrm{ad}X = \begin{pmatrix} 0 & 0 & -2 & 0 \\ 0 & 0 & 0 & 0 \\ 0 & 1 & 0 & 0 \\ 0 & 0 & 0 & 0 \end{pmatrix}, \quad \mathrm{ad}Y = \begin{pmatrix} 0 & 0 & 0 & 0 \\ 0 & 0 & 2 & 0 \\ -1 & 0 & 0 & 0 \\ 0 & 0 & 0 & 0 \end{pmatrix},$$

$$\mathrm{ad}H = \begin{pmatrix} 2 & 0 & 0 & 0 \\ 0 & -2 & 0 & 0 \\ 0 & 0 & 0 & 0 \\ 0 & 0 & 0 & 0 \end{pmatrix}, \quad \mathrm{ad}I_2 = \begin{pmatrix} 0 & 0 & 0 & 0 \\ 0 & 0 & 0 & 0 \\ 0 & 0 & 0 & 0 \\ 0 & 0 & 0 & 0 \end{pmatrix}.$$

Consider the *Killing form*

$$\kappa(a, b) := \mathrm{tr}(\mathrm{ad}(a)\mathrm{ad}(b)).$$

To get its coefficients we multiply the matrices and calculate traces of the results. In the table these traces are presented in the intersection of the corresponding row and column.

$\kappa(a,b)$	X	Y	H	I_2
X	0	4	0	0
Y	4	0	0	0
H	0	0	8	0
I	0	0	0	0

Thus the Killing form is degenerate (one row and one column is zero) and therefore the Lie algebra $g\ell(2, \mathbb{R})$ is not semisimple. Note that $g\ell(2, \mathbb{R})$ is a semidirect product of its radical $\langle I_2 \rangle$ and $\langle X, Y, H \rangle$.

Problem 11. Consider a finite dimensional Lie algebra with basis elements $\{ X_\alpha : \alpha = 1, \dots, r \}$. The commutator may be expanded in terms of the basis

$$[X_\alpha, X_\beta] = \sum_{\gamma=1}^{r} C_{\alpha\beta}^{\gamma} X_\gamma \tag{1}$$

($\alpha, \beta = 1, \dots, r$) where $C_{\alpha\beta}^{\gamma}$ are called the *structure constants*. The structure constants $C_{\alpha\beta}^{\gamma}$ in (1) are constants. The structure constants, defined by the commutation relations (1) satisfy the relations

$$C_{\alpha\beta}^{\gamma} = -C_{\beta\alpha}^{\gamma}$$

$$\sum_{\rho=1}^{r} \left(C_{\alpha\beta}^{\rho} C_{\rho\gamma}^{\delta} + C_{\beta\gamma}^{\rho} C_{\rho\alpha}^{\delta} + C_{\gamma\alpha}^{\rho} C_{\rho\beta}^{\delta} \right) = 0$$

where α, β, $\gamma = 1, \ldots, r$. This is the Third Fundamental Theorem of Lie.

Solution 11.   Since $\{ X_\alpha : \alpha = 1, \ldots, r \}$ form a basis of a Lie algebra it follows that $[X_\alpha, X_\beta] = -[X_\beta, X_\alpha]$ where

$$[X_\alpha, X_\beta] = \sum_{\rho=1}^{r} C_{\alpha\beta}^{\rho} X_\rho$$

and

$$[X_\beta, X_\alpha] = \sum_{\rho=1}^{r} C_{\beta\alpha}^{\rho} X_\rho$$

so that $C_{\alpha\beta}^{\rho} = -C_{\alpha\beta}^{\rho}$. From the Jacobi identity we obtain

$$[X_\gamma, [X_\alpha, X_\beta]] + [X_\beta, [X_\gamma, X_\alpha]] + [X_\alpha[X_\beta, X_\gamma]] = 0$$

$$\Rightarrow [X_\gamma, \sum_{\rho=1}^{r} C_{\alpha\beta}^{\rho} X_\rho] + [X_\beta, \sum_{\rho=1}^{r} C_{\gamma\alpha}^{\rho} X_\rho] + [X_\alpha, \sum_{\rho=1}^{r} C_{\beta\gamma}^{\rho} X_\rho] = 0$$

$$\Rightarrow \sum_{\rho=1}^{r} C_{\alpha\beta}^{\rho} [X_\gamma, X_\rho] + \sum_{\rho=1}^{r} C_{\gamma\alpha}^{\rho} [X_\beta, X_\rho] + \sum_{\rho=1}^{r} C_{\beta\gamma}^{\rho} [X_\alpha, X_\rho] = 0$$

$$\Rightarrow \sum_{\rho=1}^{r} C_{\alpha\beta}^{\rho} \sum_{\delta=1}^{r} C_{\gamma\rho}^{\delta} X_\delta + \sum_{\rho=1}^{r} C_{\gamma\alpha}^{\rho} \sum_{\delta=1}^{r} C_{\beta\rho}^{\delta} X_\delta + \sum_{\rho=1}^{r} C_{\beta\gamma}^{\rho} \sum_{\delta=1}^{r} C_{\alpha\rho}^{\delta} X_\delta = 0$$

$$\Rightarrow \sum_{\delta=1}^{r} (\sum_{\rho=1}^{r} C_{\alpha\beta}^{\rho} C_{\gamma\rho}^{\delta} + \sum_{\rho=1}^{r} C_{\gamma\alpha}^{\rho} C_{\beta\rho}^{\delta} + \sum_{\rho=1}^{r} C_{\beta\gamma}^{\rho} C_{\alpha\rho}^{\delta}) X_\delta = 0.$$

Since the X_δ's are linearly independent the linear combination can only be zero if

$$\sum_{\rho=1}^{r} C_{\alpha\beta}^{\rho} C_{\gamma\rho}^{\delta} + \sum_{\rho=1}^{r} C_{\gamma\alpha}^{\rho} C_{\beta\rho}^{\delta} + \sum_{\rho=1}^{r} C_{\beta\gamma}^{\rho} C_{\alpha\rho}^{\delta} = 0.$$

Problem 12.   The Lie algebra $s\ell(2, \mathbb{R})$ is spanned by the 2×2 matrices

$$h = \begin{pmatrix} 1 & 0 \\ 0 & -1 \end{pmatrix}, \quad e = \begin{pmatrix} 0 & 1 \\ 0 & 0 \end{pmatrix}, \quad f = \begin{pmatrix} 0 & 0 \\ 1 & 0 \end{pmatrix}$$

with

$$[e, f] = h, \quad [h, e] = 2e, \quad [h, f] = -2f.$$

(i) Let x be a linear combination of the basis h, e, f

$$x = c_1 h + c_2 e + c_3 f$$

where $c_1, c_2, c_3 \in \mathbb{R}$. Calculate the commutators $[h, x]$, $[e, x]$, $[f, x]$.
(ii) Show that the Lie algebra $s\ell(2, \mathbb{R})$ is simple.

Solution 12. (i) We find

$$[h, x] = 2c_2 e - 2c_3 f$$
$$[e, x] = -2c_1 e + c_3 h$$
$$[f, x] = 2c_1 f - c_2 h.$$

(ii) The Lie algebra $s\ell(2, \mathbb{R})$ is non-abelian and has no proper ideals.

Problem 13. The Lie algebra $s\ell(2, \mathbb{R})$ is simple. Study the Lie algebra $s\ell(2, \mathbb{F})$, where $\mathrm{char}\mathbb{F} = 2$.

Solution 13. The statement $\mathrm{char}(\mathbb{F}) = 2$ means that the field contains only 2 elements: 0 and 1. Thus there are $2^4 = 16$ matrices 2×2 with entries from this field. Half of these matrices have the property $\mathrm{tr}(A) = 0$. They are

$$A_1 = \begin{pmatrix} 0 & 1 \\ 0 & 0 \end{pmatrix}, \quad A_2 = \begin{pmatrix} 0 & 0 \\ 1 & 0 \end{pmatrix}, \quad A_3 = \begin{pmatrix} 0 & 1 \\ 1 & 0 \end{pmatrix}, \quad A_4 = \begin{pmatrix} 1 & 0 \\ 0 & 1 \end{pmatrix},$$

$$A_5 = \begin{pmatrix} 1 & 1 \\ 0 & 1 \end{pmatrix}, \quad A_6 = \begin{pmatrix} 1 & 0 \\ 1 & 1 \end{pmatrix}, \quad A_7 = \begin{pmatrix} 1 & 1 \\ 1 & 1 \end{pmatrix}, \quad 0_2 = \begin{pmatrix} 0 & 0 \\ 0 & 0 \end{pmatrix}.$$

Thus all these matrices are linear combinations of generators A_1, A_2, A_4 with coefficients from the field $\mathbb{F}$. The commutators are given by

$[\,,\,]$	A_1	A_2	A_4
A_1	0	A_4	0
A_2	A_4	0	0
A_4	0	0	0

Obviously A_4 commutes with the rest of the elements. Entries on the main diagonal are zero, because the table is antisymmetric. Thus we must calculate only one entry of the table

$$[A_1, A_2] = \begin{pmatrix} 1 & 0 \\ 0 & -1 \end{pmatrix} = \begin{pmatrix} 1 & 0 \\ 0 & 1 \end{pmatrix} = A_4.$$

The last equality is due to the fact that $-1 = 1$ in modulo 2 arithmetic. The Lie algebra $s\ell(2, \mathbb{R})$ is simple, but this is not true for $s\ell(2, \mathbb{F})$, since it has the proper ideal A_4.

Problem 14. Show that for a simple Lie algebra L the derived algebra $L^{(1)} = [L, L]$ is equal to L.

Solution 14.   The derived algebra $L^{(1)}$ is an ideal. Since L is simple, $L^{(1)}$ is a trivial ideal. Thus either $L^{(1)} = 0$ or $L^{(1)} = L$. Since L is non-abelian we have $L^{(1)} = L$.

Problem 15.   The Lie algebra $g\ell(n, \mathbb{R})$ can be represented by $n \times n$ matrices over $\mathbb{R}$. Consider the standard basis for the vector space of all $n \times n$ matrices over $\mathbb{R}$ consisting of the matrices M_{ij} (having 1 in the (i, j)-position and 0 elsewhere). Find the commutator

$$[M_{ij}, M_{k\ell}].$$

Solution 15.   We find

$$[M_{ij}, M_{k\ell}] = \delta_{kj} M_{i\ell} - \delta_{\ell i} M_{kj}$$

where δ_{jk} is the Kronecker delta.

Problem 16.   The semisimple Lie algebra $s\ell(3, \mathbb{R})$ has dimension 8. The standard basis is given by

$$h_1 = \begin{pmatrix} 1 & 0 & 0 \\ 0 & -1 & 0 \\ 0 & 0 & 0 \end{pmatrix}, \quad h_2 = \begin{pmatrix} 0 & 0 & 0 \\ 0 & 1 & 0 \\ 0 & 0 & -1 \end{pmatrix},$$

$$e_1 = \begin{pmatrix} 0 & 1 & 0 \\ 0 & 0 & 0 \\ 0 & 0 & 0 \end{pmatrix}, \quad e_2 = \begin{pmatrix} 0 & 0 & 0 \\ 0 & 0 & 1 \\ 0 & 0 & 0 \end{pmatrix},$$

$$f_1 = \begin{pmatrix} 0 & 0 & 0 \\ 1 & 0 & 0 \\ 0 & 0 & 0 \end{pmatrix}, \quad f_2 = \begin{pmatrix} 0 & 0 & 0 \\ 0 & 0 & 0 \\ 0 & 1 & 0 \end{pmatrix},$$

$$e_{13} = \begin{pmatrix} 0 & 0 & 1 \\ 0 & 0 & 0 \\ 0 & 0 & 0 \end{pmatrix}, \quad f_{13} = \begin{pmatrix} 0 & 0 & 0 \\ 0 & 0 & 0 \\ 1 & 0 & 0 \end{pmatrix}.$$

Find the commutator table. Find the rank of the Lie algebra.

Solution 16. By direct calculation we find the following commutation table

[,]	h_1	h_2	e_1	e_{13}	f_1	e_2	f_{13}	f_2
h_1	0	0	$2e_1$	e_{13}	$-2f_1$	$-e_2$	$-f_{13}$	f_2
h_2	0	0	$-e_1$	e_{13}	f_1	$2e_2$	$-f_{13}$	$-2f_2$
e_1	$-2e_1$	e_1	0	0	h_1	e_{13}	$-f_2$	0
e_{13}	$-e_{13}$	$-e_{13}$	0	0	$-e_2$	0	$h_1 + h_2$	e_1
f_1	$2f_1$	$-f_1$	$-h_1$	e_2	0	0	0	$-f_{13}$
e_2	e_2	$-2e_2$	$-e_{13}$	0	0	0	f_1	h_2
f_{13}	f_{13}	f_{13}	f_2	$-h_1 - h_2$	0	$-f_1$	0	0
f_2	$-f_2$	$2f_2$	0	$-e_1$	f_{13}	$-h_2$	0	0

For example we have $[h_1, h_2] = 0$. The rank of the Lie algebra is 2.

Problem 17. Give a Lie algebra which is semisimple, but not simple. Provide a matrix representation.

Solution 17. Every semisimple algebra is isomorphic to the Cartesian product of simple algebras. The Lie algebra $s\ell(2, \mathbb{R})$ is simple. Therefore the matrix algebra with a block structure

$$\begin{pmatrix} m_1 & 0_2 \\ 0_2 & m_2 \end{pmatrix}, \quad m_1, m_2 \in s\ell(2, \mathbb{R})$$

is a semisimple 6-dimensional complex Lie algebra.

Problem 18. (i) Classify all real two-dimensional Lie algebras, where $\mathbb{F} = \mathbb{R}$. Denote the basis elements by X and Y.
(ii) Find a representation of the non-abelian two-dimensional Lie algebra using differential operators.
(iii) Find a representation of the non-abelian two-dimensional Lie algebra using 2×2 matrices.

Solution 18. (i) There exists only two non-isomorphic two-dimensional Lie algebras. Let X and Y span a two-dimensional Lie algebra. Then their commutator is either 0 or $[X, Y] = Z \neq 0$. In the first case we have a two-dimensional abelian Lie algebra. In the second case we have

$$[X, Y] = Z = \alpha X + \beta Y, \quad \alpha, \beta \in \mathbb{F}$$

where α and β are not both zero. Suppose $\alpha \neq 0$. It follows that

$$[Z, \alpha^{-1} Y] = Z.$$

Hence, for all non-abelian two-dimensional Lie algebras one can choose a basis $\{Z, U\}$ such that

$$[Z, U] = Z.$$

Consequently, all non-abelian two-dimensional Lie algebras are isomorphic.
(ii) An example is the set of differential operators

$$\left\{ \frac{d}{dx}, \, x\frac{d}{dx} \right\}$$

where $Z \to d/dx$ and $U \to xd/dx$.
(iii) Since $\operatorname{tr}([Z, U]) = 0$ we have $\operatorname{tr} Z = 0$. A representation using 2×2 matrices is

$$Z = \begin{pmatrix} 0 & 1 \\ 0 & 0 \end{pmatrix}, \qquad U = \begin{pmatrix} 0 & 0 \\ 0 & 1 \end{pmatrix}.$$

Problem 19. Let I_n be the $n \times n$ identity matrix. Can one find $n \times n$ matrices A, B over $\mathbb{C}$ such that $[A, B] = I_n$?

Solution 19. No. Since $\operatorname{tr}([A, B]) = 0$ and $\operatorname{tr} I_n = n$ we cannot find such matrices. Let C be an $n \times n$ matrix over $\mathbb{C}$ with $\operatorname{tr}(C) = 0$. Show that one can find $n \times n$ matrices A, B over $\mathbb{C}$ such that $[A, B] = C$.

Problem 20. (i) Let A be an associative algebra over a field $\mathbb{F}$. We define the commutator

$$[a, b] := ab - ba$$

for all $a, b \in A$. Show that with this definition of the commutator we have a Lie algebra.
(ii) Give an example.

Solution 20. (i) The map $[a, b] = ab - ba$ is by definition bilinear and antisymmetric and satisfies the Jacobi identity.
(ii) An example is the set of all $n \times n$ matrices over $\mathbb{C}$. For any $n \times n$ matrices A, B, C we have

$$[A, [B, C]] + [C, [A, B]] + [B, [C, A]] = 0_n.$$

Problem 21. A finite-dimensional Lie algebra over a field $\mathbb{F}$ of characteristic 0 (for example $\mathbb{R}$ and $\mathbb{C}$) is called *reductive* if its adjoint representation is completely reducible. Give an example of a Lie algebra which is reductive but not semisimple.

Solution 21. An example is the Lie algebra $so(2, \mathbb{C})$. This Lie algebra is one-dimensional commutative and thus reductive but not semisimple. Note that the Lie algebras $so(n, \mathbb{C})$ with $n \geq 3$ are semisimple.

Problem 22. Consider the vector space of analytic functions $f : \mathbb{R} \to \mathbb{R}$. We define the bracket

$$[f, g] := f \frac{dg}{dx} - \frac{df}{dx} g.$$

(i) Show that this defines an infinite dimensional Lie algebra.
(ii) Consider the analytic functions

$$f(x) = \sin(x), \quad g(x) = \cos(x), \quad h(x) = 1.$$

Show that these function for a basis of a finite dimensional Lie algebra. Is the Lie algebra semisimple?
(iii) Let $m, n \in \mathbb{Z}$. Consider

$$p(x) = x^m, \qquad q(x) = x^n.$$

Find $[p, q]$.

Solution 22. (i) We have $[f, g] = -[g, f]$. For the Jacobi identity we find

$$[[f, g], h] + [[h, f], g] + [[g, h], f] = 0.$$

(ii) We find

$$[\sin(x), \cos(x)] = -\sin(x)\sin(x) - \cos(x)\cos(x) = -1$$

and

$$[\sin(x), 1] = -\cos(x), \qquad [\cos(x), 1] = \sin(x).$$

From the commutation relations we see that the Lie algebra is semisimple.
(iii) We obtain

$$[p, q] = (n - m)x^{m+n-1}.$$

Problem 23. Let $f, g, p, q : \mathbb{R} \to \mathbb{R}$ be analytic functions. Consider the differential operators

$$f(x)\frac{d}{dx} + p(x), \qquad g(x)\frac{d}{dx} + q(x).$$

Calculate the commutator

$$\left[f(x)\frac{d}{dx} + p(x), g(x)\frac{d}{dx} + q(x) \right] h(x)$$

where h is an analytic function. Discuss.

Solution 23. We have $[p, q] = 0$. Straightforward calculation yields

$$\left[f\frac{d}{dx} + p, g\frac{d}{dx} + q\right]h = \left[f\frac{d}{dx}, g\frac{d}{dx}\right]h + \left[p, g\frac{d}{dx}\right]h + \left[f\frac{d}{dx}, q\right]h + [p, q]h$$

$$= f\frac{dg}{dx}\frac{dh}{dx} - g\frac{df}{dx}\frac{dh}{dx} + f\frac{dq}{dx}h - g\frac{dp}{dx}h$$

$$= \left(f\frac{dg}{dx} - g\frac{df}{dx}\right)\frac{dh}{dx} + \left(f\frac{dq}{dx} - g\frac{dp}{dx}\right)h.$$

Thus we find the differential operator

$$\left(f\frac{dg}{dx} - g\frac{df}{dx}\right)\frac{d}{dx} + \left(f\frac{dq}{dx} - g\frac{dp}{dx}\right)$$

which is of the same form as the given differential operators. The Jacobi identity is also satisfied for such differential operators. Thus these operators form a Lie algebra.

Problem 24. Let $f, g : W \to W$ be two maps, where W is a topological vector space ($u \in W$). Assume that the *Gateaux derivative* of f and g exists, i.e.

$$f'(u)[v] := \frac{\partial f(u + \epsilon v)}{\partial \epsilon}\Big|_{\epsilon=0}, \qquad g'(u)[v] := \frac{\partial g(u + \epsilon v)}{\partial \epsilon}\Big|_{\epsilon=0}. \qquad (1)$$

The *Lie product* (or *commutator*) of f and g is defined by

$$[f, g](u) := f'(u)[g(u)] - g'(u)[f(u)]. \qquad (2)$$

Let

$$f(u) := \frac{\partial u}{\partial t} - u\frac{\partial u}{\partial x} - \frac{\partial^3 u}{\partial x^3}, \qquad g(u) := \frac{\partial u}{\partial t} - \frac{\partial^2 u}{\partial x^2}. \qquad (3)$$

Calculate the commutator $[f, g](u)$.

Solution 24. For the first term on the right-hand side of (2) we find

$$f'(u)[g(u)] \equiv \frac{\partial}{\partial \epsilon}\left[\frac{\partial}{\partial t}(u + \epsilon g(u))\right]_{\epsilon=0}$$

$$- \frac{\partial}{\partial \epsilon}\left[(u + \epsilon g(u))\frac{\partial}{\partial x}(u + \epsilon g(u))\right]_{\epsilon=0}$$

$$- \frac{\partial}{\partial \epsilon}\left[\frac{\partial^3}{\partial x^3}(u + \epsilon g(u))\right]_{\epsilon=0}.$$

Consequently,

$$f'(u)[g(u)] = \frac{\partial g(u)}{\partial t} - g(u)\frac{\partial u}{\partial x} - u\frac{\partial g(u)}{\partial x} - \frac{\partial^3 g(u)}{\partial x^3}.$$

For the second term on the right-hand side of (2) we find

$$g'(u)[f(u)] = \frac{\partial}{\partial \epsilon}\left[\frac{\partial}{\partial t}(u + \epsilon f(u))\right]_{\epsilon=0} - \frac{\partial}{\partial \epsilon}\left[\frac{\partial^2}{\partial x^2}(u + \epsilon f(u))\right]_{\epsilon=0}.$$

Consequently

$$g'(u)[f(u)] = \frac{\partial f(u)}{\partial t} - \frac{\partial^2 f(u)}{\partial x^2}.$$

Hence

$$[f,g](u) = \frac{\partial}{\partial t}(g(u) - f(u)) - g(u)\frac{\partial u}{\partial x} - u\frac{\partial g(u)}{\partial x} - \frac{\partial^3 g(u)}{\partial x^3} + \frac{\partial^2 f(u)}{\partial x^2}.$$

Inserting f and g we find

$$[f,g](u) = \frac{\partial}{\partial t}\left(-\frac{\partial^2 u}{\partial x^2} + u\frac{\partial u}{\partial x} + \frac{\partial^3 u}{\partial x^3}\right) - \left(\frac{\partial u}{\partial t} - \frac{\partial^2 u}{\partial x^2}\right)\frac{\partial u}{\partial x}$$
$$-u\left(\frac{\partial^2 u}{\partial t\partial x} - \frac{\partial^3 u}{\partial x^3}\right) - \frac{\partial^4 u}{\partial t\partial x^3} + \frac{\partial^5 u}{\partial x^5} + \frac{\partial^3 u}{\partial x^2\partial t}$$
$$-3\frac{\partial u}{\partial x}\frac{\partial^2 u}{\partial x^2} - u\frac{\partial^3 u}{\partial x^3} - \frac{\partial^5 u}{\partial x^5}.$$

It follows that

$$[f,g](u) = -2\frac{\partial u}{\partial x}\frac{\partial^2 u}{\partial x^2}.$$

Problem 25.   Let L be the real vector space $\mathbb{R}^3$. Let $\mathbf{a}, \mathbf{b} \in L$. Define

$$\mathbf{a} \times \mathbf{b} := \begin{pmatrix} a_2 b_3 - a_3 b_2 \\ a_3 b_1 - a_1 b_3 \\ a_1 b_2 - a_2 b_1 \end{pmatrix}. \tag{1}$$

Equation (1) is the *cross* or *vector product*.
(i) Show that L is a Lie algebra.
(ii) Show that in general

$$\mathbf{a} \times (\mathbf{b} \times \mathbf{c}) \neq (\mathbf{a} \times \mathbf{b}) \times \mathbf{c}.$$

Solution 25.   (i) Obviously, we have

$$(\mathbf{a} + \mathbf{b}) \times (\mathbf{c} + \mathbf{d}) = \mathbf{a} \times \mathbf{c} + \mathbf{a} \times \mathbf{d} + \mathbf{b} \times \mathbf{c} + \mathbf{b} \times \mathbf{d}.$$

From (1) we see that $\mathbf{a} \times \mathbf{b} = -\mathbf{b} \times \mathbf{a}$. Using definition (1) we obtain

$$\mathbf{a} \times (\mathbf{b} \times \mathbf{c}) + \mathbf{c} \times (\mathbf{a} \times \mathbf{b}) + \mathbf{b} \times (\mathbf{c} \times \mathbf{a}) = \mathbf{0}.$$

This equation is called the *Jacobi identity*.
(ii) Let

$$\mathbf{a} := \begin{pmatrix} 1 \\ 0 \\ 0 \end{pmatrix}, \qquad \mathbf{b} := \begin{pmatrix} 0 \\ 1 \\ 0 \end{pmatrix}, \qquad \mathbf{c} := \begin{pmatrix} 0 \\ 0 \\ 1 \end{pmatrix}$$

be the standard basis in $\mathbb{R}^3$. Then

$$\mathbf{a} \times (\mathbf{b} \times \mathbf{c}) \neq (\mathbf{a} \times \mathbf{b}) \times \mathbf{c}.$$

Problem 26.   The simple Lie algebra $s\ell(2, \mathbb{R})$ is spanned by the 2×2 matrices

$$h = \begin{pmatrix} 1 & 0 \\ 0 & -1 \end{pmatrix}, \qquad e = \begin{pmatrix} 0 & 1 \\ 0 & 0 \end{pmatrix}, \qquad f = \begin{pmatrix} 0 & 0 \\ 1 & 0 \end{pmatrix}$$

with $[e, f] = h$, $[h, e] = 2e$, $[h, f] = -2f$. Show that the bracket relations are also satisfied by the matrices

$$\rho_2(h) = \begin{pmatrix} 2 & 0 & 0 \\ 0 & 0 & 0 \\ 0 & 0 & -2 \end{pmatrix},$$

$$\rho_2(e) = \begin{pmatrix} 0 & 2 & 0 \\ 0 & 0 & 1 \\ 0 & 0 & 0 \end{pmatrix}, \qquad \rho_2(f) = \begin{pmatrix} 0 & 0 & 0 \\ 1 & 0 & 0 \\ 0 & 2 & 0 \end{pmatrix}$$

and

$$\rho_3(h) = \begin{pmatrix} 3 & 0 & 0 & 0 \\ 0 & 1 & 0 & 0 \\ 0 & 0 & -1 & 0 \\ 0 & 0 & 0 & -3 \end{pmatrix},$$

$$\rho_3(e) = \begin{pmatrix} 0 & 3 & 0 & 0 \\ 0 & 0 & 2 & 0 \\ 0 & 0 & 0 & 1 \\ 0 & 0 & 0 & 0 \end{pmatrix}, \qquad \rho_3(f) = \begin{pmatrix} 0 & 0 & 0 & 0 \\ 1 & 0 & 0 & 0 \\ 0 & 2 & 0 & 0 \\ 0 & 0 & 3 & 0 \end{pmatrix}.$$

Solution 26.   Straightforward calculation yields

$$[\rho_2(e), \rho_2(f)] = \rho_2(h), \quad [\rho_2(h), \rho_2(e)] = 2\rho_2(e), \quad [\rho_2(h), \rho_2(f)] = -2\rho_2(f)$$

and

$$[\rho_3(e), \rho_3(f)] = \rho_3(h), \quad [\rho_3(h), \rho_3(e)] = 2\rho_3(e), \quad [\rho_3(h), \rho_3(f)] = -2\rho_3(f).$$

For general n we have the $(n+1) \times (n+1)$ matrices

$$\rho_n(h) = \begin{pmatrix} n & 0 & \cdots & \cdots & 0 \\ 0 & n-2 & \cdots & \cdots & 0 \\ \vdots & \vdots & \ddots & \vdots & \vdots \\ 0 & 0 & \cdots & -n+2 & 0 \\ 0 & 0 & \cdots & 0 & -n \end{pmatrix}$$

$$\rho_n(e) = \begin{pmatrix} 0 & n & \cdots & \cdots & 0 \\ 0 & 0 & n-1 & \cdots & 0 \\ \vdots & \vdots & \ddots & \vdots & \vdots \\ 0 & 0 & \cdots & \cdots & 1 \\ 0 & 0 & \cdots & \cdots & 0 \end{pmatrix}, \quad \rho_n(f) = \begin{pmatrix} 0 & 0 & \cdots & \cdots & 0 \\ 1 & 0 & \cdots & \cdots & 0 \\ \vdots & \vdots & \ddots & \vdots & \vdots \\ 0 & 0 & \cdots & 0 & 0 \\ 0 & 0 & \cdots & n & 0 \end{pmatrix}$$

as representations.

Problem 27. Consider as a basis for the simple Lie algebra $s\ell(2, \mathbb{R})$ the invertible matrices with trace 0

$$A = \begin{pmatrix} 0 & 1 \\ -1 & 0 \end{pmatrix}, \quad B = \begin{pmatrix} 1 & 0 \\ 0 & -1 \end{pmatrix}, \quad C = \begin{pmatrix} 0 & 1 \\ 1 & 0 \end{pmatrix}.$$

(i) Find the commutators.
(ii) Show that the eigenvalues of these matrices form a group under multiplication.

Solution 27. (i) The commutators are

$$[A, B] = -2C, \quad [A, C] = 2B, \quad [B, C] = 2A.$$

(ii) The eigenvalues are ± 1 and $\pm i$. Thus they form a commutative group under multiplication.

Problem 28. The Lie algebra $su(n)$ are the $n \times n$ matrices X with the conditions $X^* = -X$, $\mathrm{tr} X = 0$. Find a basis for $su(3)$ under the condition that the elements of the basis are orthogonal to each other with respect to the *scalar product*

$$\langle X, Y \rangle := \mathrm{tr}(XY^*).$$

Solution 28. Let σ_1, σ_2, σ_3 be the Pauli spin matrices. Consider the eight 3×3 matrices

$$S_j = \begin{pmatrix} 0 & 0 \\ 0 & -i\sigma_j \end{pmatrix}, \quad j = 1, 2, 3$$

$$S_4 = \begin{pmatrix} -2i & 0 & 0 \\ 0 & i & 0 \\ 0 & 0 & i \end{pmatrix}, \quad S_5 = \begin{pmatrix} 0 & -1 & 0 \\ 1 & 0 & 0 \\ 0 & 0 & 0 \end{pmatrix},$$

$$S_6 = \begin{pmatrix} 0 & 0 & -1 \\ 0 & 0 & 0 \\ 1 & 0 & 0 \end{pmatrix}, \quad S_7 = \begin{pmatrix} 0 & i & 0 \\ i & 0 & 0 \\ 0 & 0 & 0 \end{pmatrix},$$

$$S_8 = \begin{pmatrix} 0 & 0 & i \\ 0 & 0 & 0 \\ i & 0 & 0 \end{pmatrix}.$$

Obviously, the matrices $S_1, \ldots, S_8$ satisfy the conditions given above. From the equation

$$\sum_{j=1}^{8} c_j S_j = \begin{pmatrix} 0 & 0 & 0 \\ 0 & 0 & 0 \\ 0 & 0 & 0 \end{pmatrix}$$

we find that $c_1 = c_2 = \cdots = c_8 = 0$. Thus the matrices are linearly independent. Therefore the matrices S_j ($j = 1, 2, \ldots, 8$) form a basis for the Lie algebra $su(3)$. Furthermore, a 9 element basis cannot exist since that would be a basis for the 3×3 matrices and would not satisfy the conditions above.

Problem 29. The Lie algebra $su(4)$ consists of all 4×4 matrices over $\mathbb{C}$ such that $A^* = -A$, $\mathrm{tr}A = 0$. The first condition is that the matrices are skew-hermitian. Give a basis for the Lie algebra under the condition that the elements of the basis are orthogonal to each other with respect to the scalar product

$$\langle A, B \rangle := \mathrm{tr}(AB^*).$$

Solution 29. Since $\mathrm{tr}A = 0$ the dimension of the Lie algebra is 16-1=15. We can built the basis on three diagonal matrices with trace 0 and 12 non-diagonal matrices. We find

$$g_1 = \begin{pmatrix} 0 & i & 0 & 0 \\ i & 0 & 0 & 0 \\ 0 & 0 & 0 & 0 \\ 0 & 0 & 0 & 0 \end{pmatrix}, \quad g_2 = \begin{pmatrix} 0 & 1 & 0 & 0 \\ -1 & 0 & 0 & 0 \\ 0 & 0 & 0 & 0 \\ 0 & 0 & 0 & 0 \end{pmatrix}, \quad g_3 = \begin{pmatrix} i & 0 & 0 & 0 \\ 0 & -i & 0 & 0 \\ 0 & 0 & 0 & 0 \\ 0 & 0 & 0 & 0 \end{pmatrix},$$

$$g_4 = \begin{pmatrix} 0 & 0 & i & 0 \\ 0 & 0 & 0 & 0 \\ i & 0 & 0 & 0 \\ 0 & 0 & 0 & 0 \end{pmatrix}, \quad g_5 = \begin{pmatrix} 0 & 0 & 1 & 0 \\ 0 & 0 & 0 & 0 \\ -1 & 0 & 0 & 0 \\ 0 & 0 & 0 & 0 \end{pmatrix}, \quad g_6 = \begin{pmatrix} 0 & 0 & 0 & 0 \\ 0 & 0 & i & 0 \\ 0 & i & 0 & 0 \\ 0 & 0 & 0 & 0 \end{pmatrix},$$

$$g_7 = \begin{pmatrix} 0 & 0 & 0 & 0 \\ 0 & 0 & 1 & 0 \\ 0 & -1 & 0 & 0 \\ 0 & 0 & 0 & 0 \end{pmatrix}, \quad g_8 = \frac{1}{\sqrt{3}} \begin{pmatrix} i & 0 & 0 & 0 \\ 0 & i & 0 & 0 \\ 0 & 0 & -2i & 0 \\ 0 & 0 & 0 & 0 \end{pmatrix}, \quad g_9 = \begin{pmatrix} 0 & 0 & 0 & i \\ 0 & 0 & 0 & 0 \\ 0 & 0 & 0 & 0 \\ i & 0 & 0 & 0 \end{pmatrix},$$

$$g_{10} = \begin{pmatrix} 0 & 0 & 0 & 1 \\ 0 & 0 & 0 & 0 \\ 0 & 0 & 0 & 0 \\ -1 & 0 & 0 & 0 \end{pmatrix}, \quad g_{11} = \begin{pmatrix} 0 & 0 & 0 & 0 \\ 0 & 0 & 0 & i \\ 0 & 0 & 0 & 0 \\ 0 & i & 0 & 0 \end{pmatrix}, \quad g_{12} = \begin{pmatrix} 0 & 0 & 0 & 0 \\ 0 & 0 & 0 & 1 \\ 0 & 0 & 0 & 0 \\ 0 & -1 & 0 & 0 \end{pmatrix},$$

$$g_{13} = \begin{pmatrix} 0 & 0 & 0 & 0 \\ 0 & 0 & 0 & 0 \\ 0 & 0 & 0 & i \\ 0 & 0 & i & 0 \end{pmatrix}, \quad g_{14} = \begin{pmatrix} 0 & 0 & 0 & 0 \\ 0 & 0 & 0 & 0 \\ 0 & 0 & 0 & 1 \\ 0 & 0 & -1 & 0 \end{pmatrix}, \quad g_{15} = \frac{1}{\sqrt{6}} \begin{pmatrix} i & 0 & 0 & 0 \\ 0 & i & 0 & 0 \\ 0 & 0 & i & 0 \\ 0 & 0 & 0 & -3i \end{pmatrix}.$$

Problem 30. Let X be a 2×2 matrix over $\mathbb{C}$. Let

$$L = \begin{pmatrix} -1 & 0 \\ 0 & 1 \end{pmatrix}.$$

(i) Find the condition on X such that $X^*L + LX = 0_2$. Then X is an element of the Lie algebra $u(1,1)$.
(ii) Give a basis of the Lie algebra.

Solution 30. (i) From the equation $X^*L + LX = 0_2$ we obtain the four equations

$$x_{11} + x_{11}^* = 0, \quad x_{22} + x_{22}^* = 0, \quad x_{12} - x_{12}^* = 0, \quad x_{21} - x_{21}^* = 0.$$

Let $a, b, c, d \in \mathbb{R}$ we can write X as

$$X = \begin{pmatrix} ia & c \\ d & ib \end{pmatrix}.$$

(ii) Thus a basis of the Lie algebra $u(1,1)$ is given by

$$\left\{ \begin{pmatrix} i & 0 \\ 0 & i \end{pmatrix}, \begin{pmatrix} i & 0 \\ 0 & -i \end{pmatrix}, \begin{pmatrix} 0 & 1 \\ 0 & 0 \end{pmatrix}, \begin{pmatrix} 0 & 0 \\ 1 & 0 \end{pmatrix} \right\}.$$

Is the Lie algebra semisimple?

Problem 31. There is only one simple complex Lie algebra of rank 1. If $\alpha = \lambda e_1$ is a nonzero root, $-\lambda e_1$ is the other. The canonical choice for λ is $1/\sqrt{2}$. The commutators for the basis

$$E_{-1/\sqrt{2}}, \quad H_1, \quad E_{+1/\sqrt{2}}$$

are

$$[H_1, E_{\pm 1/\sqrt{2}}] = \pm \frac{1}{\sqrt{2}} E_{\pm 1/\sqrt{2}}, \qquad [E_{+1/\sqrt{2}}, E_{-1/\sqrt{2}}] = \frac{1}{\sqrt{2}} H_1 .$$

Find a 2×2 matrix representation for this semisimple Lie algebra.

Solution 31.   Obviously the trace of the 2×2 matrices $E_{-1/\sqrt{2}}$, H_1, $E_{1/\sqrt{2}}$ must be 0. We find

$$H_1 = \frac{1}{2\sqrt{2}} \begin{pmatrix} 1 & 0 \\ 0 & -1 \end{pmatrix}, \quad E_{+1/\sqrt{2}} = \frac{1}{2} \begin{pmatrix} 0 & 1 \\ 0 & 0 \end{pmatrix}, \quad E_{-1/\sqrt{2}} = \frac{1}{2} \begin{pmatrix} 0 & 0 \\ 1 & 0 \end{pmatrix} .$$

Problem 32.   Consider the Lie algebra $g\ell(2, \mathbb{R})$. The 2×2 matrices

$$e_1 = \begin{pmatrix} 1 & 0 \\ 0 & 0 \end{pmatrix}, \quad e_2 = \begin{pmatrix} 0 & 1 \\ 0 & 0 \end{pmatrix}, \quad e_3 = \begin{pmatrix} 0 & 0 \\ 1 & 0 \end{pmatrix}, \quad e_4 = \begin{pmatrix} 0 & 0 \\ 0 & 1 \end{pmatrix}$$

form a basis of $g\ell(2, \mathbb{R})$.
(i) Find the *adjoint representation*.
(ii) Find the Killing form and the metric tensor for the Lie algebra. From these results decide whether the Lie algebra is semisimple.

Solution 32.   We have $\mathrm{ad}(e_j)e_k = [e_j, e_k]$. We find the commutators

$$[e_1, e_1] = 0_2, \quad [e_1, e_2] = e_2, \quad [e_1, e_3] = -e_3, \quad [e_1, e_4] = 0_2$$

$$[e_2, e_1] = -e_2, \quad [e_2, e_2] = 0_2, \quad [e_2, e_3] = e_1 - e_4, \quad [e_2, e_4] = e_2$$

$$[e_3, e_1] = e_3, \quad [e_3, e_2] = e_4 - e_1, \quad [e_3, e_3] = 0_2, \quad [e_3, e_4] = -e_3$$

$$[e_4, e_1] = 0_2, \quad [e_4, e_2] = -e_2, \quad [e_4, e_3] = e_3, \quad [e_4, e_4] = 0_2 .$$

It follows that

$$(e_1, e_2, e_3, e_4)\mathrm{ad}e_1 = (0_2, e_2, -e_3, 0_2)$$
$$(e_1, e_2, e_3, e_4)\mathrm{ad}e_2 = (-e_2, 0_2, e_1 - e_4, e_2)$$
$$(e_1, e_2, e_3, e_4)\mathrm{ad}e_3 = (e_3, e_4 - e_1, 0_2, -e_3)$$
$$(e_1, e_2, e_3, e_4)\mathrm{ad}e_4 = (0_2, -e_2, e_3, 0_2) .$$

This leads to the adjoint representation

$$\mathrm{ad}e_1 = \begin{pmatrix} 0 & 0 & 0 & 0 \\ 0 & 1 & 0 & 0 \\ 0 & 0 & -1 & 0 \\ 0 & 0 & 0 & 0 \end{pmatrix}, \quad \mathrm{ad}e_2 = \begin{pmatrix} 0 & 0 & 1 & 0 \\ -1 & 0 & 0 & 1 \\ 0 & 0 & 0 & 0 \\ 0 & 0 & -1 & 0 \end{pmatrix}$$

$$\text{ade}_3 = \begin{pmatrix} 0 & -1 & 0 & 0 \\ 0 & 0 & 0 & 0 \\ 1 & 0 & 0 & -1 \\ 0 & 1 & 0 & 0 \end{pmatrix}, \quad \text{ade}_4 = \begin{pmatrix} 0 & 0 & 0 & 0 \\ 0 & -1 & 0 & 0 \\ 0 & 0 & 1 & 0 \\ 0 & 0 & 0 & 0 \end{pmatrix}.$$

(ii) The Killing form is

$$\kappa(e_i, e_j) = \text{tr}(\text{ad}(e_i)\text{ad}(e_j))$$

where $i, j = 1, 2, 3, 4$. Then the *metric tensor* is

$$(g_{ij}) = (\kappa(e_i, e_j)) = \begin{pmatrix} 2 & 0 & 0 & -2 \\ 0 & 0 & 4 & 0 \\ 0 & 4 & 0 & 0 \\ -2 & 0 & 0 & 2 \end{pmatrix}.$$

Thus $\det((g_{ij})) = 0$ and therefore the Lie algebra is not semisimple. This can also be seen as follows. From the basis given above we can form the new basis

$$\begin{pmatrix} 1 & 0 \\ 0 & 1 \end{pmatrix}, \quad \begin{pmatrix} 1 & 0 \\ 0 & -1 \end{pmatrix}, \quad \begin{pmatrix} 0 & 1 \\ 0 & 0 \end{pmatrix}, \quad \begin{pmatrix} 0 & 0 \\ 1 & 0 \end{pmatrix}.$$

The identity matrix commutes with all other elements in the Lie algebra.

Problem 33. Let e_1, e_2, e_3 be the generators of a three-dimensional Lie algebra with the commutation relation

$$[e_1, e_3] = e_1, \quad [e_2, e_3] = e_1 + e_2, \quad [e_1, e_2] = 0.$$

Find the adjoint representation. Is the Lie algebra semisimple?

Solution 33. The adjoint representation is

$$\text{ade}_1 = \begin{pmatrix} 0 & 0 & 1 \\ 0 & 0 & 0 \\ 0 & 0 & 0 \end{pmatrix}, \quad \text{ade}_2 = \begin{pmatrix} 0 & 0 & 1 \\ 0 & 0 & 1 \\ 0 & 0 & 0 \end{pmatrix}, \quad \text{ade}_3 = \begin{pmatrix} -1 & -1 & 0 \\ 0 & -1 & 0 \\ 0 & 0 & 0 \end{pmatrix}.$$

No, the algebra is not semisimple. The determinant of the metric tensor is equal to 0. The Lie algebra is solvable.

Problem 34. Let I_n be the $n \times n$ unit matrix. Consider the $2n \times 2n$ matrices

$$E_1 = \frac{1}{2}\begin{pmatrix} I_n & 0_n \\ 0_n & -I_n \end{pmatrix}, \quad E_2 = \frac{1}{2}\begin{pmatrix} 0_n & I_n \\ I_n & 0_n \end{pmatrix}, \quad E_3 = \frac{1}{2}\begin{pmatrix} 0_n & I_n \\ -I_n & 0_n \end{pmatrix}.$$

Find the commutators $[E_1, E_2]$, $[E_1, E_3]$, $[E_3, E_2]$. Discuss.

Solution 34.   We obtain $[E_1, E_2] = E_3$, $[E_1, E_3] = E_2$, $[E_3, E_2] = E_1$. Thus E_1, E_2, E_3 form a basis of a simple Lie algebra.

Problem 35.   Consider the Lie algebra $u(1, 1)$. A basis using 2×2 matrices is given by

$$\tau_0 = \frac{1}{2} \begin{pmatrix} i & 0 \\ 0 & -i \end{pmatrix}, \quad \tau_1 = \frac{1}{2} \begin{pmatrix} 0 & 1 \\ 1 & 0 \end{pmatrix},$$

$$\tau_2 = \frac{1}{2} \begin{pmatrix} 0 & -i \\ i & 0 \end{pmatrix}, \quad \tau_3 = \frac{1}{2} \begin{pmatrix} i & 0 \\ 0 & i \end{pmatrix}.$$

Find the commutators. Find the 4×4 matrix

$$(A_{jk}) := \text{tr}(\tau_j \tau_k) \qquad j, k = 0, 1, 2, 3.$$

Is the matrix invertible?

Solution 35.   For the commutators we find

$$[\tau_3, \tau_0] = [\tau_3, \tau_1] = [\tau_3, \tau_2] = 0$$

and

$$[\tau_0, \tau_1] = -\tau_2, \quad [\tau_0, \tau_2] = \tau_1, \quad [\tau_1, \tau_2] = \tau_0.$$

We obtain the 4×4 diagonal matrix

$$A = \frac{1}{2} \begin{pmatrix} -1 & 0 & 0 & 0 \\ 0 & 1 & 0 & 0 \\ 0 & 0 & 1 & 0 \\ 0 & 0 & 0 & -1 \end{pmatrix}$$

which is invertible. Note that $\text{tr}(\tau_j \tau_k)$ may be considered as the inner product of the Lie algebra.

Problem 36.   Let H be a nonzero $n \times n$ hermitian matrix. Let E be a nonzero $n \times n$ matrix. Assume that $[H, E] = aE$, where $a \in \mathbb{R}$ and $a \neq 0$. Show that the matrix E cannot be hermitian.

Solution 36.   Assume that E is hermitian. Then from $[H, E] = aE$ we obtain by taking the transpose and complex conjugate of this equation

$$[E, H] = aE.$$

Adding the two equations $[H, E] = aE$, $[E, H] = aE$ we obtain $2aE = 0$. Since E is a nonzero matrix we have $a = 0$. Thus we have a contradiction and E cannot be hermitian.

Problem 37. Consider the vectors in $\mathbb{R}^2$

$$\alpha = \frac{1}{2\sqrt{3}}(1, \sqrt{3}), \qquad \beta = \frac{1}{2\sqrt{3}}(1, -\sqrt{3}).$$

Then

$$\alpha + \beta = \frac{1}{\sqrt{3}}(1, 0).$$

Find eight 3×3 matrices

$$H_1, \ H_2, \ E_\alpha, \ E_{-\alpha}, E_\beta, \ E_{-\beta}, \ E_{\alpha+\beta}, \ E_{-(\alpha+\beta)}$$

such that the following commutation relations are satisfied

$$[H_1, E_{\pm\alpha}] = \pm\frac{1}{2\sqrt{3}}E_{\pm\alpha}, \qquad [H_2, E_{\pm\alpha}] = \pm\frac{1}{2}E_{\pm\alpha}$$

$$[H_1, E_{\pm\beta}] = \pm\frac{1}{2\sqrt{3}}E_{\pm\beta}, \qquad [H_2, E_{\pm\beta}] = \pm\frac{1}{2}E_{\pm\beta}$$

$$[H_1, E_{\pm(\alpha+\beta)}] = \frac{1}{\sqrt{3}}E_{\pm(\alpha+\beta)}, \qquad [H_2, E_{\pm(\alpha+\beta)}] = 0$$

$$[E_\alpha, E_{-\alpha}] = \frac{H_1}{2\sqrt{3}} + \frac{H_2}{2}, \qquad [E_\beta, E_{-\beta}] = \frac{H_1}{2\sqrt{3}} - \frac{H_2}{2},$$

$$[E_{\alpha+\beta}, E_{-(\alpha+\beta)}] = \frac{H_1}{\sqrt{3}}, \qquad [E_\alpha, E_\beta] = \frac{1}{\sqrt{6}}E_{\alpha+\beta},$$

$$[E_\alpha, E_{\alpha+\beta}] = 0, \qquad [E_\beta, E_{\alpha+\beta}] = 0$$

$$[E_\alpha, E_{-(\alpha+\beta)}] = -\frac{1}{\sqrt{6}}E_{-\beta}, \qquad [E_\beta, E_{-(\alpha+\beta)}] = \frac{1}{\sqrt{6}}E_{-\alpha}$$

and $[H_1, H_2] = 0$. We can assume that H_1, H_2 are diagonal matrices.

Solution 37. For any $n \times n$ matrices A, B we have $\mathrm{tr}([A, B]) = 0$. Thus owing to the commutation relations all eight 3×3 matrices must have trace 0. We can also assume that the diagonal entries of the six E matrices are 0. We also have

$$E_\alpha^T = E_{-\alpha}.$$

We find

$$H_1 = \frac{1}{6}\begin{pmatrix} 0 & 0 & 0 \\ 0 & -\sqrt{3} & 0 \\ 0 & 0 & \sqrt{3} \end{pmatrix}, \qquad H_2 = \frac{1}{6}\begin{pmatrix} 2 & 0 & 0 \\ 0 & -1 & 0 \\ 0 & 0 & -1 \end{pmatrix}$$

$$E_\alpha = \frac{1}{\sqrt{6}} \begin{pmatrix} 0 & 1 & 0 \\ 0 & 0 & 0 \\ 0 & 0 & 0 \end{pmatrix}, \qquad E_{-\alpha} = \frac{1}{\sqrt{6}} \begin{pmatrix} 0 & 0 & 0 \\ 1 & 0 & 0 \\ 0 & 0 & 0 \end{pmatrix},$$

$$E_\beta = \frac{1}{\sqrt{6}} \begin{pmatrix} 0 & 0 & 0 \\ 0 & 0 & 0 \\ 1 & 0 & 0 \end{pmatrix}, \qquad E_{-\beta} = \frac{1}{\sqrt{6}} \begin{pmatrix} 0 & 0 & 1 \\ 0 & 0 & 0 \\ 0 & 0 & 0 \end{pmatrix}$$

$$E_{\alpha+\beta} = -\frac{1}{\sqrt{6}} \begin{pmatrix} 0 & 0 & 0 \\ 0 & 0 & 1 \\ 0 & 0 & 0 \end{pmatrix}, \qquad E_{-(\alpha+\beta)} = -\frac{1}{\sqrt{6}} \begin{pmatrix} 0 & 0 & 0 \\ 0 & 0 & 0 \\ 0 & 1 & 0 \end{pmatrix}.$$

Problem 38.   The Lie algebra $su(5)$ consists of all 5×5 matrices with

$$X^* = -X, \qquad \mathrm{tr}X = 0$$

for $X \in su(5)$. Give a basis of $su(5)$. The basis elements should be orthogonal with respect to the scalar product $\mathrm{tr}(X_j X_k^*)$.

Solution 38.   The basis has 25-1=24 elements. We find

$$\begin{pmatrix} 0 & i & 0 & 0 & 0 \\ i & 0 & 0 & 0 & 0 \\ 0 & 0 & 0 & 0 & 0 \\ 0 & 0 & 0 & 0 & 0 \\ 0 & 0 & 0 & 0 & 0 \end{pmatrix}, \qquad \begin{pmatrix} 0 & 1 & 0 & 0 & 0 \\ -1 & 0 & 0 & 0 & 0 \\ 0 & 0 & 0 & 0 & 0 \\ 0 & 0 & 0 & 0 & 0 \\ 0 & 0 & 0 & 0 & 0 \end{pmatrix}$$

$$\begin{pmatrix} i & 0 & 0 & 0 & 0 \\ 0 & -i & 0 & 0 & 0 \\ 0 & 0 & 0 & 0 & 0 \\ 0 & 0 & 0 & 0 & 0 \\ 0 & 0 & 0 & 0 & 0 \end{pmatrix}, \qquad \begin{pmatrix} 0 & 0 & i & 0 & 0 \\ 0 & 0 & 0 & 0 & 0 \\ i & 0 & 0 & 0 & 0 \\ 0 & 0 & 0 & 0 & 0 \\ 0 & 0 & 0 & 0 & 0 \end{pmatrix}$$

$$\begin{pmatrix} 0 & 0 & 1 & 0 & 0 \\ 0 & 0 & 0 & 0 & 0 \\ -1 & 0 & 0 & 0 & 0 \\ 0 & 0 & 0 & 0 & 0 \\ 0 & 0 & 0 & 0 & 0 \end{pmatrix}, \qquad \begin{pmatrix} 0 & 0 & 0 & 0 & 0 \\ 0 & 0 & i & 0 & 0 \\ 0 & i & 0 & 0 & 0 \\ 0 & 0 & 0 & 0 & 0 \\ 0 & 0 & 0 & 0 & 0 \end{pmatrix}$$

$$\begin{pmatrix} 0 & 0 & 0 & 0 & 0 \\ 0 & 0 & 1 & 0 & 0 \\ 0 & -1 & 0 & 0 & 0 \\ 0 & 0 & 0 & 0 & 0 \\ 0 & 0 & 0 & 0 & 0 \end{pmatrix}, \qquad \frac{1}{\sqrt{2}} \begin{pmatrix} i & 0 & 0 & 0 & 0 \\ 0 & i & 0 & 0 & 0 \\ 0 & 0 & -2i & 0 & 0 \\ 0 & 0 & 0 & 0 & 0 \\ 0 & 0 & 0 & 0 & 0 \end{pmatrix}$$

$$\begin{pmatrix} 0 & 0 & 0 & i & 0 \\ 0 & 0 & 0 & 0 & 0 \\ 0 & 0 & 0 & 0 & 0 \\ i & 0 & 0 & 0 & 0 \\ 0 & 0 & 0 & 0 & 0 \end{pmatrix}, \qquad \begin{pmatrix} 0 & 0 & 0 & 1 & 0 \\ 0 & 0 & 0 & 0 & 0 \\ 0 & 0 & 0 & 0 & 0 \\ -1 & 0 & 0 & 0 & 0 \\ 0 & 0 & 0 & 0 & 0 \end{pmatrix}$$

$$\begin{pmatrix} 0 & 0 & 0 & 0 & 0 \\ 0 & 0 & 0 & i & 0 \\ 0 & 0 & 0 & 0 & 0 \\ 0 & i & 0 & 0 & 0 \\ 0 & 0 & 0 & 0 & 0 \end{pmatrix}, \qquad \begin{pmatrix} 0 & 0 & 0 & 0 & 0 \\ 0 & 0 & 0 & 1 & 0 \\ 0 & 0 & 0 & 0 & 0 \\ 0 & -1 & 0 & 0 & 0 \\ 0 & 0 & 0 & 0 & 0 \end{pmatrix}$$

$$\begin{pmatrix} 0 & 0 & 0 & 0 & 0 \\ 0 & 0 & 0 & 0 & 0 \\ 0 & 0 & 0 & i & 0 \\ 0 & 0 & i & 0 & 0 \\ 0 & 0 & 0 & 0 & 0 \end{pmatrix}, \qquad \begin{pmatrix} 0 & 0 & 0 & 0 & 0 \\ 0 & 0 & 0 & 0 & 0 \\ 0 & 0 & 0 & 1 & 0 \\ 0 & 0 & -1 & 0 & 0 \\ 0 & 0 & 0 & 0 & 0 \end{pmatrix}$$

$$\frac{1}{\sqrt{6}}\begin{pmatrix} i & 0 & 0 & 0 & 0 \\ 0 & i & 0 & 0 & 0 \\ 0 & 0 & i & 0 & 0 \\ 0 & 0 & 0 & -3i & 0 \\ 0 & 0 & 0 & 0 & 0 \end{pmatrix}, \qquad \begin{pmatrix} 0 & 0 & 0 & 0 & i \\ 0 & 0 & 0 & 0 & 0 \\ 0 & 0 & 0 & 0 & 0 \\ 0 & 0 & 0 & 0 & 0 \\ i & 0 & 0 & 0 & 0 \end{pmatrix}$$

$$\begin{pmatrix} 0 & 0 & 0 & 0 & 1 \\ 0 & 0 & 0 & 0 & 0 \\ 0 & 0 & 0 & 0 & 0 \\ 0 & 0 & 0 & 0 & 0 \\ -1 & 0 & 0 & 0 & 0 \end{pmatrix}, \qquad \begin{pmatrix} 0 & 0 & 0 & 0 & 0 \\ 0 & 0 & 0 & 0 & i \\ 0 & 0 & 0 & 0 & 0 \\ 0 & 0 & 0 & 0 & 0 \\ 0 & i & 0 & 0 & 0 \end{pmatrix}$$

$$\begin{pmatrix} 0 & 0 & 0 & 0 & 0 \\ 0 & 0 & 0 & 0 & 1 \\ 0 & 0 & 0 & 0 & 0 \\ 0 & 0 & 0 & 0 & 0 \\ 0 & -1 & 0 & 0 & 0 \end{pmatrix}, \qquad \begin{pmatrix} 0 & 0 & 0 & 0 & 0 \\ 0 & 0 & 0 & 0 & 0 \\ 0 & 0 & 0 & 0 & i \\ 0 & 0 & 0 & 0 & 0 \\ 0 & 0 & i & 0 & 0 \end{pmatrix}$$

$$\begin{pmatrix} 0 & 0 & 0 & 0 & 0 \\ 0 & 0 & 0 & 0 & 0 \\ 0 & 0 & 0 & 0 & 1 \\ 0 & 0 & 0 & 0 & 0 \\ 0 & 0 & -1 & 0 & 0 \end{pmatrix}, \qquad \begin{pmatrix} 0 & 0 & 0 & 0 & 0 \\ 0 & 0 & 0 & 0 & 0 \\ 0 & 0 & 0 & 0 & 0 \\ 0 & 0 & 0 & 0 & i \\ 0 & 0 & 0 & i & 0 \end{pmatrix}$$

$$\begin{pmatrix} 0 & 0 & 0 & 0 & 0 \\ 0 & 0 & 0 & 0 & 0 \\ 0 & 0 & 0 & 0 & 0 \\ 0 & 0 & 0 & 0 & 1 \\ 0 & 0 & 0 & -1 & 0 \end{pmatrix}, \qquad \frac{1}{\sqrt{10}}\begin{pmatrix} i & 0 & 0 & 0 & 0 \\ 0 & i & 0 & 0 & 0 \\ 0 & 0 & i & 0 & 0 \\ 0 & 0 & 0 & i & 0 \\ 0 & 0 & 0 & 0 & -4i \end{pmatrix}.$$

These matrices are pairwise orthogonal.

Problem 39.   (i) Is the three dimensional *Heisenberg algebra* with the basis $\{x, y, z\}$ and the commutators $[x, y] = z$, $[x, z] = 0$, $[y, z] = 0$ nilpotent?
(ii) Find a matrix representation with 3×3 matrices.

Solution 39.   (i) Yes.
(ii) We have

$$
x \mapsto \begin{pmatrix} 0 & 1 & 0 \\ 0 & 0 & 0 \\ 0 & 0 & 0 \end{pmatrix}, \quad y \mapsto \begin{pmatrix} 0 & 0 & 0 \\ 0 & 0 & 1 \\ 0 & 0 & 0 \end{pmatrix}, \quad z \mapsto \begin{pmatrix} 0 & 0 & 1 \\ 0 & 0 & 0 \\ 0 & 0 & 0 \end{pmatrix}.
$$

These matrices are nilpotent matrices.

Problem 40.   If a Lie algebra is nilpotent, then the Killing form vanishes. Can we conclude that if the Killing form vanishes then the Lie algebra is nilpotent?

Solution 40.   No we cannot conclude that if the Killing form vanishes then the Lie algebra is nilpotent. Consider a three-dimensional vector space over $\mathbb{C}$ with basis a, b, c. Introducing the commutators

$$
[a, b] = b, \qquad [a, c] = ic, \qquad [b, c] = 0
$$

we can show that we have a Lie algebra. For example the Jacobi identity

$$
[a, [b, c]] + [c, [a, b]] + [b, [c, a]] = 0
$$

is satisfied. We find that $[L, L] = [L, [L, L]] = \cdots = < b, c >$. Now $[b, c] = 0$. We find that the Killing form vanishes. However the Lie algebra is not nilpotent.

Problem 41.   Show that the Lie algebra consisting of 3×3 matrices over $\mathbb{R}$ (or $\mathbb{C}$) with zeros on and below the main diagonal is nilpotent.

Solution 41.   Let

$$
A = \begin{pmatrix} 0 & a_{12} & a_{13} \\ 0 & 0 & a_{23} \\ 0 & 0 & 0 \end{pmatrix}, \qquad B = \begin{pmatrix} 0 & b_{12} & b_{13} \\ 0 & 0 & b_{23} \\ 0 & 0 & 0 \end{pmatrix}.
$$

Then the commutator $[A, B]$ is given by

$$
[A, B] = \begin{pmatrix} 0 & 0 & a_{12}b_{23} - a_{23}b_{12} \\ 0 & 0 & 0 \\ 0 & 0 & 0 \end{pmatrix}.
$$

Let

$$C = \begin{pmatrix} 0 & c_{12} & c_{13} \\ 0 & 0 & c_{23} \\ 0 & 0 & 0 \end{pmatrix}.$$

Then $[C, [A, B]] = 0$. Extend the result to higher dimensions.

Problem 42. Consider the vector space of 3×3 matrices over $\mathbb{R}$ and the two linearly independent matrices

$$A = \begin{pmatrix} 0 & 1 & 1 \\ 0 & 0 & 0 \\ 0 & 0 & 0 \end{pmatrix}, \qquad B = \begin{pmatrix} 0 & 0 & 0 \\ 0 & 0 & 1 \\ 0 & 0 & 1 \end{pmatrix}.$$

Construct a Lie algebra by calculating the commutators. Thus A and B would be the generators of the Lie algebra.

Solution 42. The commutator $[A, B]$ yields

$$[A, B] = \begin{pmatrix} 0 & 0 & 2 \\ 0 & 0 & 0 \\ 0 & 0 & 0 \end{pmatrix} = C.$$

Now

$$[A, C] = \begin{pmatrix} 0 & 0 & 0 \\ 0 & 0 & 0 \\ 0 & 0 & 0 \end{pmatrix}, \qquad [B, C] = \begin{pmatrix} 0 & 0 & -2 \\ 0 & 0 & 0 \\ 0 & 0 & 0 \end{pmatrix} = -C.$$

Thus we find a three dimensional Lie algebra with basis A, B, C. The Lie algebra is not semisimple. Is the Lie algebra nilpotent?

Problem 43. Let $z \in \mathbb{C}$. Consider the vector fields

$$V_1 = -\frac{\partial}{\partial z} - \frac{\partial}{\partial \bar{z}}, \qquad V_2 = i\frac{\partial}{\partial \bar{z}} - i\frac{\partial}{\partial z}, \qquad V_3 = \frac{2}{i}\left(z\frac{\partial}{\partial \bar{z}} - \bar{z}\frac{\partial}{\partial z} \right).$$

Find the commutators and show that we have a basis of a Lie algebra.

Solution 43. We have by direct calculation

$$[V_1, V_2] = \left(-\frac{\partial}{\partial z} - \frac{\partial}{\partial \bar{z}}\right)\left(i\frac{\partial}{\partial \bar{z}} - i\frac{\partial}{\partial z}\right) - \left(i\frac{\partial}{\partial \bar{z}} - i\frac{\partial}{\partial z}\right)\left(-\frac{\partial}{\partial z} - \frac{\partial}{\partial \bar{z}}\right) = 0$$

because all coefficients of V_1, V_2 are constant. Now

$$[V_1, V_3] = \frac{2}{i}\left(-\frac{\partial}{\partial \bar{z}} + \frac{\partial}{\partial z}\right) = 2V_2$$

$$[V_2, V_3] = 2 \left(-\frac{\partial}{\partial \bar{z}} - \frac{\partial}{\partial z} \right) = 2V_1.$$

The Jacobi identity is satisfied. Thus V_1, V_2 and V_3 form a basis of a three dimensional Lie algebra. Is the Lie algebra semisimple?

Problem 44. Consider the vector fields

$$V_{12} = x_1 \frac{\partial}{\partial x_2} - x_2 \frac{\partial}{\partial x_1}, \qquad V_{31} = x_3 \frac{\partial}{\partial x_1} - x_1 \frac{\partial}{\partial x_3}, \qquad V_{23} = x_2 \frac{\partial}{\partial x_3} - x_3 \frac{\partial}{\partial x_2}.$$

Find the commutators $[V_{12}, V_{31}]$, $[V_{31}, V_{23}]$, $[V_{23}, V_{12}]$ and show that V_{12}, V_{31}, V_{23} form a basis of a Lie algebra. Is the Lie algebra semisimple?

Solution 44. We have

$$[V_{12}, V_{31}] = V_{23}, \quad [V_{31}, V_{23}] = V_{12}, \quad [V_{23}, V_{12}] = V_{31}.$$

This is the Lie algebra $o(3)$. The Lie algebra is semisimple.

Problem 45. As an extension of the previous problem we can consider the vector fields $(n \geq 3)$

$$V_{jk} = -x_j \frac{\partial}{\partial x_k} + x_k \frac{\partial}{\partial x_j}, \qquad j = 1, 2, \ldots, n, \quad j < k$$

There are $\frac{1}{2}n(n-1)$ such vector fields. Find the commutator $[V_{jk}, V_{k\ell}]$.

Solution 45. We obtain $[V_{jk}, V_{k\ell}] = V_{\ell j}$.

Problem 46. The Lie algebra $o(2, 1)$ admits the basis

$$D_{12} = -x_1 \frac{\partial}{\partial x_2} + x_2 \frac{\partial}{\partial x_1}, \quad E_{23} = x_2 \frac{\partial}{\partial x_3} + x_3 \frac{\partial}{\partial x_2}, \quad E_{31} = x_3 \frac{\partial}{\partial x_1} + x_1 \frac{\partial}{\partial x_3}.$$

Find the commutators.

Solution 46. We obtain

$$[D_{12}, E_{23}] = -E_{31}, \quad [E_{23}, E_{31}] = D_{12}, \quad [E_{31}, D_{12}] = -E_{23}.$$

The Lie algebra is semisimple.

Problem 47. Let $n \geq 2$ and $0 \leq p \leq n$. The real semisimple Lie algebra $o(p, n-p)$ is of dimension $\frac{1}{2}n(n-1)$ and is spanned by the vector fields

$$D_{kj} = -D_{jk}, \quad E_{kj} = E_{jk}$$

$$D_{jk} = -x_j \frac{\partial}{\partial x_k} + x_k \frac{\partial}{\partial x_j}, \qquad 1 \le j < k \le p, \ p < j < k \le n$$

$$E_{jk} = x_j \frac{\partial}{\partial x_k} + x_k \frac{\partial}{\partial x_j}, \qquad 1 \le j \le p, \ p < k \le n.$$

Find the commutators.

Solution 47. Vector fields with no common subscripts commute. Otherwise we have $[D_{jk}, D_{k\ell}] = D_{\ell j}$, $[D_{jk}, E_{k\ell}] = -E_{j\ell}$, $[E_{jk}, E_{k\ell}] = D_{\ell j}$.

Problem 48. Let $z, w \in \mathbb{C}$. Consider the vector fields

$$T_1 = -\frac{w}{2i} \frac{\partial}{\partial z} - \frac{z}{2i} \frac{\partial}{\partial w}, \quad T_2 = -\frac{w}{2} \frac{\partial}{\partial z} + \frac{z}{2} \frac{\partial}{\partial w}, \quad T_3 = \frac{z}{2} \frac{\partial}{\partial z} - \frac{w}{2} \frac{\partial}{\partial w}.$$

Find the commutators $[T_1, T_2]$, $[T_2, T_3]$, $[T_3, T_1]$. Let

$$T_+ := T_1 + iT_2, \qquad T_- := T_1 - iT_2.$$

Find T_+ and T_-. Consider the function $z^m w^n$, where m is a nonnegative integer and n is a negative integer with $n < -m$. Show that $z^m w^n$ is an eigenfunction of T_3. Find the eigenvalue. Find $T_+(z^m w^n)$ and $T_-(z^m w^n)$.

Solution 48. We find $[T_1, T_2] = iT_3$, $[T_2, T_3] = -iT_1$, $[T_3, T_1] = -iT_2$. It follows that

$$T_- = iw \frac{\partial}{\partial z}, \qquad T_+ = iz \frac{\partial}{\partial w}.$$

Since

$$T_3(z^m w^n) = \left(\frac{1}{2}m - \frac{1}{2}n \right)(z^m w^n)$$

we have an eigenvalue equation with eigenvalue $\frac{1}{2}(m - n)$. We obtain

$$T_+(z^m w^n) = in(z^{m+1} w^{n-1}), \qquad T_-(z^m w^n) = im(z^{m-1} w^{n+1}).$$

Problem 49. Consider the operators

$$L_x = i \left(\cos(\phi) \cot(\alpha) \frac{\partial}{\partial \phi} + \sin(\phi) \frac{\partial}{\partial \alpha} \right)$$

$$L_y = i \left(\sin(\phi) \cot(\alpha) \frac{\partial}{\partial \phi} - \cos(\phi) \frac{\partial}{\partial \alpha} \right)$$

$$L_z = -i \frac{\partial}{\partial \phi}.$$

By calculating the commutators $[L_x, L_y]$, $[L_y, L_z]$, $[L_z, L_x]$ show that we have a basis of a Lie algebra.

Solution 49.   We obtain $[L_x, L_y] = iL_z$, $[L_y, L_z] = iL_x$, $[L_z, L_x] = iL_y$. The Jacobi identity is satisfied.

Problem 50.   Consider the linear differential operators

$$S_z = -x\frac{d}{dx} + \ell, \quad S_- = x, \quad S_+ = -x\frac{d^2}{dx^2} + 2\ell\frac{d}{dx}$$

acting on an analytic function f. Find the commutators $[S_+, S_-]$, $[S_+, S_z]$, $[S_-, S_z]$. Discuss.

Solution 50.   Since

$$S_+ S_- f = \left(-x\frac{d^2}{dx^2} + 2\ell\frac{d}{dx}\right)(xf) = -2x\frac{df}{dx} - x^2\frac{d^2 f}{dx^2} + 2\ell f + 2\ell x\frac{df}{dx}$$

and

$$S_- S_+ f = -x^2\frac{d^2 f}{dx^2} + 2\ell x\frac{df}{dx}$$

we obtain $[S_+, S_-] = 2S_z$. Analogously $[S_+, S_z] = S_+$ and $[S_-, S_z] = S_-$.

Problem 51.   Let σ_1, σ_2, σ_3 be the Pauli spin matrices with the commutation relation

$$[\sigma_1, \sigma_2] = 2i\sigma_3, \quad [\sigma_2, \sigma_3] = 2i\sigma_1, \quad [\sigma_3, \sigma_1] = 2i\sigma_2.$$

Thus we have

$$su(2) = \text{span}\{i\sigma_1, i\sigma_2, i\sigma_3\}.$$

Let $\otimes$ be the *Kronecker product*. Find the three commutators

$$[\sigma_1 \otimes I_2 + I_2 \otimes \sigma_2, \sigma_2 \otimes I_2 + I_2 \otimes \sigma_3]$$

$$[\sigma_2 \otimes I_2 + I_2 \otimes \sigma_3, \sigma_3 \otimes I_2 + I_2 \otimes \sigma_1]$$

$$[\sigma_3 \otimes I_2 + I_2 \otimes \sigma_1, \sigma_2 \otimes I_2 + I_2 \otimes \sigma_3].$$

Discuss.

Solution 51.   Using $\sigma_1\sigma_2 - \sigma_2\sigma_1 = 2i\sigma_3$ and $\sigma_2\sigma_3 - \sigma_3\sigma_2 = 2i\sigma_1$ provides

$$[\sigma_1 \otimes I_2 + I_2 \otimes \sigma_2, \sigma_2 \otimes I_2 + I_2 \otimes \sigma_3] = 2i(\sigma_3 \otimes I_2 + I_2 \otimes \sigma_1).$$

Analogously we find

$$[\sigma_2 \otimes I_2 + I_2 \otimes \sigma_3, \sigma_3 \otimes I_2 + I_2 \otimes \sigma_1] = 2i(\sigma_1 \otimes I_2 + I_2 \otimes \sigma_2)$$
$$[\sigma_3 \otimes I_2 + I_2 \otimes \sigma_1, \sigma_2 \otimes I_2 + I_2 \otimes \sigma_3] = -2i(\sigma_1 \otimes I_2 + I_2 \otimes \sigma_2).$$

Problem 52. The *Pauli spin matrices* are given by

$$\sigma_1 := \begin{pmatrix} 0 & 1 \\ 1 & 0 \end{pmatrix}, \quad \sigma_2 := \begin{pmatrix} 0 & -i \\ i & 0 \end{pmatrix}, \quad \sigma_3 := \begin{pmatrix} 1 & 0 \\ 0 & -1 \end{pmatrix}.$$

Their multiplication table is

$$\sigma_1\sigma_2 = -\sigma_2\sigma_1 = i\sigma_3, \quad \sigma_2\sigma_3 = -\sigma_3\sigma_2 = i\sigma_1, \quad \sigma_3\sigma_1 = -\sigma_1\sigma_3 = i\sigma_2.$$

Let

$$A := \sigma_1 \otimes I_2, \quad B := \sigma_2 \otimes I_2, \quad C := \sigma_3 \otimes I_2.$$

Find the commutators $[A, B]$, $[B, C]$, $[C, A]$.

Solution 52. For the commutators we find

$$[A, B] = [\sigma_1 \otimes I_2, \sigma_2 \otimes I_2] = 2i\sigma_3 \otimes I_2 = 2iC.$$

Analogously

$$[B, C] = 2iA, \qquad [C, A] = 2iB.$$

We see that the mapping $A \to \sigma_1$, $B \to \sigma_2$, $C \to \sigma_3$ preserves the commutation table. Thus the matrices A, B, C form a Lie algebra which is isomorphic to the Pauli spin matrices.

Problem 53. Consider the *Pauli spin matrices* σ_1, σ_2, σ_3 and the 4×4 matrices

$$\sigma_1 \otimes \sigma_2, \quad \sigma_2 \otimes \sigma_3, \quad \sigma_3 \otimes \sigma_2.$$

Find the commutators

$$[\sigma_1 \otimes \sigma_2, \sigma_2 \otimes \sigma_3], \quad [\sigma_2 \otimes \sigma_3, \sigma_3 \otimes \sigma_1], \quad [\sigma_3 \otimes \sigma_1, \sigma_1 \otimes \sigma_2].$$

Discuss

Solution 53. We have

$$\begin{aligned}
[\sigma_1 \otimes \sigma_2, \sigma_2 \otimes \sigma_3] &= (\sigma_1\sigma_2) \otimes (\sigma_2\sigma_3) - (\sigma_2\sigma_1) \otimes (\sigma_3\sigma_2) \\
&= (i\sigma_3) \otimes (i\sigma_1) - (-i\sigma_3) \otimes (-i\sigma_1) \\
&= -\sigma_3 \otimes \sigma_1 + \sigma_3 \otimes \sigma_1 \\
&= 0.
\end{aligned}$$

Analogously we find

$$[\sigma_2 \otimes \sigma_3, \sigma_3 \otimes \sigma_1] = 0, \quad [\sigma_3 \otimes \sigma_2, \sigma_1 \otimes \sigma_2] = 0.$$

Thus we have a commutative Lie algebra.

Problem 54. Consider the 4×4 matrices

$$S_1 = \begin{pmatrix} 0 & 1 \\ 1 & 0 \end{pmatrix} \otimes I_2, \qquad S_2 = I_2 \otimes \begin{pmatrix} 1 & 0 \\ 0 & \omega \end{pmatrix}$$

where $\omega = \exp(i\pi) = -1$. Find S_1^2, S_2^2, $S_1 S_2 S_1^{-1} S_2^{-1}$ and the commutator $[S_1, S_2]$.

Solution 54. We obtain

$$S_1^2 = I_4, \quad S_2^2 = I_4, \quad S_1 S_2 S_1^{-1} S_2^{-1} = I_4, \quad [S_1, S_2] = 0_4.$$

Problem 55. Let $\omega = \exp(2\pi i/4) \equiv \exp(\pi i/2) = i$. Consider the four 64×64 invertible matrices

$$S_1 = \begin{pmatrix} 1 & 0 & 0 & 0 \\ 0 & \omega & 0 & 0 \\ 0 & 0 & \omega^2 & 0 \\ 0 & 0 & 0 & \omega^3 \end{pmatrix} \otimes \begin{pmatrix} 1 & 0 & 0 & 0 \\ 0 & \omega^3 & 0 & 0 \\ 0 & 0 & \omega^2 & 0 \\ 0 & 0 & 0 & \omega \end{pmatrix} \otimes I_4$$

$$S_2 = I_4 \otimes \begin{pmatrix} 0 & 1 & 0 & 0 \\ 0 & 0 & 1 & 0 \\ 0 & 0 & 0 & 1 \\ 1 & 0 & 0 & 0 \end{pmatrix} \otimes I_4$$

$$S_3 = I_4 \otimes \begin{pmatrix} 1 & 0 & 0 & 0 \\ 0 & \omega & 0 & 0 \\ 0 & 0 & \omega^2 & 0 \\ 0 & 0 & 0 & \omega^3 \end{pmatrix} \otimes \begin{pmatrix} 1 & 0 & 0 & 0 \\ 0 & \omega^3 & 0 & 0 \\ 0 & 0 & \omega^2 & 0 \\ 0 & 0 & 0 & \omega \end{pmatrix}$$

$$S_4 = I_4 \otimes I_4 \otimes \begin{pmatrix} 0 & 1 & 0 & 0 \\ 0 & 0 & 1 & 0 \\ 0 & 0 & 0 & 1 \\ 1 & 0 & 0 & 0 \end{pmatrix}.$$

(i) Find S_j^4 for $j = 1, 2, 3, 4$.
(ii) Find

$$S_1 S_2 S_1^{-1} S_2^{-1}, \quad S_2 S_3 S_2^{-1} S_3^{-1}, \quad S_3 S_4 S_3^{-1} S_4^{-1}, \quad S_4 S_1 S_4^{-1} S_1^{-1}.$$

(iii) Find the commutators $[S_1, S_2]$, $[S_2, S_3]$, $[S_3, S_4]$, $[S_4, S_1]$.

Solution 55. (i) We find

$$S_1^4 = S_2^4 = S_3^4 = S_4^4 = I_{64}.$$

(ii) We obtain

$$S_1 S_2 S_1^{-1} S_2^{-1} = i I_{64}, \quad S_2 S_3 S_2^{-1} S_3^{-1} = I_{64},$$

$$S_3 S_4 S_3^{-1} S_4^{-1} = i I_{64}, \quad S_4 S_1 S_4^{-1} S_1^{-1} = I_{64}.$$

(iii) We obtain $[S_1, S_2] = [S_2, S_3] = [S_3, S_4] = [S_4, S_1] = 0_{64}$.

Problem 56. (i) Consider the non-commutative two dimensional Lie algebra with $[A, B] = A$ and

$$A = \begin{pmatrix} 0 & 1 \\ 0 & 0 \end{pmatrix}, \qquad B = \begin{pmatrix} 0 & 0 \\ 0 & 1 \end{pmatrix}.$$

Show that the matrices

$$\{ A \otimes I_2 + I_2 \otimes A, \ B \otimes I_2 + I_2 \otimes B \}$$

also form a non-commutative Lie algebra under the commutator, where $\otimes$ denotes the Kronecker product. Discuss.
(ii) We denote by $s\ell(2, \mathbb{R})$ the semisimple Lie algebra spanned by a basis $\{ E, F, H \}$ with the Lie bracket defined by

$$[H, E] = 2E, \quad [H, F] = -2F, \quad [E, F] = H.$$

Using 2×2 matrices we have the representation

$$H = \begin{pmatrix} 1 & 0 \\ 0 & -1 \end{pmatrix}, \quad F = \begin{pmatrix} 0 & 0 \\ 1 & 0 \end{pmatrix}, \quad E = \begin{pmatrix} 0 & 1 \\ 0 & 0 \end{pmatrix}.$$

Let $\otimes$ be the *Kronecker product*. Do the three 4×4 matrices

$$H \otimes I_2 + I_2 \otimes H, \quad F \otimes I_2 + I_2 \otimes F, \quad E \otimes I_2 + I_2 \otimes E$$

form a basis of a Lie algebra? Calculate the commutators.

Solution 56. (i) We have

$$[A \otimes I_2 + I_2 \otimes A, B \otimes I_2 + I_2 \otimes B] = [A, B] \otimes I_2 + I_2 \otimes [A, B]$$
$$= A \otimes I_2 + I_2 \otimes A.$$

Thus the two Lie algebras are isomorphic.

(ii) The matrices are equal to

$$A = H \otimes I_2 + I_2 \otimes H = \begin{pmatrix} 2 & 0 & 0 & 0 \\ 0 & 0 & 0 & 0 \\ 0 & 0 & 0 & 0 \\ 0 & 0 & 0 & -2 \end{pmatrix}$$

$$B = F \otimes I_2 + I_2 \otimes F = \begin{pmatrix} 0 & 0 & 0 & 0 \\ 1 & 0 & 0 & 0 \\ 1 & 0 & 0 & 0 \\ 0 & 1 & 1 & 0 \end{pmatrix}$$

$$C = E \otimes I_2 + I_2 \otimes E = \begin{pmatrix} 0 & 1 & 1 & 0 \\ 0 & 0 & 0 & 1 \\ 0 & 0 & 0 & 1 \\ 0 & 0 & 0 & 0 \end{pmatrix}.$$

By direct calculation we have $[A, B] = -2B$, $[A, C] = 2C$, $[B, C] = -A$. The Jacobi identity is satisfied.

Problem 57. Consider the Lie algebra $s\ell(2, \mathbb{R})$ with

$$x = \begin{pmatrix} 0 & 1 \\ 0 & 0 \end{pmatrix}, \qquad y = \begin{pmatrix} 0 & 0 \\ 1 & 0 \end{pmatrix}, \qquad h = \begin{pmatrix} 1 & 0 \\ 0 & -1 \end{pmatrix}$$

and the commutators $[x, h] = -2x$, $[x, y] = h$, $[y, h] = 2y$. Consider

$$h \mapsto h \otimes I_2 + I_2 \otimes h, \qquad x \mapsto x \otimes h + h \otimes x, \qquad y \mapsto y \otimes h + h \otimes y.$$

Find the three commutators

$$[h \otimes I_2 + I_2 \otimes h, x \otimes h + h \otimes x], \quad [h \otimes I_2 + I_2 \otimes h, y \otimes h + h \otimes y], \quad [x \otimes h + h \otimes x, y \otimes h + h \otimes y]$$

and thus show that we have a Lie algebra. Is this Lie algebra isomorphic to $s\ell(2, \mathbb{R})$?

Solution 57. We note that

$$h^2 = I_2, \qquad xh = -x, \qquad hx = x, \qquad yh = y, \qquad hy = -y.$$

Now

$$\begin{aligned}
[x \otimes h + h \otimes x, y \otimes h + h \otimes y] &= [x, y] \otimes h^2 + h^2 \otimes [x, y] + (xh) \otimes (hy) \\
&\quad - (hx) \otimes (yh) + (hy) \otimes (xh) - (yh) \otimes (hx) \\
&= [x, y] \otimes I_2 + I_2 \otimes [x, y] \\
&= h \otimes I_2 + I_2 \otimes h.
\end{aligned}$$

Analogously we have

$$[x \otimes h + h \otimes x, h \otimes I_2 + I_2 \otimes h] = [x,h] \otimes h + h \otimes [x,h] = -2(x \otimes h + h \otimes x)$$

$$[y \otimes h + h \otimes y, h \otimes I_2 + I_2 \otimes h] = [y,h] \otimes h + h \otimes [y,h] = 2(y \otimes h + h \otimes y).$$

Thus the two Lie algebras are isomorphic.

Problem 58. The four matrices

$$X = \begin{pmatrix} 0 & 1 \\ 0 & 0 \end{pmatrix}, \quad Y = \begin{pmatrix} 0 & 0 \\ 1 & 0 \end{pmatrix}, \quad H = \begin{pmatrix} 1 & 0 \\ 0 & -1 \end{pmatrix}, \quad I_2 = \begin{pmatrix} 1 & 0 \\ 0 & 1 \end{pmatrix}$$

form a basis of the Lie algebra $g\ell(2, \mathbb{R})$. We have

$$[X, Y] = H, \quad [H, X] = 2X, \quad [H, Y] = -2Y$$

and $[I_2, X] = [I_2, Y] = [I_2, H] = 0_2$. Can one find an invertible matrix S over $\mathbb{R}$ such that

$$SXS^{-1} = Y, \quad SYS^{-1} = X, \quad SHS^{-1} = -H ?$$

Solution 58. Yes, the matrix is

$$S = S^{-1} = \begin{pmatrix} 0 & 1 \\ 1 & 0 \end{pmatrix}.$$

Problem 59. Let A be an arbitrary $n \times n$ matrix over $\mathbb{C}$. Let T_1, T_2 be $n \times n$ matrices over $\mathbb{C}$ satisfying

$$T_1^* A + A T_1 = 0, \qquad T_2^* A + A T_2 = 0.$$

Show that

$$[T_1, T_2]^* A + A[T_1, T_2] = 0.$$

Solution 59. Using that matrix multiplication is associative we have

$$
\begin{aligned}
[T_1, T_2]^* A + A[T_1, T_2] &= ((T_1 T_2)^* - (T_2 T_1)^*)A + A(T_1 T_2 - T_2 T_1) \\
&= T_2^* T_1^* A - T_1^* T_2^* A + A T_1 T_2 - A T_2 T_1 \\
&= -T_2^* A T_1 + T_1^* A T_2 + A T_1 T_2 - A T_2 T_1 \\
&= 0.
\end{aligned}
$$

Problem 60.   Show that the dimension of a Cartan sub-Lie algebra is not in general the maximal dimension of an abelian sub-Lie algebra. Consider the Lie algebra $s\ell(2n, \mathbb{C})$ of $2n \times 2n$ matrices of trace 0.

Solution 60.   The Lie algebra $s\ell(2n, \mathbb{C})$ of $2n \times 2n$ of trace 0 has rank $2n - 1$ but has a maximal abelian sub-Lie algebra of dimension n^2 given by all matrices of the form

$$\begin{pmatrix} 0_n & X \\ 0_n & 0_n \end{pmatrix}$$

where 0_n is the $n \times n$ zero matrix and X is an arbitrary $n \times n$ matrix. The abelian sub-Lie algebra given by these matrices is not a Cartan sub-Lie algebra since it is contained in the nilpotent Lie algebra of strictly upper triangular matrices.

Problem 61.   Show that the Lie algebras $s\ell(2, \mathbb{C})$ and $so(3, \mathbb{C})$ are isomorphic.

Solution 61.   The *isomorphism* has the form

$$\begin{pmatrix} a & b \\ c & -a \end{pmatrix} \leftrightarrow \begin{pmatrix} 0 & b-c & -i(b+c) \\ c-b & 0 & 2ia \\ i(b+c) & -2ia & 0 \end{pmatrix}.$$

For $s\ell(2, \mathbb{C})$ we can choose the basis (over $\mathbb{C}$)

$$X_0 = \begin{pmatrix} 1 & 0 \\ 0 & -1 \end{pmatrix}, \qquad X_+ = \begin{pmatrix} 0 & 1 \\ 0 & 0 \end{pmatrix}, \qquad X_- = \begin{pmatrix} 0 & 0 \\ 1 & 0 \end{pmatrix}$$

with the corresponding basis for $so(3, \mathbb{C})$

$$Y_0 = \begin{pmatrix} 0 & 0 & 0 \\ 0 & 0 & 2i \\ 0 & -2i & 0 \end{pmatrix}, \quad Y_+ = \begin{pmatrix} 0 & 1 & -i \\ -1 & 0 & 0 \\ i & 0 & 0 \end{pmatrix}, \quad Y_- = \begin{pmatrix} 0 & -1 & -i \\ 1 & 0 & 0 \\ i & 0 & 0 \end{pmatrix}$$

and the commutators

$$[X_0, X_\pm] = \pm 2 X_\pm, \qquad [X_+, X_-] = X_0$$

$$[Y_0, Y_\pm] = \pm 2 Y_\pm, \qquad [Y_+, Y_-] = Y_0.$$

Problem 62.   Let L be any Lie algebra. If $x, y \in L$, define

$$\kappa(x, y) := \text{tr}(\text{ad}x\, \text{ad}y)$$

where κ is called the *Killing form* and tr(.) denotes the trace.
(i) Show that

$$\kappa([x, y], z) = \kappa(x, [y, z]).$$

(ii) If B_1 and B_2 are subspaces of a Lie algebra L, we define $[B_1, B_2]$ to be the subspace spanned by all products $[b_1, b_2]$ with $b_1 \in B_1$ and $b_2 \in B_2$. This is the set of sums

$$\sum_j [b_{1j}, b_{2j}]$$

where $b_{1j} \in B_1$ and $b_{2j} \in B_2$. Let

$$L_{(1)} := [L, L], \qquad L_{(2)} := [L, L_{(1)}], \quad \ldots \quad , L_{(k)} := [L, L_{(k-1)}].$$

A Lie algebra L is called *nilpotent* if

$$L_{(n)} = \{0\}$$

for some positive integer n. Prove that if L is nilpotent, the Killing form of L is identically zero.

Solution 62. (i) For $m \times m$ matrices we have the properties for the trace

$$\mathrm{tr}(A + B) = \mathrm{tr}A + \mathrm{tr}B, \qquad \mathrm{tr}(AB) = \mathrm{tr}(BA).$$

Moreover we apply $\mathrm{ad}[x, y] \equiv [\mathrm{ad}x, \mathrm{ad}y]$. Using these properties we obtain

$$\begin{aligned}
\kappa([x, y], z) &= \mathrm{tr}(\mathrm{ad}[x, y]\,\mathrm{ad}z) = \mathrm{tr}([\mathrm{ad}x, \mathrm{ad}y]\mathrm{ad}z) \\
&= \mathrm{tr}(\mathrm{ad}x\,\mathrm{ad}y\,\mathrm{ad}z - \mathrm{ad}y\,\mathrm{ad}x\,\mathrm{ad}z) \\
&= \mathrm{tr}(\mathrm{ad}x\,\mathrm{ad}y\,\mathrm{ad}z) - \mathrm{tr}(\mathrm{ad}y\,\mathrm{ad}x\,\mathrm{ad}z) \\
&= \mathrm{tr}(\mathrm{ad}x\,\mathrm{ad}y\,\mathrm{ad}z) - \mathrm{tr}(\mathrm{ad}x\,\mathrm{ad}z\,\mathrm{ad}y) \\
&= \mathrm{tr}(\mathrm{ad}x\,[\mathrm{ad}y, \mathrm{ad}z]) = \mathrm{tr}(\mathrm{ad}x\,\mathrm{ad}[y, z]) \\
&= \kappa(x, [y, z]).
\end{aligned}$$

(ii) For any $u \in L$ we have

$$(\mathrm{ad}x\,\mathrm{ad}y)^1(u) = [x, [y, u]] \in L^3$$

$$(\mathrm{ad}x\,\mathrm{ad}y)^2(u) = [x, [y, [x, [y, u]]]] \in L^5$$

In general

$$(\mathrm{ad}x\,\mathrm{ad}y)^k(u) \in L^{2k+1}.$$

Since L is nilpotent, say $L_{(n)} = \{0\}$ we find that for $2k + 1 \geq n$

$$(\mathrm{ad}x\,\mathrm{ad}y)^k(u) = 0 \quad \text{for all} \quad u \in L$$

or equivalently

$$(\text{ad}x\,\text{ad}y)^k = 0$$

where 0 is the zero matrix. Thus the matrix $(\text{ad}x\,\text{ad}y)$ is nilpotent. If λ is an *eigenvalue* of $(\text{ad}x\,\text{ad}y)$ with *eigenvector* $\mathbf{v}$, i.e.

$$(\text{ad}x\,\text{ad}y)\mathbf{v} = \lambda\mathbf{v}$$

then

$$\lambda^k\mathbf{v} = (\text{ad}x\,\text{ad}y)^k\mathbf{v} = 0.$$

Therefore $\lambda^k = 0$ and thus $\lambda = 0$. Thus the eigenvalues $\lambda_1, \ldots, \lambda_m$ of $(\text{ad}x\,\text{ad}y)$ are all zero, and we find

$$\kappa(x, y) = \text{tr}(\text{ad}x\,\text{ad}y) = \sum_{j=1}^{m}\lambda_j = 0$$

since the trace of a square matrix is the sum of the eigenvalues. This is true for any $x, y \in L$ and the result follows.

Problem 63.   Let $g\ell(n, \mathbb{R})$ be the Lie algebra of all $n \times n$ matrices over $\mathbb{R}$. Let

$$x \in g\ell(n, \mathbb{R}) \tag{1}$$

have n distinct eigenvalues $\lambda_1, \ldots, \lambda_n$ in $\mathbb{R}$. Prove that the eigenvalues of the $n^2 \times n^2$ matrix

$$\text{ad}x \tag{2}$$

are the n^2 scalars $\lambda_i - \lambda_j$, where $i, j = 1, 2, \ldots, n$.

Solution 63.   Since the eigenvalues $\lambda_1, \ldots, \lambda_n$ of x are real and distinct, the corresponding eigenvectors $\mathbf{v}_1, \ldots, \mathbf{v}_n$ are linearly independent (not necessarily orthogonal), and thus form a basis for $\mathbb{R}^n$. The existence of a basis of eigenvectors of x means that x can be diagonalized, i.e., there exists a matrix $R \in GL(n, \mathbb{R})$ such that

$$R^{-1}xR = \text{diag}(\lambda_1, \ldots, \lambda_n)$$

where $GL(n, \mathbb{R})$ denotes the general linear group $(GL(n, \mathbb{R}) \subset g\ell(n, \mathbb{R}))$. Now

$$\text{ad}(R^{-1}xR)(E_{ij}) \equiv [R^{-1}xR, E_{ij}] = (\lambda_i - \lambda_j)E_{ij}$$

where E_{ij} are the matrices having 1 in the (i, j) position and 0 elsewhere. Thus

$$\lambda_i - \lambda_j, \qquad i, j = 1, 2, \ldots, n$$

are eigenvalues of
$$\mathrm{ad}(R^{-1}xR).$$

Since the corresponding set of eigenvectors E_{ij} $(i,j=1,2,\ldots,n)$ is a basis of $gl(n,\mathbb{R})$ we conclude that these are the only eigenvalues of $\mathrm{ad}(R^{-1}xR)$. Now if λ is an eigenvalue of $\mathrm{ad}x$ with eigenvector $\mathbf{v}$, then

$$\mathrm{ad}(R^{-1}xR)(R^{-1}\mathbf{v}R) = R^{-1}xRR^{-1}\mathbf{v}R - R^{-1}\mathbf{v}RR^{-1}xR$$
$$= R^{-1}x\mathbf{v}R - R^{-1}\mathbf{v}xR$$
$$= R^{-1}(\mathrm{ad}x(\mathbf{v}))R.$$

Thus
$$\mathrm{ad}(R^{-1}xR)(R^{-1}\mathbf{v}R) = \lambda(R^{-1}\mathbf{v}R)$$

so that λ is an eigenvalue of $\mathrm{ad}(R^{-1}xR)$. Thus we find $\lambda = \lambda_i - \lambda_j$ for some $i,j \in \{1,\ldots,n\}$. Conversely we have

$$\mathrm{ad}(R^{-1}xR)E_{ij} = (\lambda_i - \lambda_j)E_{ij}.$$

It follows that

$$R^{-1}xRE_{ij} - E_{ij}R^{-1}xR = (\lambda_i - \lambda_j)E_{ij}.$$

Multiplying this equation from the left with R and from the right with R^{-1} yields
$$xRE_{ij}R^{-1} - RE_{ij}R^{-1}x = (\lambda_i - \lambda_j)RE_{ij}R^{-1}$$

or
$$\mathrm{ad}x(RE_{ij}R^{-1}) = (\lambda_i - \lambda_j)(RE_{ij}R^{-1}).$$

Thus $\lambda_i - \lambda_j$ is indeed an eigenvalue of $\mathrm{ad}x$ for all $i,j \in \{1,\ldots,n\}$. This proves the theorem.

Problem 64.   Consider the Lie algebra $so(4)$. Show that the Lie algebra $so(4)$ is semisimple, but not simple. Consider as a basis of $so(4)$ the six vector fields

$$V_{12} = x_2\frac{\partial}{\partial x_1} - x_1\frac{\partial}{\partial x_2}, \quad V_{23} = x_3\frac{\partial}{\partial x_2} - x_2\frac{\partial}{\partial x_3}, \quad V_{31} = x_1\frac{\partial}{\partial x_3} - x_3\frac{\partial}{\partial x_1}$$

$$W_{41} = x_1\frac{\partial}{\partial x_4} - x_4\frac{\partial}{\partial x_1}, \quad W_{42} = x_2\frac{\partial}{\partial x_4} - x_4\frac{\partial}{\partial x_2}, \quad W_{43} = x_3\frac{\partial}{\partial x_4} - x_4\frac{\partial}{\partial x_3}.$$

Solution 64.   For the commutators we find

$$[V_{12}, V_{23}] = V_{31}, \quad [V_{23}, V_{31}] = V_{12}, \quad [V_{31}, V_{12}] = V_{23}$$

and

$$[W_{41}, W_{42}] = V_{12}, \quad [W_{42}, W_{43}] = V_{23}, \quad [W_{43}, W_{41}] = V_{31}$$
$$[V_{12}, W_{41}] = W_{42}, \quad [V_{23}, W_{41}] = 0, \quad [V_{31}, W_{41}] = -W_{43}$$
$$[V_{12}, W_{42}] = -W_{41}, \quad [V_{23}, W_{42}] = W_{43}, \quad [V_{31}, W_{42}] = 0$$
$$[V_{12}, W_{43}] = 0, \quad [V_{23}, W_{43}] = -W_{42}, \quad [V_{31}, W_{43}] = W_{41} .$$

Making the linear transformation

$$X_1 = \frac{1}{2}(V_{23} + W_{41}), \quad X_2 = \frac{1}{2}(V_{31} + W_{42}), \quad X_3 = \frac{1}{2}(V_{12} + W_{43})$$

$$Y_1 = \frac{1}{2}(V_{23} - W_{41}), \quad Y_2 = \frac{1}{2}(V_{31} - W_{42}), \quad Y_3 = \frac{1}{2}(V_{12} - W_{43})$$

we obtain a new basis with the commutators ($\epsilon_{123} = +1$)

$$[X_i, X_j] = \epsilon_{ijk} X_k, \quad [Y_i, Y_j] = \epsilon_{ijk} Y_k, \quad [X_i, Y_j] = 0 .$$

The operators X_1, X_2, X_3 and Y_1, Y_2, Y_3 are separately closed under commutation, each describing the sub Lie algebra $so(3)$ of $so(4)$. Note that the Lie algebra $so(3)$ is simple. Consequently the Lie algebra $so(4)$ is the direct sum of the two $so(3)$ Lie algebras. The operators X_1, X_2, X_3 and Y_1, Y_2, Y_3 each form proper ideals in $so(4)$. Thus the Lie algebra $so(4)$ is not simple. Now these two ideals are non-abelian and therefore $so(4)$ is semisimple.

Problem 65.   Consider the semisimple Lie algebra $so(2, 1)$ whose elements X_1, X_2, X_3 satisfy the commutation relations

$$[X_1, X_2] = X_3, \quad [X_2, X_3] = -X_1, \quad [X_3, X_1] = X_2 .$$

(i) Find the adjoint representation.
(ii) Let Y_j ($j = 1, \ldots, n$) be a basis for a Lie algebra. Then $g_{jk} = \kappa(Y_j, Y_k)$ is called the *metric tensor*. From the adjoint representation find the metric tensor.

Solution 65.   (i) We obtain

$$\mathrm{ad}X_1 = \begin{pmatrix} 0 & 0 & 0 \\ 0 & 0 & 1 \\ 0 & -1 & 0 \end{pmatrix}, \quad \mathrm{ad}X_2 = \begin{pmatrix} 0 & 0 & -1 \\ 0 & 0 & 0 \\ 1 & 0 & 0 \end{pmatrix}, \quad \mathrm{ad}X_3 = \begin{pmatrix} 0 & 1 & 0 \\ = 1 & 0 & 0 \\ 0 & 0 & 0 \end{pmatrix} .$$

(ii) From (i) we obtain

$$g = \begin{pmatrix} -2 & 0 & 0 \\ 0 & 2 & 0 \\ 0 & 0 & 2 \end{pmatrix}$$

with $\det(g) = -8$. Thus in fact $so(2, 1)$ is semisimple.

Problem 66. Consider the two matrices

$$A = \begin{pmatrix} 0 & 1 \\ 0 & 0 \end{pmatrix}, \qquad B = \begin{pmatrix} 0 & 0 \\ 1 & 0 \end{pmatrix}.$$

Show that

$$[e^{tA}, e^{tB}] = t^2[A, B].$$

Solution 66. Since

$$e^{tA} = \begin{pmatrix} 1 & t \\ 0 & 1 \end{pmatrix}, \qquad e^{tB} = \begin{pmatrix} 1 & 0 \\ t & 1 \end{pmatrix}$$

we have

$$[e^{tA}, e^{tB}] = \begin{pmatrix} t^2 & 0 \\ 0 & -t^2 \end{pmatrix} = t^2 \begin{pmatrix} 1 & 0 \\ 0 & -1 \end{pmatrix} = t^2[A, B].$$

Problem 67. Let A be an $n \times n$ matrix over $\mathbb{C}$. Show that the $n \times n$ matrices R over $\mathbb{C}$ satisfying $R^*A + AR = 0$ form a Lie algebra.

Solution 67. The set of matrices R satisfying $R^*A + AR = 0$ is closed under addition and scalar multiplication. Assume that R_1, R_2 satisfy this condition and consider the commutator $[R_1, R_2] = R_1R_2 - R_2R_1$. Then we have

$$R_2^*R_1^*A - R_1^*R_2^*A = -R_2^*AR_1 + R_1^*AR_2$$
$$AR_1R_2 - AR_2R_1 = -R_1^*AR_2 + R_2^*AR_1.$$

Consequently

$$[R_1, R_2]^*A + A[R_1, R_2] = 0.$$

Problem 68. Show that the Lie algebras $so(3)$ and $su(2)$ of the Lie groups $SO(3)$ and $SU(2)$ are *isomorphic*.

Solution 68. A basis for the simple Lie algebra $so(3)$ is

$$T_1 = \begin{pmatrix} 0 & 0 & 0 \\ 0 & 0 & -1 \\ 0 & 1 & 0 \end{pmatrix}, \quad T_2 = \begin{pmatrix} 0 & 0 & 1 \\ 0 & 0 & 0 \\ -1 & 0 & 0 \end{pmatrix}, \quad T_3 = \begin{pmatrix} 0 & -1 & 0 \\ 1 & 0 & 0 \\ 0 & 0 & 0 \end{pmatrix}.$$

We have $[T_1, T_2] = T_3$, $[T_2, T_3] = T_1$, $[T_3, T_1] = T_2$. A basis for $su(2)$ is

$$\tau_1 = \frac{1}{2}\begin{pmatrix} 0 & -i \\ -i & 0 \end{pmatrix}, \quad \tau_2 = \frac{1}{2}\begin{pmatrix} 0 & -1 \\ 1 & 0 \end{pmatrix}, \quad \tau_3 = \frac{1}{2}\begin{pmatrix} -i & 0 \\ 0 & i \end{pmatrix}.$$

Then we have

$$[\tau_1, \tau_2] = \tau_3, \quad [\tau_2, \tau_3] = \tau_1, \quad [\tau_3, \tau_1] = \tau_2.$$

Thus $\tau_1 \leftrightarrow T_1$, $\tau_2 \leftrightarrow T_2$, $\tau_3 \leftrightarrow T_3$.

Problem 69.   Consider the Lie algebra $s\ell(2, \mathbb{R})$ with the basis H, X, Y and the commutators

$$[H, X] = 2X, \qquad [H, Y] = -2Y, \qquad [X, Y] = H$$

with

$$H = \begin{pmatrix} 1 & 0 \\ 0 & -1 \end{pmatrix}, \quad X = \begin{pmatrix} 0 & 1 \\ 0 & 0 \end{pmatrix}, \quad Y = \begin{pmatrix} 0 & 0 \\ 1 & 0 \end{pmatrix}.$$

Assume that $\mathbf{v} \in \mathbb{R}$ is an *eigenvector* of H, i.e. $H\mathbf{v} = \alpha\mathbf{v}$, where α is the *eigenvalue*. Calculate $H(X\mathbf{v})$. Discuss.

Solution 69.   We have

$$\begin{aligned} H(X\mathbf{v}) &= X(H\mathbf{v}) + [H, X]\mathbf{v} \\ &= X(\alpha\mathbf{v}) + 2X\mathbf{v} \\ &= (\alpha + 2)X\mathbf{v}. \end{aligned}$$

This means if $\mathbf{v}$ is an eigenvector of H with eigenvalue α, then $X\mathbf{v}$ is also an eigenvector for H with eigenvalue $\alpha + 2$.

Problem 70.   Let $S^1 := \{(x_1, x_2) : x_1^2 + x_2^2 = 1\}$. Let $\mathrm{Diff}^+(S^1)$ be the space of orientation-preserving diffeomorphism of $S^1 \to S^1$. With the product given by the composition of diffeomorphism it is an infinite dimensional Lie group. The Lie algebra is the space of vector fields on S^1. A Lie subalgebra of the complexification of the Lie algebra of $\mathrm{Diff}^+(S^1)$ is given by

$$L := \left\{ f(z)\frac{d}{dz} : f(z) \in \mathbb{C}[z, z^{-1}] \right\}.$$

This is a subspace of the space of vector fields on S^1, regarding S^1 as the unit circle in $\mathbb{C}$. We choose a basis $\{ L_n \}_{n \in \mathbb{Z}}$ of L defining

$$L_n := -z^{n+1}\frac{d}{dz}.$$

Find the commutator $[L_n, L_m]$.

Solution 70. We have

$$[L_n, L_m] = L_n L_m - L_m L_n$$
$$= (-z^{n+1}\frac{d}{dz})(-z^{m+1}\frac{d}{dz}) - (-z^{m+1}\frac{d}{dz})(-z^{n+1}\frac{d}{dz})$$
$$= (m-n)z^{n+m+1}\frac{d}{dz}$$
$$= (n-m)L_{n+m}.$$

Problem 71. Let σ_1, σ_2, σ_3 be the Pauli spin matrices. Let I_2 be the 2×2 identity matrix and 0_2 be the 2×2 zero matrix. Consider the four 4×4 matrices

$$J_1 = \frac{i}{2}\begin{pmatrix} 0_2 & \sigma_1 \\ -\sigma_1 & 0_2 \end{pmatrix}, \qquad J_2 = \frac{1}{2}\begin{pmatrix} \sigma_2 & 0_2 \\ 0_2 & \sigma_2 \end{pmatrix},$$

$$J_3 = \frac{i}{2}\begin{pmatrix} 0_2 & \sigma_3 \\ -\sigma_3 & 0_2 \end{pmatrix}, \qquad J_0 = \frac{i}{2}\begin{pmatrix} 0_2 & I_2 \\ -I_2 & 0_2 \end{pmatrix}.$$

(i) Find the commutators.
(ii) Consider the six 4×4 matrices

$$K_1 = \frac{i}{2}\begin{pmatrix} 0_2 & \sigma_3 \\ \sigma_3 & 0_2 \end{pmatrix}, \quad K_2 = \frac{i}{2}\begin{pmatrix} I_2 & 0_2 \\ 0_2 & -I_2 \end{pmatrix}, \quad K_3 = -\frac{i}{2}\begin{pmatrix} 0_2 & \sigma_1 \\ \sigma_1 & 0_2 \end{pmatrix},$$

$$Q_1 = -\frac{i}{2}\begin{pmatrix} \sigma_3 & 0_2 \\ 0_2 & -\sigma_3 \end{pmatrix}, \quad Q_2 = \frac{i}{2}\begin{pmatrix} 0_2 & I_2 \\ I_2 & 0_2 \end{pmatrix}, \quad Q_3 = \frac{i}{2}\begin{pmatrix} \sigma_1 & 0_2 \\ 0_2 & -\sigma_1 \end{pmatrix}.$$

Find the commutators. Find also the commutators $[K_j, J_k]$, $[Q_j, J_k]$, where $j, k = 1, 2, 3$ and $[K_j, J_0]$, $[Q_j, J_0]$. Thus show that we have a basis of a 10 dimensional Lie algebra.

Solution 71. (i) We obtain

$$[J_j, J_k] = i\epsilon_{jk\ell}J_\ell, \quad j, k, \ell = 1, 2, 3, \quad \epsilon_{123} = +1$$

and $[J_k, J_0] = 0$ for $k = 1, 2, 3$. Thus we have a basis of a Lie algebra.
(ii) We obtain

$$[J_j, K_k] = i\epsilon_{jk\ell}K_\ell, \qquad [J_j, Q_k] = i\epsilon_{jk\ell}Q_\ell$$

$$[K_j, K_k] = [Q_j, Q_k] = -i\epsilon_{jk\ell}J_\ell, \qquad [K_j, Q_k] = i\delta_{jk}J_0$$

$$[K_j, J_0] = iQ_j, \qquad [Q_j, J_0] = -iK_j$$

where $j, k, \ell = 1, 2, 3$. Thus $J_1, J_2, J_3, J_0, K_1, K_2, K_3, Q_1, Q_2, Q_3$ form a basis of a ten dimensional Lie algebra. This is the symplectic Lie algebra $sp(2n)$ with $n = 2$.

Problem 72. (i) Consider the matrices

$$N_1 = \begin{pmatrix} 0 & 0 & i \\ 0 & 0 & 0 \\ 0 & 0 & 0 \end{pmatrix}, \quad N_2 = \begin{pmatrix} 0 & 0 & 0 \\ 0 & 0 & i \\ 0 & 0 & 0 \end{pmatrix}, \quad L = \begin{pmatrix} 0 & -i/2 & 0 \\ i/2 & 0 & 0 \\ 0 & 0 & 0 \end{pmatrix}.$$

Show that they form a basis of a Lie algebra.
(ii) Is the Lie algebra semisimple?

Solution 72. (i) The commutators are given by

$$[N_1, N_2] = 0, \quad [N_1, L] = \frac{i}{2} N_2, \quad [N_2, L] = -\frac{i}{2} N_1.$$

Thus N_1, N_2, L satisfy closed commutation relations. Thus we have a basis of a Lie algebra. It is not necessary to the check Jacobi identity, because for matrix algebras under the commutator it is always true. It is sufficient to check that RHS's are linear combinations of basis vectors.
(ii) Since $[N_1, N_2] = 0$ the Lie algebra is not semisimple. N_1 and N_2 form a basis of a commutative Lie algebra. The adjoint representation is

$$N_1 \to \begin{pmatrix} 0 & 0 & 0 \\ 0 & 0 & -i/2 \\ 0 & 0 & 0 \end{pmatrix}, \quad N_2 \to \begin{pmatrix} 0 & 0 & i/2 \\ 0 & 0 & 0 \\ 0 & 0 & 0 \end{pmatrix}, \quad L \to \begin{pmatrix} 0 & -i/2 & 0 \\ i/2 & 0 & 0 \\ 0 & 0 & 0 \end{pmatrix}.$$

Problem 73. (i) Consider the differential operators

$$J_3 = \frac{i}{2} \left(p \frac{\partial}{\partial x} - x \frac{\partial}{\partial p} \right)$$

and

$$K_1 = \frac{i}{2} \left(p \frac{\partial}{\partial p} - x \frac{\partial}{\partial x} \right), \quad K_2 = -\frac{i}{2} \left(x \frac{\partial}{\partial p} + p \frac{\partial}{\partial x} \right).$$

Find the commutators. Show that we have a basis for a semisimple Lie algebra.
(ii) Consider the differential operators

$$\tilde{J}_3 = \frac{1}{4} \left(x^2 - \frac{d^2}{dx^2} \right)$$

and

$$\widetilde{K}_1 = -\frac{i}{4}\left(x\frac{d}{dx} + \frac{d}{dx}x\right), \quad \widetilde{K}_2 = \frac{1}{4}\left(x^2 + \frac{d^2}{dx^2}\right).$$

Find the commutators. Show that we have a basis for a semisimple Lie algebra. Are the two Lie algebras isomorphic?

Solution 73. (i) We find

$$[J_3, K_1] = iK_2, \quad [J_3, K_2] = iK_1, \quad [K_1, K_2] = iJ_3.$$

The Lie algebra is semisimple since the Killing form matrix

$$\kappa = \begin{pmatrix} 2 & 0 & 0 \\ 0 & -2 & 0 \\ 0 & 0 & -2 \end{pmatrix}$$

is invertible.
(ii) We find

$$[\widetilde{J}_3, \widetilde{K}_1] = i\widetilde{K}_2, \quad [\widetilde{J}_3, \widetilde{K}_2] = i\widetilde{K}_1, \quad [\widetilde{K}_1, \widetilde{K}_2] = i\widetilde{J}_3.$$

The two Lie algebras are isomorphic with $J_3 \leftrightarrow \widetilde{J}_3$, $K_1 \leftrightarrow \widetilde{K}_1$, $K_2 \leftrightarrow \widetilde{K}_2$.

Problem 74. Show that the vector fields

$$J_1 = -\frac{i}{2}\left(\left(p_2\frac{\partial}{\partial x_1} + p_1\frac{\partial}{\partial x_2}\right) - \left(x_2\frac{\partial}{\partial p_1} + x_1\frac{\partial}{\partial p_2}\right)\right)$$

$$J_2 = -\frac{i}{2}\left(\left(x_1\frac{\partial}{\partial x_2} - x_2\frac{\partial}{\partial x_1}\right) + \left(p_1\frac{\partial}{\partial p_2} - p_2\frac{\partial}{\partial p_1}\right)\right)$$

$$J_3 = -\frac{i}{2}\left(\left(p_1\frac{\partial}{\partial x_1} - p_2\frac{\partial}{\partial x_2}\right) - \left(x_1\frac{\partial}{\partial p_1} - x_2\frac{\partial}{\partial p_2}\right)\right)$$

$$J_0 = -\frac{i}{2}\left(\left(p_1\frac{\partial}{\partial x_1} + p_2\frac{\partial}{\partial x_2}\right) - \left(x_1\frac{\partial}{\partial p_1} + x_2\frac{\partial}{\partial p_2}\right)\right)$$

$$K_1 = -\frac{i}{2}\left(\left(p_1\frac{\partial}{\partial x_1} - p_2\frac{\partial}{\partial x_2}\right) + \left(x_1\frac{\partial}{\partial p_1} - x_2\frac{\partial}{\partial p_2}\right)\right)$$

$$K_2 = \frac{i}{2}\left(-\left(x_1\frac{\partial}{\partial x_1} + x_2\frac{\partial}{\partial x_2}\right) + \left(p_1\frac{\partial}{\partial p_1} + p_2\frac{\partial}{\partial p_2}\right)\right)$$

$$K_3 = \frac{i}{2}\left(\left(p_2\frac{\partial}{\partial x_1} + p_1\frac{\partial}{\partial x_2}\right) + \left(x_2\frac{\partial}{\partial p_1} + x_1\frac{\partial}{\partial p_2}\right)\right)$$

$$Q_1 = \frac{i}{2}\left(\left(x_1\frac{\partial}{\partial x_1} - x_2\frac{\partial}{\partial x_2}\right) - \left(p_1\frac{\partial}{\partial p_1} - p_2\frac{\partial}{\partial p_2}\right)\right)$$

$$Q_2 = -\frac{i}{2}\left(\left(p_1\frac{\partial}{\partial x_1} + p_2\frac{\partial}{\partial x_2}\right) + \left(x_1\frac{\partial}{\partial p_1} + x_2\frac{\partial}{\partial p_2}\right)\right)$$

$$Q_3 = -\frac{i}{2} \left(\left(x_2 \frac{\partial}{\partial x_1} + x_1 \frac{\partial}{\partial x_2} \right) - \left(p_2 \frac{\partial}{\partial p_1} + p_1 \frac{\partial}{\partial p_2} \right) \right)$$

form a basis of a Lie algebra under the commutator.

Solution 74. For the operators J_1, J_2, J_3, J_0 we find the commutators

$$[J_1, J_2] = iJ_3, \quad [J_3, J_1] = iJ_2, \quad [J_2, J_3] = iJ_1,$$

$$[J_0, J_1] = 0, \quad [J_0, J_2] = 0, \quad [J_0, J_3] = 0.$$

Thus we have a sub Lie algebra. Furthermore

$$[J_j, K_k] = i\epsilon_{jkl}K_l, \quad [J_j, Q_k] = i\epsilon_{jkl}Q_l,$$

$$[K_j, K_k] = [Q_j, Q_k] = -i\epsilon_{jkl}J_l, \quad [K_j, Q_k] = i\delta_{jk}J_0,$$

$$[K_j, J_0] = iQ_j, \quad [Q_j, J_0] = -iK_j$$

where $j, k, l = 1, 2, 3$. Thus J_1, J_2, J_3, J_0, K_1, K_2, K_3, Q_1, Q_2, Q_3 form a basis of a ten dimensional Lie algebra. This gives another representation of the symplectic Lie algebra $sp(2n)$ with $n = 2$.

Problem 75. Consider the spin-matrices (spin-1)

$$S_+ = \begin{pmatrix} 0 & \sqrt{2} & 0 \\ 0 & 0 & \sqrt{2} \\ 0 & 0 & 0 \end{pmatrix}, \quad S_- = \begin{pmatrix} 0 & 0 & 0 \\ \sqrt{2} & 0 & 0 \\ 0 & \sqrt{2} & 0 \end{pmatrix}, \quad S_z = \begin{pmatrix} 1 & 0 & 0 \\ 0 & 0 & 0 \\ 0 & 0 & -1 \end{pmatrix}.$$

Find the commutators and thus show that we have a basis of a Lie algebra. Show that we have a faithful representation using differential operators by

$$S_+ \to z\frac{\partial}{\partial \zeta}, \quad S_- \to \zeta\frac{\partial}{\partial z}, \quad S_z \to \frac{1}{2}\left(z\frac{\partial}{\partial z} - \zeta\frac{\partial}{\partial \zeta} \right).$$

Solution 75. We find

$$[S_+, S_-] = 2S_z, \quad [S_+, S_z] = -S_+, \quad [S_-, S_z] = S_-$$

and

$$\left[z\frac{\partial}{\partial \zeta}, \zeta\frac{\partial}{\partial z} \right] = z\frac{\partial}{\partial z} - \zeta\frac{\partial}{\partial \zeta}$$

$$\left[z\frac{\partial}{\partial \zeta}, \frac{1}{2}\left(z\frac{\partial}{\partial z} - \zeta\frac{\partial}{\partial \zeta} \right) \right] = -z\frac{\partial}{\partial \zeta}$$

$$\left[\zeta\frac{\partial}{\partial z}, \frac{1}{2}\left(z\frac{\partial}{\partial z} - \zeta\frac{\partial}{\partial \zeta} \right) \right] = \zeta\frac{\partial}{\partial z}.$$

Problem 76. Let A, B be $n \times n$ matrices over $\mathbb{C}$. Consider the expression

$$(A \otimes B - B \otimes A) \oplus (-[A, B])$$

where $\otimes$ denotes the Kronecker product, $\oplus$ denotes the direct sum and $[\,,\,]$ denotes the commutator. Find the trace of this expression. This expression plays a role for universal enveloping algebras.

Solution 76. The trace is a linear map. Thus we have

$$\begin{aligned}
\operatorname{tr}((A \otimes B - B \otimes A) \oplus (-[A, B])) &= \operatorname{tr}(A \otimes B - B \otimes A) + \operatorname{tr}(-[A, B]) \\
&= \operatorname{tr}(A \otimes B) - \operatorname{tr}(B \otimes A) - \operatorname{tr}(AB - BA) \\
&= \operatorname{tr}(A)\operatorname{tr}(B) - \operatorname{tr}(B)\operatorname{tr}(A) \\
&\quad -\operatorname{tr}(AB) + \operatorname{tr}(BA) \\
&= 0.
\end{aligned}$$

Problem 77. Consider the Lie algebra $s\ell(2, \mathbb{R})$ with the basis e_1, e_2, e_3 and the commutators

$$[e_2, e_1] = 2e_1, \quad [e_2, e_3] = -2e_3, \quad [e_1, e_3] = e_2.$$

(i) Find the adjoint representation.
(ii) Calculate $\exp(\mathrm{ad}e_1)$, $\exp(\mathrm{ad}e_2)$, $\exp(\mathrm{ad}e_3)$.

Solution 77. (i) We obtain

$$\mathrm{ad}e_1 = \begin{pmatrix} 0 & -2 & 0 \\ 0 & 0 & 1 \\ 0 & 0 & 0 \end{pmatrix}$$

$$\mathrm{ad}e_2 = \begin{pmatrix} 2 & 0 & 0 \\ 0 & 0 & 0 \\ 0 & 0 & -2 \end{pmatrix}$$

$$\mathrm{ad}e_3 = \begin{pmatrix} 0 & 0 & 0 \\ -1 & 0 & 0 \\ 0 & 2 & 0 \end{pmatrix}.$$

(ii) We find

$$\exp(\mathrm{ad}e_1) = \begin{pmatrix} 1 & -2 & -1 \\ 0 & 1 & 1 \\ 0 & 0 & 1 \end{pmatrix}$$

$$\exp(\mathrm{ad}e_2) = \begin{pmatrix} e^2 & 0 & 0 \\ 0 & 1 & 0 \\ 0 & 0 & e^{-2} \end{pmatrix}$$

$$\exp(\mathrm{ad}e_3) = \begin{pmatrix} 1 & 0 & 0 \\ -1 & 1 & 0 \\ -1 & 2 & 1 \end{pmatrix}.$$

Problem 78. The semisimple Lie algebra $s\ell(3, \mathbb{C})$ has dimension 8 and rank 2. The standard basis is given by

$$h_1 = \begin{pmatrix} 1 & 0 & 0 \\ 0 & -1 & 0 \\ 0 & 0 & 0 \end{pmatrix}, \quad h_2 = \begin{pmatrix} 0 & 0 & 0 \\ 0 & 1 & 0 \\ 0 & 0 & -1 \end{pmatrix},$$

$$e_1 = \begin{pmatrix} 0 & 1 & 0 \\ 0 & 0 & 0 \\ 0 & 0 & 0 \end{pmatrix}, \quad e_2 = \begin{pmatrix} 0 & 0 & 0 \\ 0 & 0 & 1 \\ 0 & 0 & 0 \end{pmatrix},$$

$$f_1 = \begin{pmatrix} 0 & 0 & 0 \\ 1 & 0 & 0 \\ 0 & 0 & 0 \end{pmatrix}, \quad f_2 = \begin{pmatrix} 0 & 0 & 0 \\ 0 & 0 & 0 \\ 0 & 1 & 0 \end{pmatrix},$$

$$e_{13} = \begin{pmatrix} 0 & 0 & 1 \\ 0 & 0 & 0 \\ 0 & 0 & 0 \end{pmatrix}, \quad f_{13} = \begin{pmatrix} 0 & 0 & 0 \\ 0 & 0 & 0 \\ 1 & 0 & 0 \end{pmatrix}.$$

Note that $e_1 = f_1^T$, $e_2 = f_2^T$, $e_{13} = f_{13}^T$. Find the commutation relation table.

Solution 78. We find

	h_1	h_2	e_1	e_2	e_{13}	f_1	f_2	f_{13}
h_1	0	0	$2e_1$	$-e_2$	e_{13}	$-2f_1$	f_2	$-f_{13}$
h_2	.	0	$-e_1$	$2e_2$	e_{13}	f_1	$-2f_2$	$-f_{13}$
e_1	.	.	0	e_{13}	0	h_1	0	$-f_2$
e_2	.	.	.	0	0	0	h_2	f_1
e_{13}	.	.	.	.	0	$-e_2$	e_1	$h_1 + h_2$
f_1	.	.	.	.	.	0	$-f_{13}$	0
f_2	.	.	.	.	.	.	0	0
f_{13}	.	.	.	.	.	.	.	0

Problem 79. Consider the 4×4 matrices

$$J_+ = \begin{pmatrix} 0 & 0 & 0 & 0 \\ 0 & 0 & 1 & 0 \\ 0 & 0 & 0 & 0 \\ 0 & 0 & 0 & 0 \end{pmatrix}, \quad J_- = \begin{pmatrix} 0 & 1 & 0 & 0 \\ 0 & 0 & 0 & 0 \\ 0 & 0 & 0 & 0 \\ 0 & 0 & 0 & 0 \end{pmatrix},$$

$$J_3 = \begin{pmatrix} 0 & 0 & 0 & 1 \\ 0 & 1 & 0 & 0 \\ 0 & 0 & 0 & 0 \\ 0 & 0 & 0 & 0 \end{pmatrix}, \quad E = \begin{pmatrix} 0 & 0 & 1 & 0 \\ 0 & 0 & 0 & 0 \\ 0 & 0 & 0 & 0 \\ 0 & 0 & 0 & 0 \end{pmatrix}.$$

(i) Find the commutators and thus show that we have a basis of a Lie algebra. Is the Lie algebra semisimple?

(ii) Let $\alpha, \beta, \gamma, \delta \in \mathbb{R}$. Calculate

$$\exp(\alpha J_+), \quad \exp(\beta J_-), \quad \exp(\gamma J_3), \quad \exp(\delta E).$$

Show that $\exp(\alpha J_+)\exp(\delta E) = \exp(\delta E)\exp(\alpha J_+)$.

Solution 79. (i) We obtain for the commutators

$$[J_3, J_+] = J_+, \quad [J_3, J_-] = -J_-, \quad [J_+, J_-] = -E$$

$$[E, J_+] = [E, J_-] = [E, J_3] = 0.$$

Since the matrix E commutes with J_+, J_-, J_3 we find that the Lie algebra is not semisimple.

(ii) We obtain

$$\exp(\gamma J_3) = \begin{pmatrix} 1 & 0 & 0 & \gamma \\ 0 & e^\gamma & 0 & 0 \\ 0 & 0 & 1 & 0 \\ 0 & 0 & 0 & 1 \end{pmatrix}, \quad \exp(\alpha J_+) = \begin{pmatrix} 1 & 0 & 0 & 0 \\ 0 & 1 & \alpha & 0 \\ 0 & 0 & 1 & 0 \\ 0 & 0 & 0 & 1 \end{pmatrix},$$

$$\exp(\beta J_-) = \begin{pmatrix} 1 & \beta & 0 & 0 \\ 0 & 1 & 0 & 0 \\ 0 & 0 & 1 & 0 \\ 0 & 0 & 0 & 1 \end{pmatrix}, \quad \exp(\delta E) = \begin{pmatrix} 1 & 0 & \delta & 0 \\ 0 & 1 & 0 & 0 \\ 0 & 0 & 1 & 0 \\ 0 & 0 & 0 & 1 \end{pmatrix}.$$

Direct multiplication shows that $\exp(\alpha J_+)\exp(\delta E) = \exp(\delta E)\exp(\alpha J_+)$. We find the adjoint representation

$$J_+ \rightarrow \begin{pmatrix} 0 & 0 & -1 & 0 \\ 0 & 0 & 0 & 0 \\ 0 & 0 & 0 & 0 \\ 0 & -1 & 0 & 0 \end{pmatrix}, \quad J_- \rightarrow \begin{pmatrix} 0 & 0 & 0 & 0 \\ 0 & 0 & 1 & 0 \\ 0 & 0 & 0 & 0 \\ 1 & 0 & 0 & 0 \end{pmatrix},$$

$$J_3 \rightarrow \begin{pmatrix} 1 & 0 & 0 & 0 \\ 0 & -1 & 0 & 0 \\ 0 & 0 & 0 & 0 \\ 0 & 0 & 0 & 0 \end{pmatrix}, \quad E \rightarrow \begin{pmatrix} 0 & 0 & 0 & 0 \\ 0 & 0 & 0 & 0 \\ 0 & 0 & 0 & 0 \\ 0 & 0 & 0 & 0 \end{pmatrix}.$$

Without further calculations we can conclude that the Killing's form is degenerate, since the representation of the element E is the zero matrix. The Lie algebra is not semisimple.

Problem 80.   (i) Let U, V be $n \times n$ unitary matrices such that the commutator $[U, V] = W$ is again a unitary matrix, i.e. $W^* = W^{-1}$. Show that

$$R + R^{-1} = I_n$$

where $R = U^{-1}V^{-1}UV$ is the group commutator of U^{-1} and V^{-1}.
(ii) Give an example for such matrices U, V for the case $n = 2$.

Solution 80.   (i) From $UV - VU = W$ we obtain $V^*U^* - U^*V^* = W^*$, where $U^* = U^{-1}$, $V^* = V^{-1}$, $W^* = W^{-1}$. Thus

$$(V^{-1}U^{-1} - U^{-1}V^{-1})(UV - VU) = I_n .$$

It follows that

$$U^{-1}V^{-1}UV - V^{-1}U^{-1}VU = I_n .$$

(ii) An example is

$$U = \begin{pmatrix} 0 & 1/2 + i\sqrt{3}/2 \\ 1 & 0 \end{pmatrix}, \qquad V = \begin{pmatrix} 0 & 1 \\ 1 & 0 \end{pmatrix}$$

with

$$[U, V] = \frac{1}{2}\begin{pmatrix} -1 + i\sqrt{3} & 0 \\ 0 & 1 - i\sqrt{3} \end{pmatrix} .$$

Problem 81.   Let A, B be $n \times n$ matrices over $\mathbb{C}$. Assume that $[A, B] \equiv AB - BA = 0_n$. Let $\otimes$ be the Kronecker product. Can we conclude that

$$A \otimes B - B \otimes A = 0_{n^2} ?$$

Solution 81.   No. Take for example $A = I_2$ and

$$B = \begin{pmatrix} 1 & 2 \\ 3 & 4 \end{pmatrix} .$$

Assume that $A \otimes B - B \otimes A = 0_{n^2}$. Can we conclude that $[A, B] = 0_n$?

Problem 82.   Consider the Lie algebra $s\ell(3, \mathbb{C})$. Given the roots find the structure constants. The semisimple Lie algebra $s\ell(3, \mathbb{C})$ has dimension 8 and rank 2. The root system of $s\ell(3, \mathbb{C})$ consists of 6 vectors which go from the origin to vertices of regular hexagon. Check the Jacobi identities for the resulting structure constants.

Problem 83.   Consider the semisimple Lie algebra $s\ell(n + 1, \mathbb{F})$. Let $E_{i,j}$ ($i, j \in \{1, 2, \ldots, n + 1\}$) denote the standard basis, i.e. $(n + 1) \times (n + 1)$

matrices with all entries zero except for the entry in the i-th row and j-th column which is one. We can form a *Cartan-Weyl basis* with

$$H_j := E_{j,j} - E_{j+1,j+1}, \qquad j \in \{1, 2, \ldots, n\}.$$

Show that $E_{i,j}$ are *root vectors* for $i \neq j$, i.e. there exists $\lambda_{H,i,j} \in \mathbb{F}$ such that

$$[H, E_{i,j}] = \lambda_{H,i,j} E_{i,j}$$

for all $H \in \text{span}\{ H_1, \ldots, H_n \}$.

Solution 83. Obviously $[H_j, H_k] = 0$ since H_j and H_k are diagonal. We have

$$[E_{i,j}, E_{k,l}] = E_{i,j} E_{k,l} - E_{k,l} E_{i,j} = \delta_{j,k} E_{i,l} - \delta_{l,i} E_{k,j}.$$

It follows that

$$\begin{aligned}
[H_k, E_{i,j}] &= [E_{k,k}, E_{i,j}] - [E_{k+1,k+1}, E_{i,j}] \\
&= \delta_{i,k} E_{k,j} - \delta_{j,k} E_{i,k} - \delta_{i,k+1} E_{k+1,j} + \delta_{j,k+1} E_{i,k+1}.
\end{aligned}$$

Let

$$H = \sum_{j=1}^{n} \lambda_j H_j, \qquad \lambda_1, \ldots, \lambda_n \in \mathbb{F}.$$

Then, for $i \neq j$, we find

$$\begin{aligned}
[H, E_{i,j}] &= \sum_{k=1}^{n} \lambda_k [H_k, E_{i,j}] \\
&= \sum_{k=1}^{n} \lambda_k (\delta_{i,k} E_{k,j} - \delta_{j,k} E_{i,k} - \delta_{i,k+1} E_{k+1,j} + \delta_{j,k+1} E_{i,k+1}) \\
&= \begin{cases}
(\lambda_i - \lambda_j + \lambda_{j-1}) E_{i,j} & i = 1 \\
(\lambda_i - \lambda_j - \lambda_{i-1}) E_{i,j} & j = 1 \\
(\lambda_i - \lambda_j - \lambda_{i-1} + \lambda_{j-1}) E_{i,j} & \text{otherwise}
\end{cases}
\end{aligned}$$

so that $\lambda_{H,i,j}$ is given by

$$\lambda_{H,i,j} = \begin{cases}
\lambda_i - \lambda_j + \lambda_{j-1} & i = 1 \\
\lambda_i - \lambda_j - \lambda_{i-1} & j = 1 \\
\lambda_i - \lambda_j - \lambda_{i-1} + \lambda_{j-1} & \text{otherwise}
\end{cases}.$$

Problem 84. (i) Let A, B be $n \times n$ matrices over $\mathbb{C}$. Let $z \in \mathbb{C}$. Let $\mathbf{u}, \mathbf{v} \in \mathbb{C}^n$. Calculate

$$\exp(z(A \otimes B))(\mathbf{u} \otimes \mathbf{v})$$

where $\otimes$ denotes the Kronecker product.
(ii) Simplify the result from (i) if $A^2 = I_n$ and $B^2 = I_n$.

Solution 84.   (i) Since

$$(A \otimes B)(\mathbf{u} \otimes \mathbf{v}) = (A\mathbf{u}) \otimes (B\mathbf{v})$$

we have

$$\exp(z(A \otimes B))(\mathbf{u} \otimes \mathbf{v}) = (\mathbf{u} \otimes \mathbf{v}) + \sum_{j=1}^{\infty} \frac{z^j}{j!}(A^j\mathbf{u}) \otimes (B^j\mathbf{v}).$$

(ii) If $A^2 = I_n$ and $B^2 = I_n$ we obtain

$$\exp(z(A \otimes B))(\mathbf{u} \otimes \mathbf{v}) = (\mathbf{u} \otimes \mathbf{v})\cosh(z) + (A\mathbf{u}) \otimes (B\mathbf{v})\sinh(z).$$

Problem 85.   Let A, B be $n \times n$ matrices over $\mathbb{C}$. What is the condition on A, B such that

$$[A \otimes B, B \otimes A] = 0?$$

Solution 85.   We have

$$[A{\otimes}B, B{\otimes}A] = (A{\otimes}B)(B{\otimes}A)-(B{\otimes}A)(A{\otimes}B) = (AB){\otimes}(BA)-(BA){\otimes}(AB).$$

Thus if $AB = BA$ (i.e. A and B commute) we obtain the zero matrix. This is sufficient, but is it also necessary?

Problem 86.   Let A, B, H be $n \times n$ matrices such that

$$[H, A] = 0, \qquad [H, B] = 0.$$

Show that

$$[H \otimes I_n + I_n \otimes H, A \otimes B] = 0.$$

Solution 86.   We have

$$
\begin{aligned}
[H \otimes I_n + I_n \otimes H, A \otimes B] &= [H \otimes I_n, A \otimes B] + [I_n \otimes H, A \otimes B] \\
&= (HA) \otimes B - (AH) \otimes B + A \otimes (HB) - A \otimes (BH) \\
&= [H, A] \otimes B + A \otimes [H, B] \\
&= 0.
\end{aligned}
$$

Problem 87. Let A_1, A_2, A_3 be $n \times n$ matrices over $\mathbb{C}$. The *ternary commutator* $[A_1, A_2, A_3]$ (also called the *ternutator*) is defined as

$$[A_1, A_2, A_3] := \sum_{\pi \in S_3} \operatorname{sgn}(\pi) A_{\pi(1)} A_{\pi(2)} A_{\pi(3)}$$

$$\equiv A_1 A_2 A_3 + A_2 A_3 A_1 + A_3 A_1 A_2 - A_1 A_3 A_2 - A_2 A_1 A_3 - A_3 A_2 A_1 \,.$$

(i) Let $n = 2$ and consider the Pauli spin matrices σ_x, σ_y, σ_z. Calculate the ternutator

$$[\sigma_x, \sigma_y, \sigma_z] \,.$$

(ii) Calculate

$$A_1 \otimes A_2 \otimes A_3 + A_2 \otimes A_3 \otimes A_1 + A_3 \otimes A_1 \otimes A_2 - A_1 \otimes A_3 \otimes A_2 - A_2 \otimes A_1 \otimes A_3 - A_3 \otimes A_2 \otimes A_1 \,.$$

Solution 87. (i) We obtain the diagonal matrix

$$[\sigma_x, \sigma_y, \sigma_z] = \begin{pmatrix} 6i & 0 \\ 0 & 6i \end{pmatrix} \,.$$

(ii) We obtain

$$\begin{pmatrix}
0 & 0 & 0 & 0 & 0 & 0 & 0 & 0 \\
0 & 0 & 2i & 0 & -2i & 0 & 0 & 0 \\
0 & -2i & 0 & 0 & 2i & 0 & 0 & 0 \\
0 & 0 & 0 & 0 & 0 & -2i & 2i & 0 \\
0 & 2i & -2i & 0 & 0 & 0 & 0 & 0 \\
0 & 0 & 0 & 2i & 0 & 0 & -2i & 0 \\
0 & 0 & 0 & -2i & 0 & 2i & 0 & 0 \\
0 & 0 & 0 & 0 & 0 & 0 & 0 & 0
\end{pmatrix} \,.$$

Problem 88. Let $c, x \in \mathbb{R}$. Consider the operators

$$K_1 = -ic\cos(x), \quad K_2 = -ic\sin(x), \quad K_3 = -\frac{d}{dx} \,.$$

Find the commutators and thus show that we have a basis of a Lie algebra. Is the Lie algebra semisimple?

Solution 88. We find

$$[K_3, K_1] = K_2, \quad [K_3, K_2] = -K_1, \quad [K_1, K_2] = 0 \,.$$

Thus we have a basis of a Lie algebra. The Lie algebra is not semisimple, since $[K_1, K_2] = 0$. Thus we have a two-dimensional abelian sub Lie algebra.

Problem 89. $SU(2)$ is the Lie group of all 2×2 unitary unimodular matrices, i.e., the group of all matrices of the form

$$A = \begin{pmatrix} a & b \\ -\bar{b} & \bar{a} \end{pmatrix}, \quad |a|^2 + |b|^2 = 1$$

where $a, b \in \mathbb{C}$ and $\bar{a}$ is the complex conjugate of a.
(i) Find the inverse of A.
(ii) The Lie algebra $su(2)$ of the Lie group $SU(2)$ can be identified with the vector space of all 2×2 complex skew-hermitian matrices of trace 0. A basis of $su(2)$ is given by

$$G_1 = \begin{pmatrix} 0 & i/2 \\ i/2 & 0 \end{pmatrix} = \frac{i}{2}\sigma_x,$$

$$G_2 = \begin{pmatrix} 0 & 1/2 \\ -1/2 & 0 \end{pmatrix} = \frac{i}{2}\sigma_y,$$

$$G_3 = \begin{pmatrix} -i/2 & 0 \\ 0 & i/2 \end{pmatrix} = -\frac{i}{2}\sigma_z.$$

Find the commutation relations $[G_1, G_2]$, $[G_3, G_1]$, $[G_2, G_3]$.
(iii) Find the matrix

$$M = \exp(\phi_1 G_3) \exp(\theta G_1) \exp(\phi_2 G_3).$$

Solution 89.   (i) We find the inverse matrix

$$A^{-1} = \begin{pmatrix} \bar{a} & -b \\ \bar{b} & a \end{pmatrix}.$$

(ii) We obtain $[G_1, G_2] = G_3$, $[G_3, G_1] = G_2$, $[G_2, G_3] = G_1$.
(iii) Straightforward calculation yields the unitary matrix

$$M = \begin{pmatrix} e^{-i(\phi_1+\phi_2)/2}\cos(\theta/2) & ie^{-i(\phi_1-\phi_2)/2}\sin(\theta/2) \\ ie^{i(\phi_1-\phi_2)/2}\sin(\theta/2) & e^{i(\phi_1+\phi_2)/2}\cos(\theta/2) \end{pmatrix}.$$

Problem 90.   Consider the real Lie algebra L constructed from a linear vector space with basis

$$\{ A_{ij} : i, j = 1, 2, \ldots, n \}.$$

We assume that the structure of the Lie algebra is given by the commutation relations

$$[A_{ij}, A_{k\ell}] = \delta_{kj} A_{i\ell} - \delta_{\ell i} A_{kj}$$

where δ_{ij} is the Kronecker symbol. Hence we consider the Lie algebra $g\ell(n, \mathbb{R})$. The dimension of L is n^2.
(i) Show that the Lie algebra is not simple.
(ii) Show that the mapping

$$A_{ij} \xrightarrow{\Gamma} x_i \frac{\partial}{\partial x_j}$$

is a faithful representation of the Lie algebra $g\ell(n, \mathbb{R})$.

Solution 90. (i) Let

$$N = \sum_{k=1}^{n} A_{kk} .$$

Then $[N, A_{ij}] = 0$ for all $i, j = 1, 2, \ldots, n$. Thus N is in the center of the Lie algebra.
(ii) Let $V_{ij} = x_i \partial/\partial x_j$. Then we have $[V_{ij}, V_{k\ell}] = \delta_{kj} V_{i\ell} - \delta_{\ell i} V_{kj}$. Note that the differential operator W for N is then given by

$$W = \sum_{j=1}^{n} x_j \frac{\partial}{\partial x_j} .$$

Problem 91. Consider the Lie group $O(2, \mathbb{R})$ and its Lie algebra $o(2, \mathbb{R})$. Show that $O(2, \mathbb{R})$ is not commutative. Discuss the properties of $o(2, \mathbb{R})$.

Solution 91. Let

$$A = \begin{pmatrix} -1 & 0 \\ 0 & 1 \end{pmatrix} , \qquad B = \begin{pmatrix} 0 & 1 \\ -1 & 0 \end{pmatrix} .$$

Then $A \in O(2, \mathbb{R})$, but $A \notin SO(2, \mathbb{R})$. Moreover $B \in SO(2, \mathbb{R})$. Now $AB \neq BA$. Thus $O(2, \mathbb{R})$ is not commutative. Note that $AB = -BA$. However the Lie algebra $o(2, \mathbb{R})$ is commutative. We have

$$so(2, \mathbb{R}) = o(2, \mathbb{R}) = \left\{ \begin{pmatrix} 0 & \alpha \\ -\alpha & 0 \end{pmatrix} : \alpha \in \mathbb{R} \right\}$$

which is generated by the skew-symmetric matrix

$$H = \begin{pmatrix} 0 & 1 \\ -1 & 0 \end{pmatrix} .$$

Then the Lie group $O(2, \mathbb{R})$ is the disjoint union $O(2, \mathbb{R}) = SO(2, \mathbb{R}) \cup gSO(2, \mathbb{R})$ with the matrix

$$g = \begin{pmatrix} 1 & 0 \\ 0 & -1 \end{pmatrix} .$$

Note that

$$\begin{pmatrix} 1 & 0 \\ 0 & -1 \end{pmatrix} \begin{pmatrix} \cos\alpha & \sin\alpha \\ -\sin\alpha & \cos\alpha \end{pmatrix} = \begin{pmatrix} \cos\alpha & \sin\alpha \\ \sin\alpha & -\cos\alpha \end{pmatrix}.$$

Problem 92.   The Euclidean group $E(2)$ in the plane is associated with the Lie algebra $e(2)$ defined by the commutation relations

$$[X_1, X_2] = X_3, \quad [X_1, X_3] = -X_2, \quad [X_2, X_3] = 0.$$

Show that the Lie algebra is not semisimple.

Solution 92.   The metric tensor is given by

$$g = \begin{pmatrix} -2 & 0 & 0 \\ 0 & 0 & 0 \\ 0 & 0 & 0 \end{pmatrix}.$$

Thus $\det(g) = 0$ and the Lie algebra is not semisimple. The Lie algebra contains a non-trivial abelian sub Lie algebra consisting of the elements X_2, X_3. Thus the Lie algebra cannot be reduced to a direct sum of simple Lie algebras. However we can write the Lie algebra $e(2)$ as a *semidirect sum*

$$e(2) = t(2) \oplus_s r(2)$$

where $t(2)$ is the abelian ideal and $r(2)$ the sub Lie algebra formed by X_1.

Problem 93.   The vector space of $n \times n$ matrices over $\mathbb{R}$ form a *Hilbert space* with the scalar product defined

$$\langle A, B \rangle := \operatorname{tr}(AB^*).$$

This implies a norm

$$\|A\|^2 = \operatorname{tr}(AA^*)$$

and a distance $\|A - B\|^2 = \operatorname{tr}((A - B)(A - B)^*)$. Consider the matrices

$$A = \begin{pmatrix} 0 & 1 & 0 \\ -1 & 0 & 0 \\ 0 & 0 & 0 \end{pmatrix}, \quad B = \begin{pmatrix} 0 & 0 & 0 \\ 0 & 0 & -1 \\ 0 & 1 & 0 \end{pmatrix}$$

which are elements of the simple Lie algebra $so(3)$.
(i) Are the matrices orthogonal to each other?
(ii) Find the distance between A and B.

Solution 93. (i) Yes. We have $\text{tr}(AB^*) = 0$.
(ii) We find

$$\|A - B\|^2 = \text{tr} \begin{pmatrix} 1 & 0 & -1 \\ 0 & 2 & 0 \\ -1 & 0 & 1 \end{pmatrix} = 4 \,.$$

Thus the distance is 2.

Problem 94. Let A, B be $n \times n$ matrices over $\mathbb{C}$. We define the quasi-multiplication

$$A \bullet B := \frac{1}{2}(AB + BA) \,.$$

Obviously $A \bullet B = B \bullet A$. Show that $(A^2 \bullet B) \bullet A = A^2 \bullet (B \bullet A)$. This is called the *Jordan identity*. This is an example of a *Jordan algebra*.

Solution 94. We have

$$\begin{aligned}
4(A^2 \bullet B) \bullet A &= (A^2 B + BA^2)A + A(A^2 B + BA^2) \\
&= A^2 BA + BA^3 + A^3 B + ABA^2 \\
&= A^2(BA + AB) + (BA + AB)A^2 \\
&= 4A^2 \bullet (B \bullet A) \,.
\end{aligned}$$

Problem 95. The 2×2 matrices

$$E = \begin{pmatrix} 0 & 1 \\ 0 & 0 \end{pmatrix}, \qquad F = \begin{pmatrix} 0 & 0 \\ 1 & 0 \end{pmatrix}, \qquad H = \begin{pmatrix} 1 & 0 \\ 0 & 1 \end{pmatrix}$$

form a basis for the Lie algebra $s\ell(2, \mathbb{R})$ with the commutators $[H, E] = 2E$, $[H, F] = -2F$, $[E, F] = H$. Let $k \geq 1$. Do the matrices $(m_j = j(k - j + 1))$

$$\rho_k(H) = \begin{pmatrix} k & 0 & 0 & \dots & 0 \\ 0 & k-2 & 0 & \dots & 0 \\ 0 & 0 & k-4 & \dots & 0 \\ . & . & . & \ddots & . \\ 0 & 0 & 0 & \dots & -k \end{pmatrix}$$

$$\rho_k(F) = \begin{pmatrix} 0 & 0 & 0 & \dots & . & 0 \\ 1 & 0 & 0 & \dots & . & 0 \\ 0 & 1 & 0 & \dots & . & 0 \\ . & . & . & \dots & . & . \\ 0 & 0 & 0 & \dots & 1 & 0 \end{pmatrix}, \quad \rho_k(E) = \begin{pmatrix} 0 & m_1 & 0 & \dots & 0 \\ 0 & 0 & m_2 & \dots & 0 \\ 0 & 0 & 0 & \dots & 0 \\ . & . & . & \dots & m_k \\ 0 & 0 & 0 & \dots & 0 \end{pmatrix}$$

satisfy the same commutation relations?

Solution 95.   Yes. Thus ρ_k is a representation in $\mathbb{R}^{k+1}$.

Problem 96.   The exceptional Lie algebra g_2 has the simple *roots*

$$(0, 1, -1) \qquad (1, -2, 1).$$

Find the *Cartan matrix*. Verify the properties of the Cartan matrix.

Solution 96.   From

$$A_{jk} = 2 \frac{(\alpha_j, \alpha_k)}{(\alpha_j, \alpha_j)}, \quad j, k = 1, 2$$

we find

$$A = \begin{pmatrix} 2 & -3 \\ -1 & 2 \end{pmatrix}.$$

The diagonal entries are 2. The non-diagonal entries are negative. The Cartan matrix can be written as

$$\begin{pmatrix} 2 & -3 \\ -1 & 2 \end{pmatrix} = \begin{pmatrix} 1 & 0 \\ 0 & 1/3 \end{pmatrix} \begin{pmatrix} 2 & -3 \\ -3 & 6 \end{pmatrix}$$

where the first matrix on the right-hand side is a diagonal matrix and the second matrix is symmetric. Thus the conditions of a Cartan matrix are satisfied.

Problem 97.   If $\{\, X_j \,:\, j = 1, 2, \ldots, m \,\}$ form a basis for a finite dimensional Lie algebra L then

$$g_{ij} = K(X_i, X_j)$$

is called the *metric tensor* for L. In terms of the structure constants,

$$g_{ij} = \sum_{r,s=1}^{m} C_{is}^r C_{jr}^s.$$

For a given semisimple Lie algebra L with basis $\{X_j \,:\, j = 1, \ldots, m\}$, the Casimir operator is a quantity (which is not an element of the Lie algebra) that commutes with each element of the Lie algebra. The quadratic *Casimir operator* C of a given semisimple Lie algebra L is defined by

$$C := \sum_{i=1}^{m} \sum_{j=1}^{m} (g^{ij}) X_i X_j$$

where (g^{ij}) is the inverse matrix of (g_{ij}). Note that the Casimir operator is independent of the choice of the basis. Consider the Lie algebra $so(3)$

$$L_1 := -i \left(x_2 \frac{\partial}{\partial x_3} - x_3 \frac{\partial}{\partial x_2} \right)$$

$$L_2 := -i \left(x_3 \frac{\partial}{\partial x_1} - x_1 \frac{\partial}{\partial x_3} \right)$$

$$L_3 := -i \left(x_1 \frac{\partial}{\partial x_2} - x_2 \frac{\partial}{\partial x_1} \right)$$

with the commutation relations $[L_1, L_2] = iL_3$, $[L_2, L_3] = iL_1$, $[L_3, L_1] = iL_2$. Find the quadratic Casimir operator.

Solution 97. The non-zero structure constants are $C_{12}^3 = i$, $C_{23}^1 = i$, $C_{31}^2 = i$. It follows that

$$g_{11} = \sum_{j=1}^{3} \sum_{k=1}^{3} C_{1j}^k C_{1k}^j = \sum_{j=1}^{3} (C_{1j}^2 C_{12}^j + C_{1j}^3 C_{13}^j) = C_{13}^2 C_{12}^3 + C_{12}^3 C_{13}^2 = 2.$$

Analogously $g_{22} = g_{33} = 2$. The remaining elements of the matrix (g_{jk}) are zero. It follows that

$$C = \sum_{j=1}^{3} \sum_{k=1}^{3} g^{jk} L_j L_k = g^{11} L_1^2 + g^{22} L_2^2 + g^{33} L_3^2 = \frac{1}{2}(L_1^2 + L_2^2 + L_3^2)$$

where

$$\sum_{j=1}^{3} g^{ij} g_{jk} = \sum_{j=1}^{3} g_{kj} g^{ji} = \delta_{ik}.$$

For every semisimple Lie algebra of rank r there exist r independent Casimir operators. Using the structure constants C_{ij}^k they can be found as

$$C_n = \sum_{\dots i_l, k_l \dots} C_{i_1 k_1}^{k_2} C_{i_2 k_2}^{k_3} \cdots C_{i_n k_n}^{k_1} J^{i_1} J^{i_2} \cdots J^{i_n}$$

where n takes all positive integers. Thus we arbitrarily often obtain every Casimir operator, or linear combinations of them. We have to choose r independent values for n (as small as possible). Here J_i are a basis of the semisimple Lie algebra and $J^i = \sum_j g^{ij} J_j$.

Problem 98. Let $\mathbb{Z}_2 = \{0, 1\}$ be the finite group of integers under addition modulo 2. The super-Lie algebra $g = g_0 + g_1$ is a $\mathbb{Z}_2$-graded Lie algebra with a super-Lie bracket $[,]$ which satisfies

$$[A, B] = AB - (-1)^{d(A)d(B)} BA$$

$$[A, [B, C]] = [[A, B], C] + (-1)^{d(A)d(B)}[B, [A, C]]$$

where $d(X) = 0\,(1)$ if $A \in g_0\,(g_1)$. $d(A)$ is called the degree of A. We denote by $[\,,\,]_-$ the commutator and by $[\,,\,]_+$ the anticommutator. The super-Lie algebra $s\ell(m/n)$ is defined by

$$s\ell(m/n) := \left\{ X = \begin{pmatrix} R & S \\ T & U \end{pmatrix} : \mathrm{Str}(X) := \mathrm{tr}(R) - \mathrm{tr}(U) = 0 \right\}$$

where R is an $m \times m$ matrix, S a $m \times n$ matrix, T an $n \times m$ matrix and D an $n \times n$ matrix. We assume that $m \neq n$. Consider the super-Lie algebra $s\ell(2/1)$. A basis is given by

$$E_1 = \begin{pmatrix} 0 & 1 & 0 \\ 0 & 0 & 0 \\ 0 & 0 & 0 \end{pmatrix}, \qquad E_2 = \begin{pmatrix} 0 & 0 & 0 \\ 0 & 0 & 1 \\ 0 & 0 & 0 \end{pmatrix},$$

$$F_1 = \begin{pmatrix} 0 & 0 & 0 \\ 1 & 0 & 0 \\ 0 & 0 & 0 \end{pmatrix}, \qquad F_2 = \begin{pmatrix} 0 & 0 & 0 \\ 0 & 0 & 0 \\ 0 & 1 & 0 \end{pmatrix},$$

$$G_1 = \begin{pmatrix} 0 & 0 & 1 \\ 0 & 0 & 0 \\ 0 & 0 & 0 \end{pmatrix}, \qquad G_2 = \begin{pmatrix} 0 & 0 & 0 \\ 0 & 0 & 0 \\ 1 & 0 & 0 \end{pmatrix},$$

$$H_1 = \begin{pmatrix} 1 & 0 & 0 \\ 0 & -1 & 0 \\ 0 & 0 & 0 \end{pmatrix}, \qquad H_2 = \begin{pmatrix} 0 & 0 & 0 \\ 0 & 1 & 0 \\ 0 & 0 & 1 \end{pmatrix}$$

i.e. we have four even elements E_1, F_1, H_1, H_2 and four odd elements E_2, F_2, G_1, G_2. Find the non-vanishing commutators and anticommutators.

Solution 98.   Let $a, b = 1, 2$. We find the non-vanishing (anti)commutation relations

$$[H_a, E_b]_- = K_{ab}E_b, \quad [H_a, F_b]_- = -K_{ab}F_b, \quad [E_1, F_1]_- = H_1, \quad [E_2, F_2]_+ = H_2$$

$$[H_a, G_b]_- = \Omega_{ab}G_b, \quad [E_1, E_2]_- = G_1, \quad [E_1, G_2]_- = -F_2$$

$$[E_2, G_2]_+ = F_1, \quad [F_1, F_2]_- = -G_2, \quad [F_1, G_1]_- = E_2$$

$$[F_2, G_1]_+ = E_1, \quad [G_1, G_2]_+ = H_1 + H_2$$

where

$$K = (K_{ab}) = \begin{pmatrix} 2 & -1 \\ -1 & 0 \end{pmatrix} \quad \text{and} \quad \Omega = (\Omega_{ab}) = \begin{pmatrix} 1 & -1 \\ -1 & 1 \end{pmatrix}.$$

Problem 99. Let $q > 0$. The algebra $su_q(2)$ is generated by X_+, $X_- = (X_+)^*$ and $H = H^*$ with deformed commutation relations

$$[H, X_+] = 2X_+, \quad [H, X_-] = -2X_-, \quad [X_+, X_-] = \frac{q^H - q^{-H}}{q - q^{-1}}.$$

Show by direct computation of the commutators that this algebra has the same matrix representation as the undeformed Lie algebra $su(2)$, i.e.

$$X_+ \mapsto \begin{pmatrix} 0 & 1 \\ 0 & 0 \end{pmatrix}, \quad X_- \mapsto \begin{pmatrix} 0 & 0 \\ 1 & 0 \end{pmatrix}, \quad H \mapsto \begin{pmatrix} 1 & 0 \\ 0 & -1 \end{pmatrix}.$$

Solution 99. Since

$$q^H \equiv e^{\ln(q)H} = \begin{pmatrix} e^{\ln(q)} & 0 \\ 0 & e^{-\ln(q)} \end{pmatrix} = \begin{pmatrix} q & 0 \\ 0 & 1/q \end{pmatrix}$$

and analogously

$$q^{-H} = \begin{pmatrix} 1/q & 0 \\ 0 & q \end{pmatrix}$$

we find

$$\frac{1}{q - 1/q} \left(\begin{pmatrix} q & 0 \\ 0 & 1/q \end{pmatrix} - \begin{pmatrix} 1/q & 0 \\ 0 & q \end{pmatrix} \right) = \frac{1}{q - 1/q} \begin{pmatrix} q - 1/q & 0 \\ 0 & 1/q - q \end{pmatrix}$$

$$= \begin{pmatrix} 1 & 0 \\ 0 & -1 \end{pmatrix} = H.$$

Problem 100. Let V, W be smooth vector fields defined in $\mathbb{R}^n$. Let $f, g : \mathbb{R}^n \to \mathbb{R}$ be smooth functions. Consider now the pairs (V, f), (W, g). One defines a commutator of such pairs as

$$[(V, f), (W, g)] := ([V, W], L_V g - L_W f)$$

where $L_V(.)$ denotes the *Lie derivative*. Now $[V, W]$ is a vector field again and $L_V g - L_V f$ is a smooth function again. The Lie derivative is linear and obeys the product rule. Consider $n = 2$. Let

$$V = u_2 \frac{\partial}{\partial u_1} - u_1 \frac{\partial}{\partial u_2}, \quad W = u_1 \frac{\partial}{\partial u_1} + u_2 \frac{\partial}{\partial u_2}$$

and $f(u_1, u_2) = g(u_1, u_2) = u_1^2 + u_2^2$. Calculate the commutator.

Solution 100. With $[V, W] = 0$, $L_V f = 0$, $L_W g = 2(u_1^2 + u_2^2)$ we find the commutator

$$[(V, f), (W, g)] = (0, -2(u_1^2 + u_2^2)).$$

Programming Problems

Problem 101. (i) Let L be a three dimensional Lie algebra with basis a, b, c and the commutation relations (simple Lie algebra $so(3)$)

$$[a, b] = c, \quad [b, c] = a, \quad [c, a] = b.$$

Write a SymbolicC++ program to test whether the Jacobi identity is satisfied.

(ii) Let L be a three dimensional Lie algebra with basis a, b, c and the commutation relations (solvable Lie algebra)

$$[a, b] = 0, \quad [b, c] = a + kb, \quad [c, a] = -ka + b, \quad k > 0.$$

Write a SymbolicC++ program to test whether the Jacobi identity is satisfied.

Solution 101. In Equations the commutation relations are implemented.

```
// jacobi.cpp

#include <iostream>
#include "symbolicc++.h"
using namespace std;

int main(void)
{
 Symbolic a("a"), b("b"), c("c"), k("k"), L("L");
 Symbolic x("x"), y("y"), z("z");
 Equations lie=((x,y,z,L[x+y,z]==L[x,z]+L[y,z]),
                (x,y,z,L[x,y+z]==L[x,y]+L[x,z]),
                (x,y,L[k*x,y]==k*L[x,y]),
                (x,y,L[x,k*y]==k*L[x,y]),
                (x,y,L[-x,y]==-L[x,y]),(x,y,L[x,-y]==-L[x,y]),
                (x,L[x,x]==0),(x,L[x,0]==0),(x,L[0,x]==0),
                L[b,a]==-L[a,b],L[c,b]==-L[b,c],L[a,c]==-L[c,a]);

 Symbolic j = L[a,L[b,c]] + L[c,L[a,b]] + L[b,L[c,a]];
 // Example 1
 Equations so3 = (L[a,b]==c,L[b,c]==a,L[c,a]==b);

 cout << "so3: " << so3 << endl;
 so3.insert(so3.end(),lie.begin(),lie.end());
 cout << j << " = " << j.subst_all(so3) << endl;
 // Example 2
 Equations eg = (L[a,b]==0,L[b,c]==a+k*b,L[c,a]==-k*a+b);
```

```
cout << "Another example: " << eg << endl;
eg.insert(eg.end(),lie.begin(),lie.end());
cout << j << " = " << j.subst_all(eg) << endl;
return 0;
}
```

Problem 102. Let L be a Lie algebra. One analyzes the structure of a Lie algebra L by studying its ideals. A subspace I of a Lie algebra L is called an *ideal* of L if $x \in L$ and $y \in I$ together imply $[x, y] \in I$. If L has no ideals except itself and 0, and if moreover $[L, L] \neq 0$ we call the Lie algebra L *simple*.

If L is a Lie algebra and $x \in L$, the operator $\mathrm{ad}X$ that maps y to $[x, y]$ is a linear transformation of L into itself

$$(\mathrm{ad}x)y := [x, y].$$

Then $x \to \mathrm{ad}x$ is a representation of the Lie algebra L with L itself considered as the vector space of the representation. The *Killing form* of a Lie algebra is the symmetric bilinear form

$$K(x, y) := \mathrm{tr}(\mathrm{ad}x\,\mathrm{ad}y)$$

where tr denotes the trace. If x_j $(j = 0, 1, \ldots, n-1)$ form a basis of L then

$$g_{jk} = K(x_j, x_k)$$

is called the *metric tensor* for L. Note that a Lie algebra is semisimple if and only if the matrix (g_{jk}) is nonsingular, i.e., $\det((g_{jk})) \neq 0$.

We consider the simple Lie algebra $so(3)$ with the basis elements $x(0)$, $x(1)$, $x(2)$ and the commutation relations

$$[x(0), x(1)] = x(2), \qquad [x(1), x(2)] = x(0), \qquad [x(2), x(0)] = x(1).$$

We find

$$\mathrm{ad}x(0) = \begin{pmatrix} 0 & 0 & 0 \\ 0 & 0 & -1 \\ 0 & 1 & 0 \end{pmatrix}, \quad \mathrm{ad}x(1) = \begin{pmatrix} 0 & 0 & 1 \\ 0 & 0 & 0 \\ -1 & 0 & 0 \end{pmatrix},$$

$$\mathrm{ad}x(2) = \begin{pmatrix} 0 & -1 & 0 \\ 1 & 0 & 0 \\ 0 & 0 & 0 \end{pmatrix}.$$

Solution 102. The two-dimensional array `comm(i,j)` stores the commutators.

```cpp
// adjoint.cpp

#include <iostream>
#include "symbolicc++.h"
using namespace std;

int main(void)
{
  int n=3;
  Symbolic X("X",n);
  Symbolic comm("",n,n);
  comm(0,0) = Symbolic(0), comm(0,1) = X(2); comm(0,2) = -X(1);
  comm(1,0) = -X(2); comm(1,1) = Symbolic(0); comm(1,2) = X(0);
  comm(2,0) = X(1); comm(2,1) = -X(0); comm(2,2) = Symbolic(0);
  Symbolic adX("adX",n);
  int i,j,k;
  for(i=0;i<n;i++)
  {
   Symbolic rep("",n,n);
   for(j=0;j<n;j++)
    for(k=0;k<n;k++) rep(k,j) = comm(i,j).coeff(X(k));
   adX(i) = rep;
   cout << "adX(" << i << ") = " << adX(i) << endl;
  }

  Symbolic g("",n,n);
   for(j=0;j<n;j++)
    for(k=0;k<n;k++) g(j,k) = tr(adX(j)*adX(k));
  cout << "g = " << g << endl;
  return 0;
}
```

Supplementary Problems

Problem 103. Let L be a Lie algebra. Given $H, K \subset L$. We define their bracket by

$$[H, K] := \text{span}\{\, [h, k] \,:\, h \in H,\ k \in K \,\}.$$

The derived Lie algebra of L is the ideal $[L, L]$. Show that the quotient Lie algebra $L/[L, L]$ is an abelian Lie algebra.

Problem 104. Show that the Lie algebra consisting of $n \times n$ matrices over $\mathbb{R}$ (or $\mathbb{C}$) with zeros below the main diagonal is solvable but not nilpotent.

Problem 105. (i) Let L be a finite-dimensional Lie algebra over a field $\mathbb{F}$. The linear map $\kappa : L \times L \to \mathbb{F}$ defined by

$$\kappa(a, b) := \text{tr}(\text{ad}(a)\text{ad}(b))$$

is called the Cartan-Killing form or Killing form of L, where tr denotes the trace. Show that

$$\kappa([a, b], c) = \kappa(a, [b, c]).$$

(ii) Consider the Cartan-Killing form of the finite-dimensional linear Lie algebra $g\ell(n, \mathbb{R})$. Show that

$$\kappa(X, Y) = \text{tr}((\text{ad}X)(\text{ad}Y)) = 2n\text{tr}(XY) - 2(\text{tr}X)(\text{tr}Y).$$

Problem 106. Let A, B be 2×2 matrices over $\mathbb{C}$. We define

$$A \star B := \begin{pmatrix} a_{11} & 0 & 0 & a_{12} \\ 0 & b_{11} & b_{12} & 0 \\ 0 & b_{21} & b_{22} & 0 \\ a_{21} & 0 & 0 & a_{22} \end{pmatrix}.$$

Consider the Lie algebra $s\ell(2, \mathbb{R})$ with the basis

$$h = \begin{pmatrix} 1 & 0 \\ 0 & -1 \end{pmatrix}, \quad e = \begin{pmatrix} 0 & 1 \\ 0 & 0 \end{pmatrix}, \quad f = \begin{pmatrix} 0 & 0 \\ 1 & 0 \end{pmatrix}.$$

Find $h \star h$, $e \star e$, $f \star f$. Calculate the commutators

$$[e \star e, f \star f], \quad [h \star h, e \star e], \quad [h \star h, f \star f].$$

Discuss.

Problem 107.   Do the vector fields

$$\frac{\partial}{\partial x_j}, \quad x_k \frac{\partial}{\partial x_j}, \quad x_\ell x_k \frac{\partial}{\partial x_j}, \qquad j, k, \ell = 1, \ldots, n$$

form a Lie algebra under the commutator? If so classify the Lie algebra. Consider first the case $n = 1$.

Problem 108.   Let $s = 0, 1/2, 1, 3/2, \ldots$. The spin s generators of the Lie algebra $s\ell(2, \mathbb{C})$ at site k are given by

$$S_{+,k} = z_k^2 \frac{\partial}{\partial z_k} - 2s z_k, \quad S_{-,k} = -\frac{\partial}{\partial z_k}, \quad S_{3,k} = z_k \frac{\partial}{\partial z_k} - s.$$

Find the commutators.

Problem 109.   Let $k \geq 1$. Do the vector fields

$$\frac{\partial}{\partial x}, \quad \frac{\partial}{\partial u}, \quad x \frac{\partial}{\partial u}, \quad \ldots, \quad x^k \frac{\partial}{\partial u}, \quad x \frac{\partial}{\partial x} + (k+1) u \frac{\partial}{\partial u}$$

form a basis of a Lie algebra under the commutator? If so classify the Lie algebra.

Problem 110.   The infinitesimal generators (vector fields) of the six dimensional Lie algebra $so(4)$ can be written in terms of the variables (x_1, x_2, x_3, x_4) as

$$X_1 = x_3 \frac{\partial}{\partial x_2} - x_2 \frac{\partial}{\partial x_3}, \quad X_2 = x_1 \frac{\partial}{\partial x_3} - x_3 \frac{\partial}{\partial x_1}, \quad X_3 = x_2 \frac{\partial}{\partial x_1} - x_1 \frac{\partial}{\partial x_2},$$

$$Y_1 = x_1 \frac{\partial}{\partial x_4} - x_4 \frac{\partial}{\partial x_1}, \quad Y_2 = x_2 \frac{\partial}{\partial x_4} - x_4 \frac{\partial}{\partial x_2}, \quad Y_3 = x_3 \frac{\partial}{\partial x_4} - x_4 \frac{\partial}{\partial x_3}.$$

(i) Find the commutators.
(ii) Show that the Lie algebra is semisimple, but not simple.
(iii) Find elements of the compact Lie group $SO(4)$ by calculating

$$\exp(\alpha_1 X_1 + \alpha_2 X_2 + \alpha_3 X_3 + \beta_1 Y_1 + \beta_2 Y_2 + \beta_3 Y_3)$$

and

$$\exp(\alpha_1 X_1) \exp(\alpha_2 X_2) \exp(\alpha_3 X_3) \exp(\beta_1 Y_1) \exp(\beta_2 Y_2) \exp(\beta_3 Y_3)$$

where $\alpha_j, \beta_j \in \mathbb{R}$.

Chapter 4

Applications

Problem 1. An *icosahedron* is a regular polyhedron with 20 identical equilateral triangular faces, 30 edges and 12 vertices. Let $\tau = \frac{1}{2}(1 + \sqrt{5})$ be the *golden mean number*. The Cartesian coordinates defining the vertices of an icosahedron with edge length 2, centered at the origin are given by

$$(0, \pm 1, \pm \tau), \quad (\pm 1, \pm \tau, 0), \quad (\pm \tau, 0, \pm 1).$$

A discrete subgroup of the compact Lie group $SO(3)$ of order 60 is called the *icosahedral group*, which is the symmetry group of the icosahedron. It is isomorphic to the group of even permutations A_5 of five elements. Its double cover $2A_5$ of 120 elements (binary icosahedral group) can be represented by quaternions or equivalently by 2×2 unitary matrices of determinant one. Let

$$E_1 = \begin{pmatrix} 0 & -i \\ -i & 0 \end{pmatrix}, \quad E_2 = \begin{pmatrix} 0 & -1 \\ 1 & 0 \end{pmatrix}, \quad E_3 = \begin{pmatrix} -i & 0 \\ 0 & i \end{pmatrix}.$$

The 120 elements of the binary icosahedral group can be generated by the elements (2×2 matrices)

$$A = \frac{1}{2}(\tau I_2 - \sigma E_1 + E_3), \quad B = \frac{1}{2}(I_2 - \sigma E_2 + \tau E_3)$$

where $\sigma = \frac{1}{2}(1 - \sqrt{5}) = -1/\tau$. Let $C = E_3$. Find A^5, B^3, C^2 and ABC.

Solution 1. We have

$$A = \frac{1}{2}\begin{pmatrix} \tau - i & \sigma i \\ \sigma i & \tau + i \end{pmatrix}, \quad B = \frac{1}{2}\begin{pmatrix} 1 - i\tau & \sigma \\ -\sigma & 1 + i\tau \end{pmatrix}.$$

Thus we find $A^5 = B^3 = C^2 = ABC = -I_2$.

211

Problem 2.   Consider the Hilbert space $\mathcal{H} = \mathbb{C}^4$ and the hermitian matrices

$$H = \begin{pmatrix} 2 & -1 & 0 & -1 \\ -1 & 2 & -1 & 0 \\ 0 & -1 & 2 & -1 \\ -1 & 0 & -1 & 2 \end{pmatrix}$$

$$A = \begin{pmatrix} 1 & 0 & 0 & 0 \\ 0 & 0 & 0 & 0 \\ 0 & 0 & -1 & 0 \\ 0 & 0 & 0 & 0 \end{pmatrix}, \qquad B = \begin{pmatrix} 0 & 0 & 0 & 0 \\ 0 & 1 & 0 & 0 \\ 0 & 0 & 0 & 0 \\ 0 & 0 & 0 & -1 \end{pmatrix}.$$

Using the *Kronecker product* we consider the 16×16 matrix

$$K = \frac{1}{2}(H \otimes I_4 + I_4 \otimes H) - \lambda A \otimes A - \mu B \otimes B$$

acting in the Hilbert space $\mathbb{C}^{16}$, where $\lambda, \mu \geq 0$. We consider K as a Hamilton operator.
(i) Consider the permutation matrix

$$R = \begin{pmatrix} 0 & 1 & 0 & 0 \\ 0 & 0 & 1 & 0 \\ 0 & 0 & 0 & 1 \\ 1 & 0 & 0 & 0 \end{pmatrix}.$$

Thus the inverse of R exists. Calculate

$$(R \otimes R)^{-1} K (R \otimes R)$$

and the commutator $[R \otimes R, K]$. Discuss the case $\lambda = \mu$.
(ii) Consider the permutation matrices

$$S = \begin{pmatrix} 0 & 0 & 1 & 0 \\ 0 & 1 & 0 & 0 \\ 1 & 0 & 0 & 0 \\ 0 & 0 & 0 & 1 \end{pmatrix}, \qquad T = \begin{pmatrix} 1 & 0 & 0 & 0 \\ 0 & 0 & 0 & 1 \\ 0 & 0 & 1 & 0 \\ 0 & 1 & 0 & 0 \end{pmatrix}.$$

Find the commutators $[S \otimes S, K]$ and $[T \otimes T, K]$. Discuss.

Solution 2.   (i) We find for $\lambda = \mu$ that

$$(R \otimes R)^{-1} K (R \otimes R) = K, \qquad [R \otimes R, K] = 0.$$

(ii) We find $[S \otimes S, K] = 0$ and $[T \otimes T, K] = 0$ for any $\mu, \lambda \geq 0$.

Problem 3. Let R be a 4×4 matrix and I_2 be the 2×2 unitary matrix. Let R be of the form

$$R = \begin{pmatrix} a & 0 & 0 & 0 \\ 0 & b & c & 0 \\ 0 & d & e & 0 \\ 0 & 0 & 0 & f \end{pmatrix} \equiv (a) \oplus \begin{pmatrix} b & c \\ d & e \end{pmatrix} \oplus (f)$$

where $a, b, c, d, e, f \in \mathbb{R}$ and $\oplus$ denotes the direct sum. Find the conditions on a, b, c, d, e, f such that

$$(R \otimes I_2)(I_2 \otimes R)(R \otimes I_2) = (I_2 \otimes R)(R \otimes I_2)(I_2 \otimes R).$$

Assume that $a \neq 0$.

Solution 3. For the left-hand side we have

$$(R \otimes I_2)(I_2 \otimes R)(R \otimes I_2) =$$

$$(a^3) \oplus \begin{pmatrix} a^2b & abc & ac^2 \\ abd & b^2e + acd & bce + ace \\ ad^2 & bde + ade & cde + ae^2 \end{pmatrix} \oplus \begin{pmatrix} b^2f + bcd & bcf + bce & c^2f \\ bdf + bde & cdf + be^2 & cef \\ d^2f & def & ef^2 \end{pmatrix} \oplus (f^3).$$

For the right-hand side we have

$$(I_2 \otimes R)(R \otimes I_2)(I_2 \otimes R) =$$

$$(a^3) \oplus \begin{pmatrix} ab^2 + bcd & abc + bce & ac^2 \\ abd + bde & acd + be^2 & ace \\ ad^2 & ade & a^2e \end{pmatrix} \oplus \begin{pmatrix} bf^2 & bcf & c^2f \\ bdf & b^2e + cdf & bce + cef \\ d^2f & bde + def & cde + e^2f \end{pmatrix} \oplus (f^3).$$

Thus we find the following seven conditions

$$b(cd + ab - a^2) = 0$$
$$b(cd + bf - f^2) = 0$$
$$e(cd + ae - a^2) = 0$$
$$e(cd + ef - f^2) = 0$$
$$bce = 0$$
$$bde = 0$$
$$be(b - e) = 0.$$

Thus we obtain the two R matrices with $b = 0$

$$R = \begin{pmatrix} a & 0 & 0 & 0 \\ 0 & 0 & c & 0 \\ 0 & d & a - cd/a & 0 \\ 0 & 0 & 0 & a \end{pmatrix}, \quad R = \begin{pmatrix} a & 0 & 0 & 0 \\ 0 & 0 & c & 0 \\ 0 & d & a - cd/a & 0 \\ 0 & 0 & 0 & -cd/a \end{pmatrix}.$$

Problem 4.   Consider the alternating group A_4. Find the conjugacy classes. Find the character table.

Solution 4.   The alternating group A_4 has four conjugacy classes. The number of elements in these classes are 1, 4, 4, 3. There are three one-dimensional representations and one irreducible three dimensional representation. The character table is

	1	4	4	3
A_4	1	(123)	(132)	(12)(34)
A_1	1	1	1	1
A_2	1	ω	ω^2	1
A_3	1	ω^2	ω	1
B	3	0	0	-1

where $\omega = e^{2\pi i/3}$.

Problem 5.   (i) Let I_2 be the 2×2 identity matrix and σ_x, σ_y, σ_z be the Pauli spin matrices. The *Pauli group* $\mathcal{P}_1$ consisting of 16 elements is given by

$$\mathcal{P}_1 = \{\, \pm I_2,\ \pm i I_2,\ \pm \sigma_x,\ \pm i \sigma_x,\ \pm \sigma_y,\ \pm i \sigma_y,\ \pm \sigma_z,\ \pm i \sigma_z \,\}$$

and matrix multiplication as composition. Is there a subgroup with 8 elements?

(ii) Consider the group of order 8 consisting of the *quaternions*

$$\{\, \pm 1,\ \pm I,\ \pm J,\ \pm K \,\}$$

under multiplication. A matrix representation of the quaternions is given by

$$1 \mapsto \begin{pmatrix} 1 & 0 \\ 0 & 1 \end{pmatrix} = I_2$$

$$I \mapsto -i \begin{pmatrix} 0 & 1 \\ 1 & 0 \end{pmatrix} = -i\sigma_x$$

$$J \mapsto -i \begin{pmatrix} 0 & -i \\ i & 0 \end{pmatrix} = -i\sigma_y$$

$$K \mapsto -i \begin{pmatrix} 1 & 0 \\ 0 & -1 \end{pmatrix} = -i\sigma_z.$$

Find the conjugacy classes of this group. Find the character table.

Solution 5.   (i) One such subgroup is

$$\{\, \pm I_2,\ \pm \sigma_x,\ \pm i \sigma_y,\ \pm \sigma_z \,\}.$$

(ii) There are five conjugacy classes. They are

$$\{\,I_2\,\}, \quad \{\,-I_2\,\}, \quad \{\,i\sigma_x, \,-i\sigma_x\,\}, \quad \{\,i\sigma_y, \,-i\sigma_y\,\}, \quad \{\,i\sigma_z, \,-i\sigma_z\,\}.$$

Thus there are five irreducible representations. Four of these are one-dimensional and one is two-dimensional.

Problem 6. The n-qubit *Pauli group* is defined by

$$\mathcal{P}_n = \{\,I_2, \sigma_x, \sigma_y, \sigma_z\,\}^{\otimes n} \otimes \{\,\pm 1, \pm i\,\} \tag{1}$$

where σ_x, σ_y, σ_z are the 2×2 Pauli matrices and I_2 is the 2×2 identity matrix. The dimension of the Hilbert space under consideration is dim $\mathcal{H} = 2^n$. Thus each element of the Pauli group $\mathcal{P}_n$ is (up to an overall phase ± 1, $\pm i$) a Kronecker product of Pauli matrices and 2×2 identity matrices acting on n qubits. The order of the group is given by 2^{2n+2}.
(i) Find all subgroups of the Pauli group $\mathcal{P}_1$.
(ii) Find all conjugacy classes of the Pauli group $\mathcal{P}_1$.

Solution 6. (i) From the multiplication table of the Pauli group we see that P_1 has a 8 elements subgroup Q_8.

Q_8	I_2	$-I_2$	$i\sigma_x$	$i\sigma_y$	$i\sigma_z$	$-i\sigma_x$	$-i\sigma_y$	$-i\sigma_z$
I_2	I_2	$-I_2$	$i\sigma_x$	$i\sigma_y$	$i\sigma_z$	$-i\sigma_x$	$-i\sigma_y$	$-i\sigma_z$
$-I_2$	$-I_2$	I_2	$-i\sigma_x$	$-i\sigma_y$	$-i\sigma_z$	$i\sigma_x$	$i\sigma_y$	$i\sigma_z$
$i\sigma_x$	$i\sigma_x$	$-i\sigma_x$	$-I_2$	$-i\sigma_z$	$i\sigma_y$	I_2	$i\sigma_z$	$-i\sigma_y$
$i\sigma_y$	$i\sigma_y$	$-i\sigma_y$	$i\sigma_z$	$-I_2$	$-i\sigma_x$	$-i\sigma_z$	I_2	$i\sigma_x$
$i\sigma_z$	$i\sigma_z$	$-i\sigma_z$	$-i\sigma_y$	$i\sigma_x$	$-I_2$	$i\sigma_y$	$-i\sigma_x$	I_2
$-i\sigma_x$	$-i\sigma_x$	$i\sigma_x$	I_2	$i\sigma_z$	$-i\sigma_y$	$-I_2$	$-i\sigma_z$	$i\sigma_y$
$-i\sigma_y$	$-i\sigma_y$	$i\sigma_y$	$-i\sigma_z$	I_2	$i\sigma_x$	$i\sigma_z$	$-I_2$	$-i\sigma_x$
$-i\sigma_z$	$-i\sigma_z$	$i\sigma_z$	$i\sigma_y$	$-i\sigma_x$	I_2	$-i\sigma_y$	$i\sigma_x$	$-I_2$

The mapping

$$I \to -i\sigma_x, \quad J \to -i\sigma_y, \quad K \to -i\sigma_x$$

provides an homomorphism with the quaternion group Q_8. It follows that

$$(-i\sigma_x)(-i\sigma_x) = -I_2 \Rightarrow i^2 = -1$$

$$(-i\sigma_x)(-i\sigma_y) = -i\sigma_z \Rightarrow ij = k.$$

Q_8 is a normal subgroup of $\mathcal{P}_1$, since for all $g \notin Q_8$ we have $gqg^{-1} \in Q_8$. The table also shows that there are 3 cyclic subgroups of Q_8 - they are, of course, also subgroups of $\mathcal{P}_1$
- 4-th order cyclic subgroup with generator $i\sigma_x$
- 4-th order cyclic subgroup with generator $i\sigma_y$

- 4-th order cyclic subgroup with generator $i\sigma_z$.

Each of these subgroups is normal in Q_8. Consider the first subgroup, let $g \neq \pm i\sigma_x$. Then $g = a\sigma_y$ or $g = a\sigma_z$ ($a = \pm i$). In the first case

$$(a\sigma_y)i\sigma_x(a\sigma_y)^{-1} = \sigma_y i(\sigma_x\sigma_y) = i\sigma_y i\sigma_z = -i\sigma_x \in N$$

and the same for $g = a\sigma_z$. In $\mathcal{P}_1$ we can extend the cyclic subgroup with generator $i\sigma_x$, adding elements $\pm iI_2, \pm\sigma_x$. This new subgroup is not cyclic, but it is also normal (in $\mathcal{P}_1$). The argumentation is the same. This is a maximal normal subgroup, because it does not belong to any other proper normal subgroup. To prove this, we simply add some element to subgroup, then add all its products etc. The result of such extension is $\mathcal{P}_1$. We need these considerations to build 1-dimensional representations of $\mathcal{P}_1$. Representation sends elements of some normal subgroup to unity and all another elements of $\mathcal{P}_1$ to -1.

(ii) The conjugacy classes of the Pauli group $\mathcal{P}_1$ are

1) $\{I_2\}$ - 1 element of order 1
2) $\{-I_2\}$ - 1 element of order 2
3) $\{iI_2\}$ - 1 element of order 4
4) $\{-iI_2\}$ - 1 element of order 4
5) $\{\sigma_x, -\sigma_x\}$ - 2 elements of order 2
6) $\{i\sigma_x, -i\sigma_x\}$ - 2 elements of order 4
7) $\{\sigma_y, -\sigma_y\}$ - 2 elements of order 2
8) $\{i\sigma_y, -i\sigma_y\}$ - 2 elements of order 4
9) $\{\sigma_z, -\sigma_z\}$ - 2 elements of order 2
10) $\{i\sigma_z, -i\sigma_z\}$ - 2 elements of order 4.

Thus there is 10 irreducible representations of $\mathcal{P}_1$. The sum of squares of dimensions of representations is equal to 16. The solution is unique: There are 8 one-dimensional representations and 2 two-dimensional representations.

Problem 7. The n-qubit *Clifford group* $\mathcal{C}_n$ is the normalizer of the Pauli group. A $2^n \times 2^n$ unitary matrix U acting on n qubits is an element of the Clifford group iff

$$UMU^* \in \mathcal{P}_n \quad \text{for each} \quad M \in \mathcal{P}_n .$$

This means the unitary matrix U acting by conjugation takes a Kronecker product of Pauli matrices to Kronecker product of Pauli matrices. An element of the Clifford group is defined as this action by conjugation, so that the overall phase of the unitary matrix U is not relevant. In other words the Clifford group is the group of all matrices that leave the Pauli group invariant.

(i) Show that the *Hadamard matrix*

$$H = \frac{1}{\sqrt{2}} \begin{pmatrix} 1 & 1 \\ 1 & -1 \end{pmatrix}$$

is a member of the Clifford group $\mathcal{C}_1$.

(ii) Show that the matrix

$$U = \begin{pmatrix} 1 & 0 \\ 0 & i \end{pmatrix}$$

is an element of the Clifford group $\mathcal{C}_1$.

(iii) Show that the 4×4 matrix

$$U_{XOR} = \begin{pmatrix} 1 & 0 & 0 & 0 \\ 0 & 1 & 0 & 0 \\ 0 & 0 & 0 & 1 \\ 0 & 0 & 1 & 0 \end{pmatrix}$$

is an element of the Clifford group $\mathcal{C}_2$.

Solution 7. (i) We have $H = H^* = H^{-1}$ and

$$H\sigma_x H^{-1} = \sigma_z, \quad H\sigma_y H^{-1} = -\sigma_y, \quad H\sigma_z H^{-1} = \sigma_x \,.$$

Consider conjugations of elements of the Pauli group $\mathcal{P}_1$ by H with respect to

$$HIH = I, \quad H(-I)H = -I, \quad H(iI)H = iI, \quad H(-iI)H = -iI$$

$$H\sigma_x H = \sigma_z, \quad H(-\sigma)H = -\sigma_z, \quad H(i\sigma_x)H = i\sigma_z, \quad H(-i\sigma_x)H = -i\sigma_z$$

$$H\sigma_y H = -\sigma_y, \quad H(-\sigma_y)H = \sigma_y, \quad Hi\sigma_y)H = -i\sigma_y, \quad H(-i\sigma_y)H = \sigma_y,$$

$$H\sigma_z H = \sigma_x, \quad H(-\sigma_z)H = -\sigma_x, \quad H(i\sigma_z)H = i\sigma_x, \quad H(-iZ)H = -i\sigma_x \,.$$

Thus for all 16 elements M of the Pauli group $\mathcal{P}_1$ we have

$$HMH^{-1} \in \mathcal{P}_1.$$

Thus H is a member of the Clifford group $\mathcal{C}_1$.

(ii) We have

$$U(\pm\sigma_x)U^{-1} = \pm\sigma_y, \qquad U(\pm i\sigma_x)U^{-1} = \pm i\sigma_y,$$

$$U(\pm\sigma_y)U^{-1} = \mp\sigma_x, \qquad U(\pm\sigma_z)U^{-1} = \mp i\sigma_x \,.$$

Thus U belongs to the 1-qubit Clifford group.

(iii) Note that $U_{XOR} = U_{XOR}^*$ and U_{XOR} is a permutation matrix. We have to test $U_{XOR}MU_{XOR}^* \in \mathcal{P}_2$ for all 64 elements M in $\mathcal{P}_2$. For example

$$U_{XOR}(\sigma_x \otimes \sigma_y)U_{XOR} = \sigma_y \otimes \sigma_z.$$

For the other 63 calculation one would apply computer algebra which shows that U_{XOR} is an element of the Clifford group $\mathcal{C}_2$.

Problem 8. Consider the system of linear equations $Ax = b$, where A is an invertible 3×3. Assume that the linear system is invariant under $x_1 \leftrightarrow x_2$, i.e.

$$A\mathbf{x} = A\mathbf{x}|_{x_1 \leftrightarrow x_2} = \mathbf{b}.$$

Can we conclude that $x_1 = x_2$?

Solution 8. Let

$$A = \begin{pmatrix} a_{11} & a_{12} & a_{13} \\ a_{21} & a_{22} & a_{23} \\ a_{31} & a_{32} & a_{33} \end{pmatrix}.$$

From the invariance condition we find the three equations

$$(a_{11} - a_{12})(x_1 - x_2) = 0$$
$$(a_{21} - a_{22})(x_1 - x_2) = 0$$
$$(a_{31} - a_{32})(x_1 - x_2) = 0.$$

These equations are satisfied if $x_1 = x_2$. If $a_{11} = a_{12}$, $a_{21} = a_{22}$, $a_{31} = a_{32}$ we find that the matrix A is not invertible.

Problem 9. The system of partial differential equations of the system of *chiral field equations* can be written as

$$\frac{\partial \mathbf{u}}{\partial t} + \frac{\partial \mathbf{u}}{\partial x} - \mathbf{u} \times (J\mathbf{v}) = 0$$
$$\frac{\partial \mathbf{v}}{\partial t} - \frac{\partial \mathbf{v}}{\partial x} - \mathbf{v} \times (J\mathbf{u}) = 0$$

where $\mathbf{u} = (u_1, u_2, u_3)^T$, $\mathbf{v} = (v_1, v_2, v_3)^T$, $\mathbf{u}^2 = \mathbf{v}^2 = 1$, $J = \mathrm{diag}(j_1, j_2, j_3)$ is a 3×3 diagonal matrix and $\times$ denotes the vector product. Consider the linear mapping $M : \mathbb{R}^3 \to so(3)$

$$M(\mathbf{u}) = \begin{pmatrix} 0 & u_3 & -u_2 \\ -u_3 & 0 & u_1 \\ u_2 & -u_1 & 0 \end{pmatrix}$$

where $so(3)$ is the simple Lie algebra of the 3×3 skew-symmetric matrices. Rewrite the system of partial differential equations using $M(\mathbf{u})$.

Solution 9. First we note that $[M(\mathbf{u}), M(\mathbf{v})] = -M(\mathbf{u} \times \mathbf{v})$. Thus

$$\frac{\partial M(\mathbf{u})}{\partial t} + \frac{\partial M(\mathbf{u})}{\partial x} + [M(\mathbf{u}), M(J\mathbf{v})] = 0$$

$$\frac{\partial M(\mathbf{v})}{\partial t} - \frac{\partial M(\mathbf{v})}{\partial x} + [M(\mathbf{v}), M(J\mathbf{u})] = 0$$

with $\mathbf{u}^2 = \mathbf{v}^2 = 1$.

Problem 10. The *Landau-Lifshitz equation* is given by

$$\frac{\partial \mathbf{u}}{\partial t} = \mathbf{u} \times \frac{\partial^2 \mathbf{u}}{\partial x^2} + \mathbf{u} \times (K\mathbf{u})$$

where $\mathbf{u} = (u_1, u_2, u_3)^T$, $\mathbf{u}^2 = 1$, $K = \mathrm{diag}(k_1, k_2, k_3)$ is a 3×3 diagonal matrix and $\times$ denotes the vector product. Consider the linear mapping $M : \mathbb{R}^3 \to so(3)$

$$M(\mathbf{u}) = \begin{pmatrix} 0 & u_3 & -u_2 \\ -u_3 & 0 & u_1 \\ u_2 & -u_1 & 0 \end{pmatrix}$$

where $so(3)$ is the simple Lie algebra of the 3×3 skew-symmetric matrices. Rewrite the Landau-Lifshitz equation using $M(\mathbf{u})$.

Solution 10. First we note that $[M(\mathbf{u}), M(\mathbf{v})] = -M(\mathbf{u} \times \mathbf{v})$. Thus we find

$$\frac{\partial M(\mathbf{u})}{\partial t} + [M(\mathbf{u}), M(\mathbf{u})_{xx}] + [M(\mathbf{u}), M(K\mathbf{u})] = 0$$

with $\mathbf{u}^2 = 1$ and $M(\mathbf{u})_{xx} \equiv \partial^2 M / \partial x^2$.

Problem 11. Let $\{ X_1, \ldots, X_n \}$ be a basis of a Lie algebra L and $\{ C_{ij}^k \}$ be the *structure constants* over this basis. Then we can represent L in the linear space $C^\infty(L^*)$ by the linear differential operators

$$\hat{X}_i = -\sum_{j=1}^n \sum_{k=1}^n C_{ij}^k x_k \frac{\partial}{\partial x_j}$$

where $[X_i, X_j] = \sum_{k=1}^n C_{ij}^k X_k$ $(1 \le i < j \le n)$. The differential operators satisfy

$$[\hat{X}_i, \hat{X}_j] = \sum_{k=1}^n C_{ij}^k \hat{X}_k .$$

Therefore they constitute a representation of the Lie algebra L. An analytic function $f \in C^\infty(L^*)$ is called an *invariant* of L if and only if it is a solution of the system

$$\{ \hat{X}_i f = 0, \quad 1 \le i \le n \} .$$

Polynomial solutions of the system correspond to classical Casimir invariants after symmetrization. Apply this to the simple Lie algebra $so(3)$ with the basis $\{\, X_1,\, X_2,\, X_3 \,\}$ and the commutation relations

$$[X_1, X_2] = X_3, \qquad [X_2, X_3] = X_1, \qquad [X_3, X_1] = X_2.$$

Solution 11.   According to the formula

$$\hat{X}_i = -\sum_{j=1}^{n} \sum_{k=1}^{n} C_{ij}^{k} x_k \frac{\partial}{\partial x_j}$$

we express $\hat{X}_i$ in the following way

$$\hat{X}_1 = x_3 \frac{\partial}{\partial x_2} - x_2 \frac{\partial}{\partial x_3}, \quad \hat{X}_2 = x_1 \frac{\partial}{\partial x_3} - x_3 \frac{\partial}{\partial x_1}, \quad \hat{X}_3 = x_2 \frac{\partial}{\partial x_1} - x_1 \frac{\partial}{\partial x_2}.$$

These operators are linearly dependent, since

$$x_1 \hat{X}_1 + x_2 \hat{X}_2 + x_3 \hat{X}_3 = 0.$$

Let $f : \mathbb{R}^3 \to \mathbb{R}$

$$f(x_1, x_2, x_3) = x_1^2 + x_2^2 + x_3^2.$$

Then

$$\hat{X}_1 f = 0, \quad \hat{X}_2 f = 0, \quad \hat{X}_3 f = 0.$$

Thus f is an invariant.

Problem 12.   Let $n \geq 2$. The discrete *Weyl algebra* is generated by the $n \times n$ matrices $\{\, U, V, I_n \}$ over the complex numbers satisfying

$$UV = \omega VU, \qquad U^n = V^n = I_n$$

where $\omega := \exp(2\pi i/n)$.
(i) Find U, V for $n = 2$.
(ii) Find U, V for $n = 3$.

Solution 12.   (i) With $e^{i\pi} = -1$ we find

$$U = \begin{pmatrix} 1 & 0 \\ 0 & -1 \end{pmatrix}, \quad V = \begin{pmatrix} 0 & 1 \\ 1 & 0 \end{pmatrix}.$$

(ii) With $\omega = e^{2i\pi/3}$ we find

$$U = \begin{pmatrix} 1 & 0 & 0 \\ 0 & \omega & 0 \\ 0 & 0 & \omega^2 \end{pmatrix}, \quad V = \begin{pmatrix} 0 & 0 & 1 \\ 1 & 0 & 0 \\ 0 & 1 & 0 \end{pmatrix}.$$

Problem 13. A basis for the Lie algebra $su(N)$, for odd N, may be built from two unitary unimodular $N \times N$ matrices

$$
g = \begin{pmatrix} 1 & 0 & 0 & \cdots & 0 \\ 0 & \omega & 0 & \cdots & 0 \\ 0 & 0 & \omega^2 & \cdots & 0 \\ \vdots & \vdots & \vdots & \ddots & \vdots \\ 0 & 0 & 0 & \cdots & \omega^{N-1} \end{pmatrix}, \qquad h = \begin{pmatrix} 0 & 1 & 0 & \cdots & 0 \\ 0 & 0 & 1 & \cdots & 0 \\ \vdots & \vdots & \vdots & \ddots & \vdots \\ 0 & 0 & 0 & \cdots & 1 \\ 1 & 0 & 0 & \cdots & 0 \end{pmatrix}
$$

where ω is a primitive Nth root of unity, i.e. with period not smaller than N, here taken to be $\exp(4\pi i/N)$. We obviously have

$$ hg = \omega gh. \tag{1} $$

(i) Find g^N and h^N.

(ii) Find $\operatorname{tr} g$.

(iii) Let $\mathbf{m} = (m_1, m_2)$, $\mathbf{n} = (n_1, n_2)$ and define $\mathbf{m} \times \mathbf{n} := m_1 n_2 - m_2 n_1$, where $m_1 = 0, 1, \ldots, N-1$, $m_2 = 0, 1, \ldots, N-1$. The complete set of unitary unimodular $N \times N$ matrices

$$ J_{m_1, m_2} := \omega^{m_1 m_2/2} g^{m_1} h^{m_2} $$

suffice to span the Lie algebra $su(N)$, where $J_{0,0} = I_N$. Find $J^*_{(m_1. m_2)}$.

(iv) Calculate the matrix product $J_{\mathbf{m}} J_{\mathbf{n}}$.

(v) Find the commutator $[J_{\mathbf{m}}, J_{\mathbf{n}}]$.

Solution 13. (i) Since $\omega^N = 1$ we find $g^N = I_N$. We also obtain $h^N = I_N$.

(ii) Since

$$ 1 + \omega + \omega^2 + \cdots + \omega^{N-1} = 0 $$

we find $\operatorname{tr} g = 0$.

(iii) Obviously we have

$$ J^*_{(m_1, m_2)} = J_{(-m_1, -m_2)}. $$

(iv) Using equation (1) we find

$$ J_{\mathbf{m}} J_{\mathbf{n}} = \omega^{\mathbf{n} \times \mathbf{m}/2} J_{\mathbf{m}+\mathbf{n}}. $$

(v) Using the result from (iv) we obtain

$$ [J_{\mathbf{m}}, J_{\mathbf{n}}] = -2i \sin\left(\frac{2\pi}{N} \mathbf{m} \times \mathbf{n}\right) J_{\mathbf{m}+\mathbf{n}}. $$

Problem 14. Let $t \in \mathbb{R}$. Let $A(t)$ be a smooth $n \times n$ matrix-valued function. Then we have

$$ \frac{d}{dt} e^{A(t)} = e^{A(t)} \left(\frac{I_n - \exp(-\operatorname{ad}_{A(t)})}{\operatorname{ad}_{A(t)}} \left(\frac{dA}{dt} \right) \right) $$

where $\mathrm{ad}Y(X) := [Y, X]$. Let

$$A(t) = \begin{pmatrix} 0 & t \\ -t & 0 \end{pmatrix} = t \begin{pmatrix} 0 & 1 \\ -1 & 0 \end{pmatrix}.$$

Calculate the derivative $de^{A(t)}/dt$ using this expression.

Solution 14. We have

$$\frac{dA}{dt} = \begin{pmatrix} 0 & 1 \\ -1 & 0 \end{pmatrix}.$$

Now

$$\frac{I_n - \exp(-\mathrm{ad}_{A(t)})}{\mathrm{ad}_{A(t)}} = \frac{I_n - (I_n - \mathrm{ad}_{A(t)} + \frac{1}{2!}\mathrm{ad}_{A(t)} - \cdots)}{\mathrm{ad}_{A(t)}}$$

$$= (I_n - \frac{1}{2!}\mathrm{ad}_{A(t)} + \cdots)$$

and

$$\mathrm{ad}_{A(t)}\frac{dA}{dt} = [A(t), dA/dt]$$

$$= \begin{pmatrix} 0 & t \\ -t & 0 \end{pmatrix}\begin{pmatrix} 0 & 1 \\ -1 & 0 \end{pmatrix} - \begin{pmatrix} 0 & 1 \\ -1 & 0 \end{pmatrix}\begin{pmatrix} 0 & t \\ -t & 0 \end{pmatrix}$$

$$= \begin{pmatrix} 0 & 0 \\ 0 & 0 \end{pmatrix}.$$

Thus

$$\frac{d}{dt}e^{A(t)} = e^{A(t)}\begin{pmatrix} 0 & 1 \\ -1 & 0 \end{pmatrix} = \begin{pmatrix} \cos t & \sin t \\ -\sin t & \cos t \end{pmatrix}\begin{pmatrix} 0 & 1 \\ -1 & 0 \end{pmatrix}$$

$$= \begin{pmatrix} -\sin t & \cos t \\ -\cos t & -\sin t \end{pmatrix}.$$

Problem 15. Consider the one-dimensional Hamilton operator

$$\hat{H} = -\frac{\hbar^2}{2}\frac{d^2}{dx^2} + V(x)$$

where the potential V is a smooth function of x. We want to find a linear differential operator $\hat{I}$ that commutes with the operator $\hat{H}$. We make the ansatz for the differential operator $\hat{I}$ as

$$\hat{I} = -i\hbar^3\frac{d^3}{dx^3} + \frac{i}{2}\hbar\left(\frac{df}{dx} + f(x)\frac{d}{dx}\right).$$

Find the conditions on V and f such that $[\hat{H}, \hat{I}] = 0$.

Solution 15. The condition $[\hat{H}, \hat{I}] = 0$ provides the following two differential equations for V and f

$$\frac{df}{dx} - 3\frac{dV}{dx} = 0, \qquad \hbar^2\left(\frac{d^3f}{dx^3} - 4\frac{d^3V}{dx^3}\right) + 4f\frac{dV}{dx} = 0.$$

Thus $f(x) = 3V(x) + c_1$. Inserting this into the second equation provides the nonlinear third order differential equation for V

$$\hbar^2\frac{d^3V}{dx^3} + 4(3V + c_1)\frac{dV}{dx} = 0.$$

This third order differential equation can be integrated twice. If we introduce the *Weierstrass* $\wp$ *function* we have

$$V(x) = \hbar^2\wp(x) - c_1, \qquad f(x) = 3\hbar^2\wp(x).$$

The Weierstrass $\wp$ function satisfies

$$\left(\frac{d\wp}{dx}\right)^2 = 4\wp^3(x) - g_2\wp - g_3.$$

Problem 16. Consider the Hamilton function

$$H(\mathbf{p}, \mathbf{q}) = \frac{1}{2}(p_1^2 + p_2^2) - (Aq_1^2 + Bq_2^2 + Cq_1^2q_2 + Dq_2^3)$$

where $\mathbf{q} = (q_1, q_2) \in \mathbb{R}^2$ and $\mathbf{p} = (p_1, p_2) \in \mathbb{R}^2$ are the position vector and the momentum vector of the particle, respectively. The Hamilton function is a constant of motion, i.e. $H(\mathbf{p}, \mathbf{q}) = E$. Consider the real analytic diffeomorphism

$$r = |\mathbf{q}|$$
$$\theta = \arctan(q_2/q_1)$$
$$u = \dot{r} = (q_1p_1 + q_2p_2)/|\mathbf{q}|$$
$$v = r\dot{\theta} = (q_1p_2 - q_2p_1)/|\mathbf{q}|$$

which is a *McGehee-type transformation* of the second kind. The equations of motion take the form

$$\dot{r} = u$$
$$\dot{\theta} = v/r$$
$$\dot{u} = v^2/r + 2(A\cos^2\theta + B\sin^2\theta)r + 3(C\cos^2\theta + D\sin^2\theta)r^2\sin\theta$$
$$\dot{v} = -uv/r + 2(B - A)r\sin\theta\cos\theta + (3(D - C)\sin^2\theta + C)r^2\cos\theta$$

and the energy takes the form

$$\frac{1}{2}(u^2 + v^2) - (A\cos^2\theta + B\sin^2\theta)r^2 - (C\cos^2\theta + D\sin^2\theta)r^3\sin\theta = E.$$

Find the discrete symmetries of the vector field given by the autonomous system of differential equations.

Solution 16.   The vector field has four discrete symmetries

$$\begin{aligned}
S_0 &= (r, \theta, u, v, t) \quad \text{identity} \\
S_1 &= (r, \theta, -u, -v, -t), \\
S_2 &= (r, \pi - \theta, -u, v, -t), \\
S_3 &= (r, \pi - \theta, u, -v, t).
\end{aligned}$$

These discrete symmetries form an Abelian group G isomorphic to Klein's group.

Problem 17.   Consider the eigenvalue problem (Schrödinger equation)

$$\left(-\frac{\hbar^2}{2m}\frac{d^2}{dx^2} + V_0(A\cosh(ax) - 1)^2\right)\psi(x) = E\psi(x)$$

where V_0 is positive (dimension energy), $0 < A < 1$ and a is positive with dimension inverse length. Introducing the dimensionless parameters S and B

$$V_0 = \frac{\hbar^2 a^2}{8m}(2S+1)^2, \qquad A = \frac{B}{(2S+1)}, \qquad -\frac{1}{2} < S < \infty, \quad B > 0$$

and the dimensionless length $\zeta = ax$ we find with $\widetilde{\psi}(\zeta(x)) = \psi(x)$ the eigenvalue equation in the form

$$\hat{H}\widetilde{\psi}(\zeta) = \epsilon\widetilde{\psi}(\zeta)$$

with the Hamilton operator

$$\hat{H} = -\frac{d^2}{d\zeta^2} + \frac{1}{4}B^2\sinh^2\zeta - B(S+1/2)\cosh\zeta$$

and

$$\epsilon = \frac{2m}{\hbar^2 a^2}E - \frac{1}{4}(2S+1)^2\left(1 + \left(\frac{B}{2S+1}\right)^2\right).$$

Consider the differential operators

$$S_1 = S \cosh \zeta - \frac{B}{2} \sinh^2 \zeta - \sinh \zeta \frac{d}{d\zeta}$$

$$S_2 = i \left(-S \sinh \zeta + \frac{B}{2} \sinh \zeta \cosh \zeta + \cosh \zeta \frac{d}{d\zeta} \right)$$

$$S_3 = \frac{B}{2} \sinh \zeta + \frac{d}{d\zeta} .$$

(i) Find the commutators and thus show that they form a basis of a Lie algebra.
(ii) Show that the Hamilton operator can be expressed with the operators S_1, S_2, S_3.
(iii) Find $S_1^2 + S_2^2 + S_3^2$.

Solution 17. (i) We obtain for the commutators

$$[S_j, S_k] = i \sum_{\ell=1}^{3} \epsilon_{jk\ell} S_\ell .$$

(ii) Since

$$S_3^2 = \frac{B^2}{4} \sinh^2 \zeta + \frac{B}{2} \cosh \zeta + B \sinh \zeta \frac{d}{d\zeta} + \frac{d^2}{d\zeta^2}$$

we obtain

$$\hat{H} = -S_3^2 - BS_1 .$$

(iii) Using $\cosh^2 \zeta - \sinh^2 \zeta = 1$ we find $S_1^2 + S_2^2 + S_3^2 = S(S+1)I$, where I is the identity operator.

Problem 18. (i) Consider the Lie algebra $so(2,3)$. The ten generators $L_1, L_2, L_3, D_1, D_2, D_3, F_1, F_2, F_3$ and V satisfy the commutation relations

$$[L_j, L_k] = i\epsilon_{jk\ell} L_\ell$$
$$[L_j, V] = 0$$
$$[L_j, D_k] = i\epsilon_{jk\ell} D_\ell$$
$$[L_j, F_k] = i\epsilon_{jk\ell} F_\ell$$
$$[D_j, F_k] = -i\delta_{jk} V$$
$$[D_j, D_k] = -i\epsilon_{jk\ell} L_\ell$$
$$[F_j, F_k] = -i\epsilon_{jk\ell} L_\ell$$
$$[D_j, V] = -iF_j$$
$$[F_j, V] = iD_j$$

where $j, k = 1, 2, 3$ and $\epsilon_{123} = \epsilon_{312} = \epsilon_{231} = +1$. Find the adjoint representation of $so(2,3)$.

(ii) Find the *Killing form* $\kappa(x,y) := \text{tr}(\text{ad}(x)\text{ad}(y))$. Then show that the Lie algebra is semisimple.

(iii) Find the two quadratic *Casimir operators* of $so(2,3)$. The Casimir operators commute with all elements of the Lie algebra.

Solution 18.   (i) From the commutation relations we find the adjoint representation

$$\text{ad}(L_1) = \begin{pmatrix} 0 & 0 & 0 & 0 & 0 & 0 & 0 & 0 & 0 & 0 \\ 0 & 0 & -i & 0 & 0 & 0 & 0 & 0 & 0 & 0 \\ 0 & i & 0 & 0 & 0 & 0 & 0 & 0 & 0 & 0 \\ 0 & 0 & 0 & 0 & 0 & 0 & 0 & 0 & 0 & 0 \\ 0 & 0 & 0 & 0 & 0 & -i & 0 & 0 & 0 & 0 \\ 0 & 0 & 0 & 0 & i & 0 & 0 & 0 & 0 & 0 \\ 0 & 0 & 0 & 0 & 0 & 0 & 0 & 0 & 0 & 0 \\ 0 & 0 & 0 & 0 & 0 & 0 & 0 & 0 & -i & 0 \\ 0 & 0 & 0 & 0 & 0 & 0 & 0 & i & 0 & 0 \\ 0 & 0 & 0 & 0 & 0 & 0 & 0 & 0 & 0 & 0 \end{pmatrix}$$

$$\text{ad}(L_2) = \begin{pmatrix} 0 & 0 & i & 0 & 0 & 0 & 0 & 0 & 0 & 0 \\ 0 & 0 & 0 & 0 & 0 & 0 & 0 & 0 & 0 & 0 \\ -i & 0 & 0 & 0 & 0 & 0 & 0 & 0 & 0 & 0 \\ 0 & 0 & 0 & 0 & 0 & i & 0 & 0 & 0 & 0 \\ 0 & 0 & 0 & 0 & 0 & 0 & 0 & 0 & 0 & 0 \\ 0 & 0 & 0 & -i & 0 & 0 & 0 & 0 & 0 & 0 \\ 0 & 0 & 0 & 0 & 0 & 0 & 0 & 0 & i & 0 \\ 0 & 0 & 0 & 0 & 0 & 0 & 0 & 0 & 0 & 0 \\ 0 & 0 & 0 & 0 & 0 & 0 & -i & 0 & 0 & 0 \\ 0 & 0 & 0 & 0 & 0 & 0 & 0 & 0 & 0 & 0 \end{pmatrix}$$

$$\text{ad}(L_3) = \begin{pmatrix} 0 & -i & 0 & 0 & 0 & 0 & 0 & 0 & 0 & 0 \\ i & 0 & 0 & 0 & 0 & 0 & 0 & 0 & 0 & 0 \\ 0 & 0 & 0 & 0 & 0 & 0 & 0 & 0 & 0 & 0 \\ 0 & 0 & 0 & 0 & -i & 0 & 0 & 0 & 0 & 0 \\ 0 & 0 & 0 & i & 0 & 0 & 0 & 0 & 0 & 0 \\ 0 & 0 & 0 & 0 & 0 & 0 & 0 & 0 & 0 & 0 \\ 0 & 0 & 0 & 0 & 0 & 0 & 0 & -i & 0 & 0 \\ 0 & 0 & 0 & 0 & 0 & 0 & i & 0 & 0 & 0 \\ 0 & 0 & 0 & 0 & 0 & 0 & 0 & 0 & 0 & 0 \\ 0 & 0 & 0 & 0 & 0 & 0 & 0 & 0 & 0 & 0 \end{pmatrix}$$

$$\mathrm{ad}(D_1) = \begin{pmatrix} 0 & 0 & 0 & 0 & 0 & 0 & 0 & 0 & 0 & 0 \\ 0 & 0 & 0 & 0 & 0 & i & 0 & 0 & 0 & 0 \\ 0 & 0 & 0 & 0 & -i & 0 & 0 & 0 & 0 & 0 \\ 0 & 0 & 0 & 0 & 0 & 0 & 0 & 0 & 0 & 0 \\ 0 & 0 & -i & 0 & 0 & 0 & 0 & 0 & 0 & 0 \\ 0 & i & 0 & 0 & 0 & 0 & 0 & 0 & 0 & 0 \\ 0 & 0 & 0 & 0 & 0 & 0 & 0 & 0 & 0 & -i \\ 0 & 0 & 0 & 0 & 0 & 0 & 0 & 0 & 0 & 0 \\ 0 & 0 & 0 & 0 & 0 & 0 & 0 & 0 & 0 & 0 \\ 0 & 0 & 0 & 0 & 0 & 0 & -i & 0 & 0 & 0 \end{pmatrix}$$

$$\mathrm{ad}(D_2) = \begin{pmatrix} 0 & 0 & 0 & 0 & 0 & -i & 0 & 0 & 0 & 0 \\ 0 & 0 & 0 & 0 & 0 & 0 & 0 & 0 & 0 & 0 \\ 0 & 0 & 0 & i & 0 & 0 & 0 & 0 & 0 & 0 \\ 0 & 0 & i & 0 & 0 & 0 & 0 & 0 & 0 & 0 \\ 0 & 0 & 0 & 0 & 0 & 0 & 0 & 0 & 0 & 0 \\ -i & 0 & 0 & 0 & 0 & 0 & 0 & 0 & 0 & 0 \\ 0 & 0 & 0 & 0 & 0 & 0 & 0 & 0 & 0 & 0 \\ 0 & 0 & 0 & 0 & 0 & 0 & 0 & 0 & 0 & -i \\ 0 & 0 & 0 & 0 & 0 & 0 & 0 & 0 & 0 & 0 \\ 0 & 0 & 0 & 0 & 0 & 0 & 0 & -i & 0 & 0 \end{pmatrix}$$

$$\mathrm{ad}(D_3) = \begin{pmatrix} 0 & 0 & 0 & 0 & i & 0 & 0 & 0 & 0 & 0 \\ 0 & 0 & 0 & -i & 0 & 0 & 0 & 0 & 0 & 0 \\ 0 & 0 & 0 & 0 & 0 & 0 & 0 & 0 & 0 & 0 \\ 0 & -i & 0 & 0 & 0 & 0 & 0 & 0 & 0 & 0 \\ i & 0 & 0 & 0 & 0 & 0 & 0 & 0 & 0 & 0 \\ 0 & 0 & 0 & 0 & 0 & 0 & 0 & 0 & 0 & 0 \\ 0 & 0 & 0 & 0 & 0 & 0 & 0 & 0 & 0 & 0 \\ 0 & 0 & 0 & 0 & 0 & 0 & 0 & 0 & 0 & 0 \\ 0 & 0 & 0 & 0 & 0 & 0 & 0 & 0 & 0 & -i \\ 0 & 0 & 0 & 0 & 0 & 0 & 0 & 0 & -i & 0 \end{pmatrix}$$

$$\mathrm{ad}(F_1) = \begin{pmatrix} 0 & 0 & 0 & 0 & 0 & 0 & 0 & 0 & 0 & 0 \\ 0 & 0 & 0 & 0 & 0 & 0 & 0 & 0 & i & 0 \\ 0 & 0 & 0 & 0 & 0 & 0 & 0 & -i & 0 & 0 \\ 0 & 0 & 0 & 0 & 0 & 0 & 0 & 0 & 0 & i \\ 0 & 0 & 0 & 0 & 0 & 0 & 0 & 0 & 0 & 0 \\ 0 & 0 & 0 & 0 & 0 & 0 & 0 & 0 & 0 & 0 \\ 0 & 0 & 0 & 0 & 0 & 0 & 0 & 0 & 0 & 0 \\ 0 & 0 & -i & 0 & 0 & 0 & 0 & 0 & 0 & 0 \\ 0 & i & 0 & 0 & 0 & 0 & 0 & 0 & 0 & 0 \\ 0 & 0 & 0 & i & 0 & 0 & 0 & 0 & 0 & 0 \end{pmatrix}$$

$$\mathrm{ad}(F_2) = \begin{pmatrix} 0 & 0 & 0 & 0 & 0 & 0 & 0 & 0 & -i & 0 \\ 0 & 0 & 0 & 0 & 0 & 0 & 0 & 0 & 0 & 0 \\ 0 & 0 & 0 & 0 & 0 & 0 & i & 0 & 0 & 0 \\ 0 & 0 & 0 & 0 & 0 & 0 & 0 & 0 & 0 & 0 \\ 0 & 0 & 0 & 0 & 0 & 0 & 0 & 0 & 0 & i \\ 0 & 0 & 0 & 0 & 0 & 0 & 0 & 0 & 0 & 0 \\ 0 & 0 & i & 0 & 0 & 0 & 0 & 0 & 0 & 0 \\ 0 & 0 & 0 & 0 & 0 & 0 & 0 & 0 & 0 & 0 \\ -i & 0 & 0 & 0 & 0 & 0 & 0 & 0 & 0 & 0 \\ 0 & 0 & 0 & 0 & i & 0 & 0 & 0 & 0 & 0 \end{pmatrix}$$

$$\mathrm{ad}(F_3) = \begin{pmatrix} 0 & 0 & 0 & 0 & 0 & 0 & 0 & i & 0 & 0 \\ 0 & 0 & 0 & 0 & 0 & 0 & -i & 0 & 0 & 0 \\ 0 & 0 & 0 & 0 & 0 & 0 & 0 & 0 & 0 & 0 \\ 0 & 0 & 0 & 0 & 0 & 0 & 0 & 0 & 0 & 0 \\ 0 & 0 & 0 & 0 & 0 & 0 & 0 & 0 & 0 & 0 \\ 0 & 0 & 0 & 0 & 0 & 0 & 0 & 0 & 0 & i \\ 0 & -i & 0 & 0 & 0 & 0 & 0 & 0 & 0 & 0 \\ i & 0 & 0 & 0 & 0 & 0 & 0 & 0 & 0 & 0 \\ 0 & 0 & 0 & 0 & 0 & 0 & 0 & 0 & 0 & 0 \\ 0 & 0 & 0 & 0 & 0 & i & 0 & 0 & 0 & 0 \end{pmatrix}$$

$$\mathrm{ad}(V) = \begin{pmatrix} 0 & 0 & 0 & 0 & 0 & 0 & 0 & 0 & 0 & 0 \\ 0 & 0 & 0 & 0 & 0 & 0 & 0 & 0 & 0 & 0 \\ 0 & 0 & 0 & 0 & 0 & 0 & 0 & 0 & 0 & 0 \\ 0 & 0 & 0 & 0 & 0 & 0 & -i & 0 & 0 & 0 \\ 0 & 0 & 0 & 0 & 0 & 0 & 0 & -i & 0 & 0 \\ 0 & 0 & 0 & 0 & 0 & 0 & 0 & 0 & -i & 0 \\ 0 & 0 & 0 & i & 0 & 0 & 0 & 0 & 0 & 0 \\ 0 & 0 & 0 & 0 & i & 0 & 0 & 0 & 0 & 0 \\ 0 & 0 & 0 & 0 & 0 & i & 0 & 0 & 0 & 0 \\ 0 & 0 & 0 & 0 & 0 & 0 & 0 & 0 & 0 & 0 \end{pmatrix} .$$

(ii) For the Killing form

$$\kappa(x, y) = \mathrm{tr}(\mathrm{ad}(x)\mathrm{ad}(y))$$

we find the invertible diagonal matrix

$$\kappa = \mathrm{diag}(6\ 6\ 6\ -6\ -6\ -6\ -6\ -6\ -6\ 6) .$$

We see that B is nondegenerate diagonal matrix. Thus the Lie algebra is semisimple.

(iii) The quadratic Casimir operators are

$$C_1 = L^2 + V^2 + D^2 + L^2, \qquad C_2 = L^2 + V^2 - D^2 - F^2 .$$

Problem 19. Any unitary $2^n \times 2^n$ matrix U can be decomposed as

$$U = \begin{pmatrix} U_1 & 0 \\ 0 & U_2 \end{pmatrix} \begin{pmatrix} C & S \\ -S & C \end{pmatrix} \begin{pmatrix} U_3 & 0 \\ 0 & U_4 \end{pmatrix}$$

where U_1, U_2, U_3, U_4 are $2^{n-1} \times 2^{n-1}$ unitary matrices and C and S are the $2^{n-1} \times 2^{n-1}$ diagonal matrices

$$C = \mathrm{diag}(\cos \alpha_1, \cos \alpha_2, \ldots, \cos \alpha_{2^n/2})$$
$$S = \mathrm{diag}(\sin \alpha_1, \sin \alpha_2, \ldots, \sin \alpha_{2^n/2})$$

where $\alpha_j \in \mathbb{R}$. This decomposition is called the *cosine-sine decomposition* of U. Let V, W be $2^n \times 2^n$ unitary matrices. Assume we have the cosine-sine decomposition of V and W. Let $\otimes$ be the Kronecker product. Do we have a cosine-sine decomposition of $V \otimes W$?

Solution 19. No we do not have a cosine-sine decomposition of $V \otimes W$.

Problem 20. Consider the linear operators L and M defined by

$$L\psi(x,t,\lambda) = \left(i\frac{\partial}{\partial x} + U(x,t,\lambda) \right) \psi(x,t,\lambda)$$

$$M\psi(x,t,\lambda) = \left(i\frac{\partial}{\partial t} + V(x,t,\lambda) \right) \psi(x,t,\lambda).$$

Find the condition on L and M such that $[L, M] = 0$, where $[\,,\,]$ denotes the commutator. The potentials $U(x,t,\lambda)$ and $V(x,t,\lambda)$ (square matrices) are typically chosen as elements of some semisimple Lie algebra.

Solution 20. We obtain the matrix-valued differential equation

$$i\frac{\partial}{\partial x}V - i\frac{\partial}{\partial t}U + [U, V] = 0.$$

Problem 21. Consider the time-independent Schrödinger equation for the wavefunction ψ of a three-particle system in one space dimension

$$-\frac{\hbar^2}{2m}\Delta\psi + \left(\sum_{j \leq k=1}^{3} (\alpha(x_j - x_k)^2 + \beta(x_j - x_k)^4) \right) \psi = E\psi$$

where Δ is the *Laplacian*

$$\Delta := \frac{\partial^2}{\partial x_1^2} + \frac{\partial^2}{\partial x_2^2} + \frac{\partial^2}{\partial x_3^2}.$$

This means we consider three particles of equal mass m moving in a single spatial dimension with the potential V given by

$$V(x_j, x_k) = \alpha(x_j - x_k)^2 + \beta(x_j - x_k)^4, \qquad \beta > 0.$$

Let

$$X := \frac{1}{3}(x_1 + x_2 + x_3)$$

be the *center of mass* of the system. We introduce the coordinate system X, ρ, θ

$$r_1 = x_1 - X = \sqrt{\frac{2}{3}}\rho\cos(\theta - 2\pi/3)$$

$$r_2 = x_2 - X = \sqrt{\frac{2}{3}}\rho\cos(\theta + 2\pi/3)$$

$$r_3 = x_3 - X = \sqrt{\frac{2}{3}}\rho\cos(\theta).$$

(i) Find

$$\sum_{j=1}^{3} r_j, \qquad \sum_{j=1}^{3} r_j^2, \qquad \sum_{j=1}^{3} r_j^4$$

and

$$\sum_{j \leq k=1}^{3} (x_j - x_k)^2, \qquad \sum_{j \leq k=1}^{3} (x_j - x_k)^4.$$

(ii) Rewrite the metric tensor field

$$g = dx_1 \otimes dx_1 + dx_2 \otimes dx_2 + dx_3 \otimes dx_3$$

using the coordinates X, ρ, θ.

(iii) Rewrite the Laplacian Δ using the coordinates X, ρ, θ.

Solution 21. (i) We obtain

$$\sum_{j=1}^{3} r_j = 0, \qquad \sum_{j=1}^{3} r_j^2 = \rho^2, \qquad \sum_{j=1}^{3} r_j^4 = \frac{1}{2}\rho^4.$$

We obtain

$$\sum_{j \leq k=1}^{3} (x_j - x_k)^2 = \frac{1}{2}\sum_{j,k=1}^{3} (r_j - r_k)^2 = \frac{1}{2}\sum_{j,k=1}^{3} (r_j^2 + r_k^2 - 2r_j r_k) = 3\sum_{j=1}^{3} r_j^2 = 3\rho^2$$

and

$$\sum_{j \le k=1}^{3} (x_j - x_k)^4 = \frac{1}{2} \sum_{j,k=1}^{3} (r_j - r_k)^4$$

$$= \frac{1}{2} \sum_{j,k=1}^{3} (r_j^4 + r_k^4 - 4r_j^3 r_k - 4r_j r_k^3 + 6r_j^2 r_k^2)$$

$$= 3 \sum_{j=1}^{3} r_j^4 + 3 \left(\sum_{j=1}^{3} r_j^2 \right)^2 = \frac{9}{2} \rho^4 .$$

(iii) We find the metric tensor field

$$g = 3dX \otimes dX + d\rho \otimes d\rho + \rho^2 d\theta \otimes d\theta .$$

(iv) The Laplace operator takes the form

$$\Delta = \frac{1}{3} \frac{\partial^2}{\partial X^2} + \frac{1}{\rho} \frac{\partial}{\partial \rho} \left(\rho \frac{\partial}{\partial \rho} \right) + \frac{1}{\rho^2} \frac{\partial^2}{\partial \theta^2} .$$

Thus the wave function $\psi(X, \rho, \theta)$ is separable since the potential only depends on ρ, i.e. $\psi(X, \rho, \theta) = U(\rho)W(\theta)Y(X)$.

Problem 22. Let $n \ge 2$. The *braid group* $\mathcal{B}_n$ on n strings is given by the generators $\sigma_1, \sigma_2, \ldots, \sigma_{n-1}$ and the defining relations

$$\sigma_i \sigma_{i+1} \sigma_i = \sigma_{i+1} \sigma_i \sigma_{i+1} \quad \text{for} \quad 1 \le i \le n-2$$

and

$$\sigma_i \sigma_j = \sigma_j \sigma_i \quad \text{for} \quad |i - j| \ge 2 \quad \text{and} \quad 1 \le i, j \le n-1 .$$

Let $\mathbb{Z}[t, t^{-1}]$ be the Laurent polynomial ring over the integer numbers. Let $n \ge 4$. We define the reduced *Burau representation*

$$R : \mathcal{B}_n \to GL(n-1, \mathbb{Z}[t, t^{-1}])$$

by

$$R(bb') = R(b)R(b')$$

$$R(\sigma_1) = \begin{pmatrix} -t & 1 \\ 0 & 1 \end{pmatrix} \oplus I_{n-3}, \qquad R(\sigma_{n-1}) = I_{n-3} \oplus \begin{pmatrix} 1 & 0 \\ t & -t \end{pmatrix}$$

$$R(\sigma_i) = I_{i-2} \oplus \begin{pmatrix} 1 & 0 & 0 \\ t & -t & 1 \\ 0 & 0 & 1 \end{pmatrix} \oplus I_{n-i-2}$$

where $\oplus$ denotes the direct sum and I_{i-2} is the $i-2 \times i-2$ unit matrix. Consider the case $n = 4$.

(i) Give $R(\sigma_1)$, $R(\sigma_2)$, $R(\sigma_3)$.

(ii) Show that

$$R(\sigma_1)R(\sigma_2)R(\sigma_1) = R(\sigma_2)R(\sigma_1)R(\sigma_2).$$

(iii) Show that

$$R(\sigma_1)R(\sigma_3) = R(\sigma_3)R(\sigma_1).$$

(iv) Find the inverse matrices $R^{-1}(\sigma_1)$, $R^{-1}(\sigma_2)$, $R^{-1}(\sigma_3)$.

Solution 22.   (i) We have the 3×3 matrices

$$R(\sigma_1) = \begin{pmatrix} -t & 1 & 0 \\ 0 & 1 & 0 \\ 0 & 0 & 1 \end{pmatrix}, \quad R(\sigma_2) = \begin{pmatrix} 1 & 0 & 0 \\ t & -t & 1 \\ 0 & 0 & 1 \end{pmatrix}, \quad R(\sigma_3) = \begin{pmatrix} 1 & 0 & 0 \\ 0 & 1 & 0 \\ 0 & t & -t \end{pmatrix}.$$

(ii) We find the *braid like relation*

$$R(\sigma_1)R(\sigma_2)R(\sigma_1) = \begin{pmatrix} 0 & -t & 1 \\ -t^2 & 0 & 1 \\ 0 & 0 & 1 \end{pmatrix} = R(\sigma_2)R(\sigma_1)R(\sigma_2).$$

(iii) We have

$$R(\sigma_1)R(\sigma_3) = \begin{pmatrix} -t & 1 & 0 \\ 0 & 1 & 0 \\ 0 & t & -t \end{pmatrix} = R(\sigma_3)R(\sigma_1).$$

(iv) For the inverse matrices we find

$$R^{-1}(\sigma_1) = \begin{pmatrix} -1/t & 1/t & 0 \\ 0 & 1 & 0 \\ 0 & 0 & 1 \end{pmatrix}$$

$$R^{-1}(\sigma_2) = \begin{pmatrix} 1 & 0 & 0 \\ 1 & -1/t & 1/t \\ 0 & 0 & 1 \end{pmatrix}$$

$$R^{-1}(\sigma_3) = \begin{pmatrix} 1 & 0 & 0 \\ 0 & 1 & 0 \\ 0 & 1 & -1/t \end{pmatrix}.$$

Problem 23.   Let $X_1, X_2, \ldots, X_n$ be the basis of a finite dimensional Lie algebra. Let $\epsilon \in \mathbb{R}$. We define

$$f_{jk}(\epsilon) := \exp(\epsilon X_j) X_k \exp(-\epsilon X_j), \qquad j, k = 1, 2, \ldots, n.$$

Find the derivative $df_{jk}(\epsilon)/d\epsilon$ and express it using the structure constants.

Solution 23. We have

$$\frac{df_{jk}}{d\epsilon} = \exp(\epsilon X_j)[X_j, X_k]\exp(-\epsilon X_j) = e^{\epsilon X_j}\sum_{\ell=1}^{n}C_{jk}^{\ell}X_{\ell}e^{-\epsilon X_j} = \sum_{\ell=1}^{n}C_{jk}^{\ell}f_{j\ell}.$$

This is a system of linear differential equations with constant coefficients.

Problem 24. Consider the simple Lie algebra $so(3)$ with the basis $\{X_1, X_2, X_3\}$, where $[X_1, X_2] = X_3$, $[X_2, X_3] = X_1$, $[X_3, X_1] = X_2$. Let $\epsilon \in \mathbb{R}$. Calculate

$$f_{12}(\epsilon) = e^{\epsilon X_1}X_2e^{-\epsilon X_1}$$
$$f_{23}(\epsilon) = e^{\epsilon X_2}X_3e^{-\epsilon X_2}$$
$$f_{31}(\epsilon) = e^{\epsilon X_3}X_1e^{-\epsilon X_3}$$
$$f_{21}(\epsilon) = e^{\epsilon X_2}X_1e^{-\epsilon X_2}$$
$$f_{32}(\epsilon) = e^{\epsilon X_3}X_2e^{-\epsilon X_3}$$
$$f_{13}(\epsilon) = e^{\epsilon X_1}X_3e^{-\epsilon X_1}.$$

Use the technique of parameter differentiation and solve the initial value problem of the autonomous system of differential equations.

Solution 24. Differentiation with respect to ϵ yields

$$\frac{df_{12}(\epsilon)}{d\epsilon} = e^{\epsilon X_1}[X_1, X_2]e^{-\epsilon X_1} = e^{\epsilon X_1}X_3e^{-\epsilon X_1} = f_{13}(\epsilon)$$

$$\frac{df_{23}(\epsilon)}{d\epsilon} = e^{\epsilon X_2}[X_2, X_3]e^{-\epsilon X_2} = e^{\epsilon X_2}X_1e^{-\epsilon X_2} = f_{21}(\epsilon)$$

$$\frac{df_{31}(\epsilon)}{d\epsilon} = e^{\epsilon X_3}[X_3, X_1]e^{-\epsilon X_3} = e^{\epsilon X_3}X_2e^{-\epsilon X_3} = f_{32}(\epsilon)$$

$$\frac{df_{21}(\epsilon)}{d\epsilon} = e^{\epsilon X_2}[X_2, X_1]e^{-\epsilon X_2} = -e^{\epsilon X_2}X_3e^{-\epsilon X_2} = -f_{23}(\epsilon)$$

$$\frac{df_{32}(\epsilon)}{d\epsilon} = e^{\epsilon X_3}[X_3, X_2]e^{-\epsilon X_3} = -e^{\epsilon X_3}X_1e^{-\epsilon X_3} = -f_{31}(\epsilon)$$

$$\frac{df_{13}(\epsilon)}{d\epsilon} = e^{\epsilon X_1}[X_1, X_3]e^{-\epsilon X_1} = -e^{\epsilon X_1}X_2e^{-\epsilon X_1} = -f_{12}(\epsilon).$$

Thus we have the autonomous system of linear differential equations

$$\begin{pmatrix} df_{12}/d\epsilon \\ df_{23}/d\epsilon \\ df_{31}/d\epsilon \\ df_{21}/d\epsilon \\ df_{32}/d\epsilon \\ df_{13}/d\epsilon \end{pmatrix} = \begin{pmatrix} 0 & 0 & 0 & 0 & 0 & 1 \\ 0 & 0 & 0 & 1 & 0 & 0 \\ 0 & 0 & 0 & 0 & 1 & 0 \\ 0 & -1 & 0 & 0 & 0 & 0 \\ 0 & 0 & -1 & 0 & 0 & 0 \\ -1 & 0 & 0 & 0 & 0 & 0 \end{pmatrix} \begin{pmatrix} f_{12} \\ f_{23} \\ f_{31} \\ f_{21} \\ f_{32} \\ f_{13} \end{pmatrix}.$$

We denote the 6×6 matrix on the right-hand side by A. Then $A^2 = -I_6$. Therefore $A^3 = -A$, $A^4 = I_6$. Thus $\exp(\epsilon A) = I_6 \cos \epsilon + A \sin \epsilon$. Therefore the solution of the initial value problem is

$$
\begin{pmatrix} f_{12}(\epsilon) \\ f_{23}(\epsilon) \\ f_{31}(\epsilon) \\ f_{21}(\epsilon) \\ f_{32}(\epsilon) \\ f_{13}(\epsilon) \end{pmatrix} = (I_6 \cos \epsilon + A \sin \epsilon) \begin{pmatrix} f_{12} \\ f_{23} \\ f_{31} \\ f_{21} \\ f_{32} \\ f_{13} \end{pmatrix}.
$$

It follows that

$$
\begin{aligned}
e^{\epsilon X_1} X_2 e^{-\epsilon X_1} &= X_2 \cos \epsilon + X_3 \sin \epsilon \\
e^{\epsilon X_2} X_3 e^{-\epsilon X_2} &= X_3 \cos \epsilon + X_1 \sin \epsilon \\
e^{\epsilon X_3} X_1 e^{-\epsilon X_3} &= X_1 \cos \epsilon + X_2 \sin \epsilon \\
e^{\epsilon X_2} X_1 e^{-\epsilon X_2} &= X_1 \cos \epsilon - X_3 \sin \epsilon \\
e^{\epsilon X_3} X_2 e^{-\epsilon X_3} &= X_2 \cos \epsilon - X_1 \sin \epsilon \\
e^{\epsilon X_1} X_3 e^{-\epsilon X_1} &= X_3 \cos \epsilon - X_2 \sin \epsilon.
\end{aligned}
$$

Problem 25.   Consider the simple Lie algebra $su(1,1)$ with the basis $\{K_0, K_+, K_-\}$, where

$$
[K_0, K_+] = K_+, \qquad [K_0, K_-] = -K_-, \qquad [K_-, K_+] = 2K_0.
$$

Let $\epsilon \in \mathbb{R}$. Calculate

$$
\begin{aligned}
f_{0+}(\epsilon) &= e^{\epsilon K_0} K_+ e^{-\epsilon K_0} \\
f_{0-}(\epsilon) &= e^{\epsilon K_0} K_- e^{-\epsilon K_0} \\
f_{+-}(\epsilon) &= e^{\epsilon K_+} K_- e^{-\epsilon K_+} \\
f_{+0}(\epsilon) &= e^{\epsilon K_+} K_0 e^{-\epsilon K_+} \\
f_{-0}(\epsilon) &= e^{\epsilon K_-} K_0 e^{-\epsilon K_-} \\
f_{-+}(\epsilon) &= e^{\epsilon K_-} K_+ e^{-\epsilon K_-}.
\end{aligned}
$$

Use the technique of parameter differentiation and solve the initial value problem of the autonomous system of differential equations.

Solution 25.   Differentiation with respect to ϵ yields

$$
\frac{df_{0+}}{d\epsilon} = e^{\epsilon K_0}[K_0, K_+]e^{-\epsilon K_0} = e^{\epsilon K_0} K_+ e^{-\epsilon K_0} = f_{0+}(\epsilon)
$$

$$
\frac{df_{0-}}{d\epsilon} = e^{\epsilon K_0}[K_0, K_-]e^{-\epsilon K_0} = -e^{\epsilon K_0} K_- e^{-\epsilon K_0} = -f_{0-}(\epsilon)
$$

$$\frac{df_{+-}}{d\epsilon} = e^{\epsilon K_+}[K_+, K_-]e^{-\epsilon K_+} = -2e^{\epsilon K_+}K_0 e^{-\epsilon K_+} = -2f_{+0}(\epsilon)$$

$$\frac{df_{+0}}{d\epsilon} = e^{\epsilon K_+}[K_+, K_0]e^{-\epsilon K_+} = -e^{\epsilon K_+}K_+ e^{-\epsilon K_+} = -K_+$$

$$\frac{df_{-0}}{d\epsilon} = e^{\epsilon K_-}[K_-, K_0]e^{-\epsilon K_-} = e^{\epsilon K_-}K_- e^{-\epsilon K_-} = K_-$$

$$\frac{df_{-+}}{d\epsilon} = e^{\epsilon K_-}[K_-, K_+]e^{-\epsilon K_-} = 2e^{\epsilon K_-}K_0 e^{-\epsilon K_-} = 2f_{-0}(\epsilon).$$

The initial value problem $f_{0+}(0) = K_+$, $f_{0-}(0) = K_-$, $f_{+-}(0) = K_-$, $f_{+0}(0) = K_0$, $f_{-0}(0) = K_0$, $f_{-+}(0) = K_+$ can be easily solved. We obtain

$$f_{0+}(\epsilon) = K_+ e^{\epsilon}$$
$$f_{0-}(\epsilon) = K_- e^{-\epsilon}$$
$$f_{+0}(\epsilon) = -K_+\epsilon + K_0$$
$$f_{-0}(\epsilon) = K_-\epsilon + K_0$$
$$f_{+-}(\epsilon) = K_+\epsilon^2 - 2K_0\epsilon + K_-$$
$$f_{-+}(\epsilon) = K_-\epsilon^2 + 2K_0\epsilon + K_+.$$

It follows that

$$e^{\epsilon K_0}K_+ e^{-\epsilon K_0} = K_+ e^{\epsilon}$$
$$e^{\epsilon K_0}K_- e^{-\epsilon K_0} = K_- e^{-\epsilon}$$
$$e^{\epsilon K_+}K_- e^{-\epsilon K_+} = K_+\epsilon^2 - 2K_0\epsilon + K_-$$
$$e^{\epsilon K_+}K_0 e^{-\epsilon K_+} = -K_+\epsilon + K_0$$
$$e^{\epsilon K_-}K_0 e^{-\epsilon K_-} = K_-\epsilon + K_0$$
$$e^{\epsilon K_-}K_+ e^{-\epsilon K_-} = K_-\epsilon^2 + 2K_0\epsilon + K_+.$$

Problem 26. Consider the Lie algebra $s\ell(2, \mathbb{R})$ with the basis H, X, Y and the commutators $[H, X] = 2X$, $[H, Y] = -2Y$, $[X, Y] = H$, where

$$X = \begin{pmatrix} 0 & 1 \\ 0 & 0 \end{pmatrix}, \quad Y = \begin{pmatrix} 0 & 0 \\ 1 & 0 \end{pmatrix}, \quad H = \begin{pmatrix} 1 & 0 \\ 0 & -1 \end{pmatrix}.$$

Let $\epsilon \in \mathbb{R}$. Calculate

$$f_{XY}(\epsilon) = e^{\epsilon X}Y e^{-\epsilon X}$$
$$f_{XH}(\epsilon) = e^{\epsilon X}H e^{-\epsilon X}$$
$$f_{YX}(\epsilon) = e^{\epsilon Y}X e^{-\epsilon Y}$$
$$f_{HX}(\epsilon) = e^{\epsilon H}X e^{-\epsilon H}$$
$$f_{YH}(\epsilon) = e^{\epsilon Y}H e^{-\epsilon Y}$$
$$f_{HY}(\epsilon) = e^{\epsilon H}Y e^{-\epsilon H}$$

using the technique of parameter differentiation. Note that the "initial conditions" are $f_{YX}(0) = f_{HX}(0) = X$, $f_{XY}(0) = f_{HY}(0) = Y$, $f_{XH}(0) = f_{YH}(0) = H$.

Solution 26. Using the commutation relations we find

$$\frac{df_{XY}}{d\epsilon} = e^{\epsilon X}[X, Y]e^{-\epsilon X} = f_{XH}$$

$$\frac{df_{XH}}{d\epsilon} = e^{\epsilon X}[X, H]e^{-\epsilon X} = -2X$$

$$\frac{df_{YX}}{d\epsilon} = e^{\epsilon Y}[Y, X]e^{-\epsilon Y} = -f_{YH}$$

$$\frac{df_{HX}}{d\epsilon} = e^{\epsilon H}[H, X]e^{-\epsilon H} = 2f_{HX}$$

$$\frac{df_{YH}}{d\epsilon} = e^{\epsilon Y}[Y, H]e^{-\epsilon Y} = 2Y$$

$$\frac{df_{HY}}{d\epsilon} = e^{\epsilon H}[H, Y]e^{-\epsilon H} = -2f_{HY}.$$

The solution of the initial value problem is

$$f_{XY}(\epsilon) = -\epsilon^2 X + \epsilon H + Y$$
$$f_{XH}(\epsilon) = -2\epsilon X + H$$
$$f_{YX}(\epsilon) = -\epsilon^2 Y - \epsilon H + X$$
$$f_{HX}(\epsilon) = e^{2\epsilon} X$$
$$f_{YH}(\epsilon) = 2\epsilon Y + H$$
$$f_{HY}(\epsilon) = e^{-2\epsilon} Y.$$

Problem 27. Consider the non-abelian finite dimensional Lie algebra L over $\mathbb{C}$ with the basis $\{X_1, \ldots, X_n\}$ and the commutators

$$[X_j, X_k] = \sum_{m=1}^{n} C_{jk}^m X_m, \qquad j, k = 1, 2, \ldots, n. \tag{1}$$

Consider the exponential operators

$$\hat{O}_I(\boldsymbol{\alpha}) = \exp\left(\sum_{k=1}^{n} \alpha_k X_k\right) \tag{2}$$

and

$$\hat{O}_{II}(\boldsymbol{\beta}) = \prod_{k=1}^{n} \exp(\beta_k X_k). \tag{3}$$

Let $\epsilon \in \mathbb{R}$. Consider

$$\hat{O}_I(\epsilon\boldsymbol{\alpha}) = \exp\left(\epsilon \sum_{k=1}^{n} \alpha_k X_k\right) \tag{4}$$

and

$$\hat{O}_{II}(\boldsymbol{\beta}(\epsilon)) = \prod_{k=1}^{n} \exp(\beta_k(\epsilon) X_k). \tag{5}$$

Impose the condition

$$\hat{O}_I(\epsilon\boldsymbol{\alpha}) = \hat{O}_{II}(\boldsymbol{\beta}(\epsilon)) \tag{6}$$

and find the system of differential equations for $\beta_1(\epsilon), \beta_2(\epsilon), \ldots, \beta_n(\epsilon)$.

Solution 27. From the condition (6) using differentiation with respect to ϵ we obtain

$$\frac{d\hat{O}_{II}(\boldsymbol{\beta}(\epsilon))}{d\epsilon} \hat{O}_{II}^{-1}(\boldsymbol{\beta}(\epsilon)) = \frac{d\hat{O}_I(\epsilon\boldsymbol{\alpha})}{d\epsilon} \hat{O}_I^{-1}(\epsilon\boldsymbol{\alpha}).$$

Thus

$$\frac{d\hat{O}_{II}(\boldsymbol{\beta}(\epsilon))}{d\epsilon} \hat{O}_{II}^{-1}(\boldsymbol{\beta}(\epsilon)) = \sum_{k=1}^{n} \alpha_k X_k.$$

Now

$$\frac{d\hat{O}_{II}(\boldsymbol{\beta}(\epsilon))}{d\epsilon} \hat{O}_{II}^{-1}(\boldsymbol{\beta}(\epsilon)) = \frac{d\beta_1}{d\epsilon} X_1 + \frac{d\beta_2}{d\epsilon} e^{\beta_1 X_1} X_2 e^{-\beta_1 X_1}$$

$$+ \frac{d\beta_3}{d\epsilon} e^{\beta_1 X_1} e^{\beta_2 X_2} X_3 e^{-\beta_2 X_2} e^{-\beta_1 X_1}$$

$$\vdots$$

$$+ \frac{d\beta_k}{d\epsilon} e^{\beta_1 X_1} \cdots e^{\beta_k X_{k-1}} X_k e^{-\beta_k X_{k-1}} \cdots e^{-\beta_1 X_1}$$

$$\vdots$$

$$+ \frac{d\beta_n}{d\epsilon} e^{\beta_1 X_1} \cdots e^{\beta_{n-1} X_{n-1}} X_n e^{-\beta_{n-1} X_{n-1}} \cdots e^{-\beta_1 X_1}.$$

Problem 28. Apply the technique from the previous problem to the two-dimensional Lie algebra with basis X_1, X_2 and $[X_1, X_2] = X_1$ using parameter differentiation.

Solution 28. Let $\epsilon \in \mathbb{R}$. We obtain

$$\frac{d\hat{O}_{II}(\boldsymbol{\beta}(\epsilon))}{d\epsilon} = e^{\beta_1(\epsilon) X_1} X_1 e^{\beta_2(\epsilon) X_2} \frac{d\beta_1}{d\epsilon} + e^{\beta_1(\epsilon) X_1} e^{\beta_2(\epsilon) X_2} X_2 \frac{d\beta_2}{d\epsilon}.$$

Since
$$\hat{O}_{II}^{-1}(\beta(\epsilon)) = e^{-\beta_2(\epsilon)X_2} e^{-\beta_1(\epsilon)X_1}$$

we have

$$\frac{d\hat{O}_{II}(\beta(\epsilon))}{d\epsilon} \hat{O}_{II}^{-1}(\beta(\epsilon)) = X_1 \frac{d\beta_1}{d\epsilon} + e^{\beta_1(\epsilon)X_1} X_2 e^{-\beta_1(\epsilon)X_1} \frac{d\beta_2}{d\epsilon} \,.$$

From the commutation relation $[X_1, X_2] = X_1$ we have

$$e^{\beta_1(\epsilon)X_1} X_2 e^{-\beta_1(\epsilon)X_1} = X_2 + \beta_1(\epsilon)X_1 \,.$$

Therefore

$$X_1 \frac{d\beta_1}{d\epsilon} + (X_2 + \beta_1(\epsilon)X_1)\frac{d\beta_2}{d\epsilon} = \alpha_1 X_1 + \alpha_2 X_2 \,.$$

Comparing coefficients of X_1 and X_2 yields

$$\frac{d\beta_1}{d\epsilon} + \beta_1 \frac{d\beta_2}{d\epsilon} = \alpha_1, \qquad \frac{d\beta_2}{d\epsilon} = \alpha_2 \,.$$

Consequently

$$\frac{d\beta_1}{d\epsilon} = -\beta_1\alpha_2 + \alpha_1, \qquad \frac{d\beta_2}{d\epsilon} = \alpha_2$$

with the initial conditions $\beta_1(0) = 0$, $\beta_2(0) = 0$. The solution is

$$\beta_1(\epsilon) = e^{-\alpha_2\epsilon} + \alpha_1\epsilon - 1, \qquad \beta_2(\epsilon) = \alpha_2\epsilon \,.$$

Thus

$$\exp(\epsilon(\alpha_1 X_1 + \alpha_2 X_2)) = \exp((e^{-\alpha_2\epsilon} + \alpha_1\epsilon - 1)X_1)\exp(\epsilon\alpha_2 X_2) \,.$$

With $\epsilon = 1$ we have

$$\exp(\alpha_1 X_1 + \alpha_2 X_2) = \exp((e^{-\alpha_2} + \alpha_1 - 1)X_1)\exp(\alpha_2 X_2) \,.$$

Problem 29. Let A, H be $n \times n$ hermitian matrices, where H plays the role of the Hamilton operator. The *Heisenberg equation of motion* is given by

$$\frac{dA(t)}{dt} = \frac{i}{\hbar}[H, A(t)] \,.$$

with $A = A(t = 0) = A(0)$ and the solution of the initial value problem

$$A(t) = e^{iHt/\hbar} A e^{-iHt/\hbar} \,.$$

Let E_j $(j = 1, 2, \ldots, n^2)$ be an orthonormal basis in the *Hilbert space* $\mathcal{H}$ of the $n \times n$ matrices with scalar product

$$\langle X, Y \rangle := \operatorname{tr}(XY^*), \qquad X, Y \in \mathcal{H}.$$

Now $A(t)$ can be expanded using this orthonormal basis as

$$A(t) = \sum_{j=1}^{n^2} c_j(t) E_j$$

and H can be expanded as

$$H = \sum_{j=1}^{n^2} h_j E_j.$$

Find the time evolution for the coefficients $c_j(t)$, i.e. dc_j/dt, where $j = 1, 2, \ldots, n^2$.

Solution 29. We have

$$\frac{dA(t)}{dt} = \sum_{j=1}^{n^2} \frac{dc_j}{dt} E_j.$$

Inserting this equation and the expansion for H into the Heisenberg equation of motion we arrive at

$$\sum_{j=1}^{n^2} \frac{dc_j}{dt} E_j = \frac{i}{\hbar} \sum_{k=1}^{n^2} \sum_{k=1}^{n^2} h_k c_j(t) [E_k, E_j].$$

Taking the scalar product of the left and right-hand side of this equation with E_ℓ gives

$$\sum_{j=1}^{n^2} \frac{dc_j(t)}{dt} \operatorname{tr}(E_j E_\ell^*) = \frac{i}{\hbar} \sum_{k=1}^{n^2} \sum_{k=1}^{n^2} h_k c_j(t) \operatorname{tr}(([E_k, E_j]) E_\ell^*)$$

where $\ell = 1, 2, \ldots, n^2$. Since $\operatorname{tr}(E_j E_\ell^*) = \delta_{j\ell}$ we obtain

$$\frac{dc_\ell}{dt} = \frac{i}{\hbar} \sum_{k=1}^{n^2} \sum_{j=1}^{n^2} h_k c_j(t) \operatorname{tr}(E_k E_j E_\ell^* - E_j E_k E_\ell^*)$$

where $\ell = 1, 2, \ldots, n^2$.

Problem 30.   Let A, B be $n \times n$ matrices over $\mathbb{C}$. Then we have

$$e^A B e^{-A} = B + [A, B] + \frac{1}{2!}[A, [A, B]] + \frac{1}{3!}[A, [A, [A, B]]] + \cdots \quad (1)$$

and

$$e^A B e^A = B + [A, B]_+ + \frac{1}{2!}[A, [A, B]_+]_+ + \frac{1}{3!}[A[A, [A, B]_+]_+]_+ + \cdots \quad (2)$$

where $[,]$ is the commutator and $[,]_+$ denotes the anticommutator. From (2) we find

$$e^A B e^{-A} = \left(B + [A, B]_+ + \frac{1}{2!}[A, [A, B]_+]_+ + \frac{1}{3!}[A[A, [A, B]_+]_+]_+ + \cdots \right) e^{-2A} \quad (3)$$

$$e^A B e^{-A} = e^{2A} \left(B - [A, B]_+ + \frac{1}{2!}[A, [A, B]_+]_+ - \frac{1}{3!}[A[A, [A, B]_+]_+]_+ + \cdots \right). \quad (4)$$

Let σ_1 and σ_3 be the Pauli spin matrices

$$\sigma_1 = \begin{pmatrix} 0 & 1 \\ 1 & 0 \end{pmatrix}, \qquad \sigma_3 = \begin{pmatrix} 1 & 0 \\ 0 & -1 \end{pmatrix}.$$

Calculate

$$\widetilde{\sigma}_1 = e^{i\sigma_3\phi/2} \sigma_1 e^{-i\sigma_3\phi/2}$$

where $\phi \in \mathbb{R}$.

Solution 30.   Since the anticommutator of σ_1 and σ_3 vanishes, i.e. $[\sigma_1, \sigma_3]_+ = 0$ we utilize equation (3) (or (4)) to calculate $\widetilde{\sigma}_1$. We obtain

$$e^{i\sigma_3\phi/2} \sigma_1 e^{-i\sigma_3\phi/2} = \sigma_1 e^{-i\sigma_3\phi}$$

$$= \begin{pmatrix} 0 & 1 \\ 1 & 0 \end{pmatrix} \begin{pmatrix} e^{-i\phi} & 0 \\ 0 & e^{i\phi} \end{pmatrix}$$

$$= \begin{pmatrix} 0 & e^{i\phi} \\ e^{-i\phi} & 0 \end{pmatrix}.$$

Problem 31.   Consider the Lie algebra $s\ell(2, \mathbb{R})$ with the basis $\{H, X, Y\}$ and the commutators

$$[H, X] = 2X, \quad [H, Y] = -2Y, \quad [X, Y] = H$$

where

$$X = \begin{pmatrix} 0 & 1 \\ 0 & 0 \end{pmatrix}, \quad Y = \begin{pmatrix} 0 & 0 \\ 1 & 0 \end{pmatrix}, \quad H = \begin{pmatrix} 1 & 0 \\ 0 & -1 \end{pmatrix}.$$

(i) Let $\alpha, \beta, \gamma \in \mathbb{R}$. Calculate $\exp(\alpha H)$, $\exp(\beta X)$, $\exp(\gamma Y)$.

(ii) Calculate $\exp(\alpha H)\exp(\beta X)\exp(\gamma Y)$.

Solution 31. (i) Since H is a diagonal matrix we have

$$\exp\left(\alpha\begin{pmatrix} 1 & 0 \\ 0 & -1 \end{pmatrix}\right) = \begin{pmatrix} e^\alpha & 0 \\ 0 & e^{-\alpha} \end{pmatrix}.$$

Since

$$\begin{pmatrix} 0 & 1 \\ 0 & 0 \end{pmatrix}\begin{pmatrix} 0 & 1 \\ 0 & 0 \end{pmatrix} = \begin{pmatrix} 0 & 0 \\ 0 & 0 \end{pmatrix}, \qquad \begin{pmatrix} 0 & 0 \\ 1 & 0 \end{pmatrix}\begin{pmatrix} 0 & 0 \\ 1 & 0 \end{pmatrix} = \begin{pmatrix} 0 & 0 \\ 0 & 0 \end{pmatrix}$$

we find

$$\exp\left(\beta\begin{pmatrix} 0 & 1 \\ 0 & 0 \end{pmatrix}\right) = \begin{pmatrix} 1 & \beta \\ 0 & 1 \end{pmatrix}, \qquad \exp\left(\gamma\begin{pmatrix} 0 & 0 \\ 1 & 0 \end{pmatrix}\right) = \begin{pmatrix} 1 & 0 \\ \gamma & 1 \end{pmatrix}.$$

(ii) Using the result from (i) we obtain

$$\begin{pmatrix} e^\alpha & 0 \\ 0 & e^{-\alpha} \end{pmatrix}\begin{pmatrix} 1 & \beta \\ 0 & 1 \end{pmatrix}\begin{pmatrix} 1 & 0 \\ \gamma & 1 \end{pmatrix} = \begin{pmatrix} e^\alpha + \beta\gamma e^\alpha & e^\alpha\beta \\ e^{-\alpha}\gamma & e^{-\alpha} \end{pmatrix}.$$

The determinant of this matrix is obviously 1.

Problem 32. Let $\alpha, \beta, \gamma \in \mathbb{R}$ and

$$X = \begin{pmatrix} 0 & 1 \\ 0 & 0 \end{pmatrix}, \quad Y = \begin{pmatrix} 0 & 0 \\ 1 & 0 \end{pmatrix}, \quad H = \begin{pmatrix} 1 & 0 \\ 0 & -1 \end{pmatrix}.$$

Consider the matrix

$$M(\alpha, \beta, \gamma) = \alpha H + \beta X + \gamma Y = \begin{pmatrix} \alpha & \beta \\ \gamma & -\alpha \end{pmatrix}$$

with trace 0. Thus M is an element of $s\ell(2, \mathbb{R})$. Calculate

$$\exp(M(\alpha, \beta, \gamma)).$$

Solution 32. We have

$$M^2 = \begin{pmatrix} \alpha^2 + \beta\gamma & 0 \\ 0 & \alpha^2 + \beta\gamma \end{pmatrix} = (\alpha^2 + \beta\gamma)I_2$$

and

$$M^3 = \begin{pmatrix} \alpha(\alpha^2 + \beta\gamma) & \beta(\alpha^2 + \beta\gamma) \\ \gamma(\alpha^2 + \beta\gamma) & -\alpha(\alpha^2 + \beta\gamma) \end{pmatrix}.$$

We set $s := \sqrt{\alpha^2 + \beta\gamma}$. Then we have

$$e^M = \begin{pmatrix} \cosh(\sqrt{s}) + \frac{\alpha}{\sqrt{s}}\sinh(\sqrt{s}) & \frac{\beta}{\sqrt{s}}\sinh(\sqrt{s}) \\ \frac{\gamma}{\sqrt{s}}\sinh(\sqrt{s}) & \cosh(\sqrt{s}) - \frac{\alpha}{\sqrt{s}}\sinh(\sqrt{s}) \end{pmatrix}.$$

Since $\mathrm{tr}(M) = 0$, the determinant of e^M is obviously 1.

Problem 33. Let I_n be the $n \times n$ unit matrix and 0_n be the $n \times n$ zero matrix. Show that the $2n \times 2n$ matrices

$$\tau_1 = \begin{pmatrix} 0_n & I_n \\ I_n & 0_n \end{pmatrix}, \quad \tau_2 = \begin{pmatrix} 0_n & -iI_n \\ iI_n & 0_n \end{pmatrix}, \quad \tau_3 = \begin{pmatrix} I_n & 0_n \\ 0_n & -I_n \end{pmatrix}$$

form a basis of a Lie algebra under the commutator.

Solution 33. Consider the commutators of the τ_i matrices

$$[\tau_1, \tau_2] = 2i\tau_3, \quad [\tau_1, \tau_3] = -2i\tau_2, \quad [\tau_2, \tau_3] = 2i\tau_1.$$

The Jacobi identity is always satisfied for $n \times n$ matrices. Thus the matrices τ_1, τ_2, τ_3 form a basis of a Lie algebra under the commutator. Is the Lie algebra simple?

Problem 34. Let z_1, z_2 be complex variables. Show that differential operators (vector fields)

$$J_1 = \frac{1}{2}\left(z_1\frac{\partial}{\partial z_2} + z_2\frac{\partial}{\partial z_1}\right), \quad J_2 = \frac{1}{2}i\left(z_1\frac{\partial}{\partial z_2} - z_2\frac{\partial}{\partial z_1}\right),$$

$$J_3 = \frac{1}{2}\left(z_2\frac{\partial}{\partial z_2} - z_1\frac{\partial}{\partial z_1}\right)$$

form a basis of a Lie algebra under the commutator.

Solution 34. We obtain

$$[J_1, J_2] = \frac{1}{2}i\left(z_2\frac{\partial}{\partial z_2} - z_1\frac{\partial}{\partial z_1}\right) = iJ_3$$

$$[J_1, J_3] = \frac{1}{2}\left(z_1\frac{\partial}{\partial z_2} - z_2\frac{\partial}{\partial z_1}\right) = -iJ_2$$

$$[J_2, J_3] = \frac{1}{2}i\left(z_1\frac{\partial}{\partial z_2} + z_2\frac{\partial}{\partial z_1}\right) = iJ_1.$$

The Jacobi identity is satisfied for all vector fields. Thus the vector fields J_k form a basis of a simple Lie algebra under the commutator.

Problem 35. Let $j = 0, 1/2, 1, 3/2, \ldots$. Show that the differential operators

$$J_+ = 2jz - z^2 \frac{\partial}{\partial z}, \quad J_- = \frac{\partial}{\partial z}, \quad J_z = z \frac{\partial}{\partial z} - j$$

form a basis of a Lie algebra under the commutator.

Solution 35. We obtain

$$[J_+, J_-] = 2J_z, \quad [J_+, J_z] = -J_+, \quad [J_-, J_z] = J_- .$$

The Jacobi identity is also satisfied.

Problem 36. Let $j = 0, 1/2, 1, 3/2, \ldots$. Show that the differential operators

$$J_+ = z \left(j - z \frac{\partial}{\partial z} \right), \quad J_- = \frac{1}{z} \left(j + z \frac{\partial}{\partial z} \right), \quad J_z = z \frac{\partial}{\partial z}$$

form a basis of a Lie algebra under the commutator.

Solution 36. We obtain

$$[J_+, J_-] = 2J_z, \quad [J_+, J_z] = -J_+, \quad [J_-, J_z] = J_- .$$

The Jacobi identity is also satisfied.

Problem 37. Consider $C^\infty(\mathbb{R}^{2n})$ functions which depend on $2n$ variables

$$(\mathbf{p}, \mathbf{q}) = (q_1, \ldots, q_n, p_1, \ldots, p_n) .$$

These variables are known as the *phase space variables* where q_j denotes the position and p_j the momentum. The commutator of such two functions f and g is defined by the *Poisson bracket*

$$[f, g] := \sum_{j=1}^{n} \left(\frac{\partial f}{\partial q_j} \frac{\partial g}{\partial p_j} - \frac{\partial f}{\partial p_j} \frac{\partial g}{\partial q_j} \right) .$$

The vector space of $C^\infty(\mathbb{R}^{2n})$ functions $f(\mathbf{p}, \mathbf{q})$ form an infinite dimensional Lie algebra under the Poisson bracket. Consider the case with $n = 2$ and the 10 functions

$$f_{12} = q_1 p_2 - q_2 p_1$$

$$f_{13} = \frac{1}{\sqrt{2p_0}} \left(\left(\frac{1}{2} \mathbf{p}^2 - p_0 \right) q_1 - (\mathbf{q} \cdot \mathbf{p}) p_1 \right)$$

$$f_{23} = \frac{1}{\sqrt{2p_0}} \left(\left(\frac{1}{2} \mathbf{p}^2 - p_0 \right) q_2 - (\mathbf{q} \cdot \mathbf{p}) p_2 \right)$$

$$f_{14} = \frac{1}{\sqrt{2p_0}} \left(\left(\frac{1}{2}\mathbf{p}^2 + p_0 \right) q_1 - (\mathbf{q} \cdot \mathbf{p})p_1 \right)$$

$$f_{24} = \frac{1}{\sqrt{2p_0}} \left(\left(\frac{1}{2}\mathbf{p}^2 + p_0 \right) q_2 - (\mathbf{q} \cdot \mathbf{p})p_2 \right)$$

$$f_{34} = (\mathbf{q} \cdot \mathbf{p})$$

$$f_{15} = rp_1$$

$$f_{25} = rp_2$$

$$f_{35} = \frac{1}{\sqrt{2p_0}} \left(\frac{1}{2}\mathbf{p}^2 - p_0 \right) r$$

$$f_{45} = \frac{1}{\sqrt{2p_0}} \left(\frac{1}{2}\mathbf{p}^2 + p_0 \right) r$$

where p_0 is a constant and

$$\mathbf{p}^2 := p_1^2 + p_2^2, \quad r := \sqrt{q_1^2 + q_2^2}, \quad \mathbf{q} \cdot \mathbf{p} := q_1 p_1 + q_2 p_2 \,.$$

(i) Calculate the Poisson brackets and thus show that we have a basis for the Lie algebra $so(3, 2)$.
(ii) Find sub Lie algebras.

Solution 37.   (i) Let

$$g_{11} = g_{22} = g_{33} = 1, \quad g_{44} = g_{55} = -1, \quad g_{ij} = 0, \ i \neq j \,.$$

Then we have for the non-zero commutators

$$[f_{ij}, f_{ik}] = g_{ii} f_{jk}, \quad f_{ij} = -f_{ji}, \quad i, j, k = 1, \ldots, 5 \,.$$

The Lie algebra plays a role for the *Kepler problem*.
(ii) Lie sub algebras are given by $so(3)$ with f_{12}, f_{13}, f_{23} and by $so(1, 2)$ with f_{34}, f_{35}, f_{45}.

Problem 38.   Consider $C^\infty(\mathbb{R}^{2n})$ functions which depend on the $2n$ variables

$$(\mathbf{q}, \mathbf{p}) = (q_1, \ldots, q_n, p_1, \ldots, p_n) \,.$$

These variables are known as the phase space variables where q_j denotes the position and q_j denotes the momentum. The commutator of two such functions f and g is defined by the *Poisson bracket*

$$[f(\mathbf{p}, \mathbf{q}), g(\mathbf{p}, \mathbf{q})] = \sum_{j=1}^{n} \left(\frac{\partial f}{\partial q_j} \frac{\partial g}{\partial p_j} - \frac{\partial f}{\partial p_j} \frac{\partial g}{\partial q_j} \right) \,.$$

The vector space of $C^\infty(\mathbb{R}^{2n})$ functions $f(\mathbf{p}, \mathbf{q})$ form a Lie algebra under the Poisson bracket.
(i) Let

$$f(\mathbf{p}, \mathbf{q}) = \frac{1}{2}p_1^2 + \frac{1}{2}p_2^2 + \frac{1}{12}(q_1^4 + q_2^4) + \frac{1}{2}q_1^2 q_2^2$$

and

$$g(\mathbf{p}, \mathbf{q}) = 3p_1 p_2 + q_1 q_2 (q_1^2 + q_2^2).$$

Find $[f, g]$. Discuss.
(ii) Consider the *Toda lattice* with Hamilton function

$$H(\mathbf{p}, \mathbf{q}) = \frac{1}{2}(p_1^2 + p_2^2 + p_3^2) + e^{q_1 - q_2} + e^{q_2 - q_3} + e^{q_3 - q_1}.$$

Let

$$I_1(\mathbf{p}, \mathbf{q}) = p_1 + p_2 + p_3$$
$$I_2(\mathbf{p}, \mathbf{q}) = p_1 p_2 + p_2 p_3 + p_3 p_1 - e^{q_1 - q_2} - e^{q_2 - q_3} - e^{q_3 - q_1}$$
$$I_3(\mathbf{p}, \mathbf{q}) = p_1 p_2 p_3 - p_1 e^{q_2 - q_3} - p_2 e^{q_3 - q_1} - p_3 e^{q_1 - q_2}.$$

Find the Poisson brackets. Discuss.

Solution 38. (i) We obtain

$$[f(\mathbf{p}, \mathbf{q}), g(\mathbf{p}, \mathbf{q})] = 0.$$

The function f can be considered as a Hamilton function, where $p_1^2/2 + p_2^2/2$ is the kinetic term. Since $[f, g] = 0$ we say that g is a first integral of the Hamilton system.
(ii) We obtain

$$[H, I_1] = [H, I_2] = [H, I_3] = 0.$$

Thus I_1, I_2, I_3 are first integrals of the Hamilton system. We find $[I_1, I_2] = 0$, $[I_2, I_3] = 0$, $[I_3, I_1] = 0$. Thus the system of first integrals is in involution. Thus the Toda system is *Liouville integrable*.

Problem 39. The *Moyal bracket* (also called *sine bracket*) is defined as

$$[f(\mathbf{p}, \mathbf{q}), g(\mathbf{p}, \mathbf{q})]_{MB} := f(\mathbf{p}, \mathbf{q}) \frac{2}{\hbar} \sin\left(\frac{\hbar}{2}(\overleftarrow{\partial_{\mathbf{q}}} \cdot \overrightarrow{\partial_{\mathbf{p}}} - \overleftarrow{\partial_{\mathbf{p}}} \cdot \overrightarrow{\partial_{\mathbf{q}}})\right) g(\mathbf{p}, \mathbf{q})$$

where

$$\partial_{\mathbf{q}} \cdot \partial_{\mathbf{p}} := \sum_{j=1}^{n} \frac{\partial}{\partial q_j} \frac{\partial}{\partial p_j}, \qquad \partial_{\mathbf{p}} \cdot \partial_{\mathbf{q}} := \sum_{j=1}^{n} \frac{\partial}{\partial p_j} \frac{\partial}{\partial q_j}.$$

The Moyal bracket reduces to the Poisson bracket when $\hbar \to 0$. The Moyal bracket is antisymmetric in its arguments f and g. It satisfies the Jacobi identity. The Moyal bracket defines an infinite dimensional Lie algebra. Consider the Hamilton function

$$H(\mathbf{p}, \mathbf{q}) = \frac{1}{2}p_1^2 + \frac{1}{2}p_2^2 + \frac{16}{3}q_2^3 + q_1q_2^2 + \mu q_1^{-2}.$$

Let

$$I(\mathbf{p}, \mathbf{q}) = p_1^4 + 4(q_1^2 q_2 + \mu q_1^{-2})p_1^2 - \frac{4}{3}q_1^3 p_1 p_2 - \frac{4}{3}q_1^4 q_2^2 + \frac{8}{3}\mu q_2 - \frac{2}{9}q_1^6 + 4\mu^2 q_1^{-4}.$$

Find the Moyal bracket. Discuss.

Solution 39.   For the Poisson bracket we obtain $[H, I] = 0$. The Moyal bracket $[H, I]_{MB}$ does not vanish, but we have an $\hbar^2$ term. We can define

$$\widetilde{I} = I + \hbar^2 J(q_1, q_2).$$

Then the condition that $[H, \widetilde{I}]_{MB}$ vanish provides the conditions

$$\frac{\partial J}{\partial q_2} = 0 \quad \text{and} \quad \frac{\partial J}{\partial q_1} = 24q_1^{-5}\mu.$$

Thus $[H, \widetilde{I}]_{MB} = 0$, where

$$\widetilde{I} = I - 6\hbar^2 \mu q_1^{-4}.$$

Problem 40.   Consider the *Hénon-Heiles Hamilton function*

$$H(p_1, p_2, x_1, x_2) = \frac{1}{2}(p_1^2 + p_2^2 + x_1^2 + x_2^2) + cx_1^2 x_2 - \frac{1}{3}x_2^3$$

with the equations of motion

$$\frac{d^2 x_1}{dt^2} = -x_1 - 2cx_1 x_2, \qquad \frac{d^2 x_2}{dt^2} = -x_2 - cx_1^2 + x_2^2.$$

Show that the equations of motions are invariant for $c = 1$ under the $2\pi/3$ rotation

$$\begin{pmatrix} x_1' \\ x_2' \end{pmatrix} = \begin{pmatrix} 1/2 & \sqrt{3}/2 \\ -\sqrt{3}/2 & 1/2 \end{pmatrix} \begin{pmatrix} x_1 \\ x_2 \end{pmatrix}.$$

Solution 40.   This rotation of space induces the same rotation of momentum. Thus

$$\begin{pmatrix} p_1' \\ p_2' \end{pmatrix} = \begin{pmatrix} 1/2 & \sqrt{3}/2 \\ -\sqrt{3}/2 & 1/2 \end{pmatrix} \begin{pmatrix} p_1 \\ p_2 \end{pmatrix}.$$

The expression for the Hamilton function consists of 3 terms. The first two

$$\frac{1}{2}(p_1^2 + p_2^2), \qquad \frac{1}{2}(x_1^2 + x_2^2)$$

are invariants of any rotation, because rotations preserve the Euclidean metric. The last term

$$x_1^2 x_2 - \frac{1}{3}x_2^3$$

is equal to the imaginary part of the function $z^3/3$ (where $z = x_1 + ix_2$). Rotation of the complex plane by $2\pi/3$ yields $z' = e^{i2\pi/3}z$. Thus z^3 is an invariant. Its imaginary part is also invariant. Thus the Hamilton function is invariant as a sum of three invariant terms. The equations of motion are also invariant under rotation by $2\pi/3$.

Problem 41. Let $\mathbf{x}$, $\mathbf{y}$, $\mathbf{z}$ be unit vectors in the Euclidean space $\mathbb{R}^2$. Find the maximum of the function

$$f(\mathbf{x}, \mathbf{y}, \mathbf{z}) = \|\mathbf{x} - \mathbf{y}\|^2 + \|\mathbf{y} - \mathbf{z}\|^2 + \|\mathbf{z} - \mathbf{x}\|^2$$

with respect to the constraints

$$\|\mathbf{x}\| = \|\mathbf{y}\| = \|\mathbf{z}\| = 1$$

i.e., $\mathbf{x}$, $\mathbf{y}$, $\mathbf{z}$ are unit vectors. Give an interpretation of the result. Solve the problem using the *Lagrange multiplier method*. Discuss the discrete symmetries of the problem.

Solution 41. Let λ_x, λ_y, λ_z be the Lagrange's multipliers. Then from

$$L(\mathbf{x}, \mathbf{y}, \mathbf{z}) = \|\mathbf{x} - \mathbf{y}\|^2 + \|\mathbf{y} - \mathbf{z}\|^2 + \|\mathbf{z} - \mathbf{x}\|^2 - \lambda_x\|\mathbf{x}\|^2 - \lambda_y\|\mathbf{y}\|^2 - \lambda_z\|\mathbf{z}\|^2$$

we obtain the 9 equations

$$2\mathbf{x} - \mathbf{y} - \mathbf{z} - \lambda_x\mathbf{x} = 0$$
$$2\mathbf{y} - \mathbf{z} - \mathbf{x} - \lambda_y\mathbf{y} = 0$$
$$2\mathbf{z} - \mathbf{x} - \mathbf{y} - \lambda_z\mathbf{z} = 0.$$

Thus the vectors $\mathbf{x}$, $\mathbf{y}$, $\mathbf{z}$ are linearly dependent. This means that they lie in the plane, which goes through the origin. Since either $\mathbf{x} = \alpha\mathbf{y}$ or $\mathbf{x} = \beta\mathbf{z}$ for some $\alpha, \beta \in \mathbb{R}$ we find

$$\lambda_x = \lambda_y = \lambda_z = 3, \qquad \mathbf{x} + \mathbf{y} + \mathbf{z} = \mathbf{0}.$$

The three vectors $\mathbf{x}$, $\mathbf{y}$, $\mathbf{z}$ have equal lengths and their sum is the zero vector. Consider the triangle with sides $\mathbf{x}$, $\mathbf{y}$, $\mathbf{z}$. The triangle is equilateral.

Thus all angles between vectors are equal. Three vectors **x**, **y**, **z** are vertices of another equilateral triangle. Its sides are equal to $\sqrt{3}$ and the maximum value of the function f is

$$(\sqrt{3})^2 + (\sqrt{3})^2 + (\sqrt{3})^2 = 9.$$

Problem 42.   The D_4 symmetry consists of four rotations each of 90 degrees and four reflection mirrors with each angle between them being 45 degrees. Consider the system of second order ordinary differential equations

$$\frac{d^2 x_1}{dt^2} = -x_1 + (\mu - x_1^2 - x_2^2)\frac{dx_1}{dt}$$

$$\frac{d^2 x_2}{dt^2} = -x_2 + (\mu - x_2^2 - x_1^2)\frac{dx_2}{dt}$$

where $\mu \in \mathbb{R}$ is a real bifurcation parameter. Show that as μ varies through zero, there is a *Hopf bifurcation* with D_4 symmetry.

Solution 42.   We rewrite this system as the system of 4 first order equations

$$\frac{dx_1}{dt} = v_1$$

$$\frac{dx_2}{dt} = v_2$$

$$\frac{dv_1}{dt} = -x_1 + (\mu - x_1^2 - x_2^2)v_1$$

$$\frac{dv_2}{dt} = -x_2 + (\mu - x_2^2 - x_1^2)v_2.$$

This system has a fixed point at $(x_1^*, x_2^*, v_1^*, v_2^*) = (0, 0, 0, 0)$. The Jacobian matrix at the fixed point is

$$J = \begin{pmatrix} 0 & 0 & 1 & 0 \\ 0 & 0 & 0 & 1 \\ -1 & 0 & \mu & 0 \\ 0 & -1 & 0 & \mu \end{pmatrix}.$$

Its characteristic polynomial is

$$\lambda^4 - 2\mu\lambda^3 + \lambda^2(\mu^2 + 2) - 2\mu\lambda + 1 = 0.$$

We see that when $\mu = 0$, the characteristic equation has two double imaginary roots in the points $\pm i$. The general roots are also double

$$\lambda = \frac{1}{2}(\mu \pm \sqrt{\mu^2 - 4}).$$

When $\mu > 0$ the real parts of the roots are also positive and the fixed point is a repeller. When $\mu < 0$ the real parts of the roots are also negative and the fixed point is an attractor. Thus by $\mu = 0$ the system suffers a Hopf bifurcation and we expect the presence of a limit cycle. This limit cycle is the unit circle. All the rest of the paths are invariants of D_4, because the dynamical system has the symmetries

$$x_1' = x_2, \qquad x_2' = x_1$$

$$x_1' = -x_1, \qquad x_2 = x_2.$$

All elements of D_4 are products of these generators. Thus as μ varies through zero, there is a Hopf bifurcation with D_4 symmetry.

Problem 43. Let $\sigma \in \mathbb{R}$. Show that the differential operators

$$J_+ = -t\left(w\frac{\partial}{\partial w} - t\frac{\partial}{\partial t} + \sigma\right), \quad J_- = -\frac{1}{t}\left((w-1)\frac{\partial}{\partial w} + t\frac{\partial}{\partial t} + \sigma\right)$$

$$J_3 = t\frac{\partial}{\partial t}$$

form a basis of semisimple Lie algebra. The *Casimir operator* is given by

$$C = J_3^2 + J_+J_- - J_3.$$

Calculate C.

Solution 43. We find

$$[J_3, J_\pm] = \pm J_\pm, \qquad [J_+, J_-] = 2J_3.$$

The Casimir operator is

$$C = w(w-1)\frac{\partial^2}{\partial w^2} + t\frac{\partial^2}{\partial w \partial t} + (\sigma+1)(2w-1)\frac{\partial}{\partial w} + \sigma(\sigma+1).$$

Thus $[C, J_3] = [C, J_+] = [C, J_-] = 0$.

Problem 44. Let $\mu \in \mathbb{R}$ and

$$\sigma_z = \begin{pmatrix} 1 & 0 \\ 0 & -1 \end{pmatrix}, \qquad I_2 = \begin{pmatrix} 1 & 0 \\ 0 & 1 \end{pmatrix}.$$

Let $x \in \mathbb{R}$. Show that the operators in a product space

$$T_+ = e^x\left(i\frac{d}{dx} + \frac{1}{2} + \mu\right) \otimes \sigma_z, \quad T_- = e^{-x}\left(-i\frac{d}{dx} + \frac{1}{2} + \mu\right) \otimes \sigma_z$$

$$T_3 = i\frac{d}{dx} \otimes I_2$$

form a basis of a Lie algebra under the commutator, where $\otimes$ denotes the tensor product.

Solution 44.   Note that $\sigma_z^2 = I_2$. Thus we obtain

$$[T_+, T_-] = 2iT_3, \quad [T_+, T_3] = -iT_+, \quad [T_-, T_3] = iT_- .$$

The Jacobi identity is also satisfied

$$[T_+, [T_-, T_3]] + [T_3, [T_+, T_-]] + [T_-, [T_3, T_+]] = 0 .$$

Problem 45.   Consider the Hamilton function

$$H(p, q) = \frac{p^2}{2m} + K\frac{q^4}{4}$$

for the *anharmonic oscillator*. The equations of motion are given by

$$\frac{dp}{dt} = -\frac{\partial H}{\partial q} = -Kq^3, \qquad \frac{dq}{dt} = \frac{\partial H}{\partial p} = \frac{p}{m}$$

with the initial values $p(0)$ and $q(0)$.
(i) Show that H is a first integral.
(ii) Solve the initial value problem using *Jacobi elliptic functions*

Solution 45.   (i) We have

$$\frac{dH}{dt} = \frac{p}{m}\frac{dp}{dt} + Kq^3\frac{dq}{dt} = \frac{p}{m}(-Kq^3) + Kq^3\frac{p}{m} = 0 .$$

Thus H is a first integral and represents the (total) energy E of the system. This means $H(t) = H(0) = E$.
(ii) We have to integrate

$$\frac{dp}{-Kq^3} = \frac{dq}{p/m} = \frac{dt}{1}$$

where we use the first integral given in (i). Let (energy)

$$E = \frac{p^2(0)}{2m} + \frac{Kq^4(0)}{4}$$

and

$$\omega = \left(\frac{K}{2m}\right)^{1/2}, \qquad C = \left(\frac{4E}{K}\right)^{1/4} .$$

We obtain as the solution for the initial value problem

$$q(t) = \frac{p(0)/(m\omega)\mathrm{sn}(\omega t, i) + q(0)\mathrm{cn}(\omega t, i)\mathrm{dn}(\omega t, i)}{1 + (q(0)/C)^2\mathrm{sn}^2(\omega t, i)}$$

$$p(t) = \frac{p(0)\mathrm{cn}(\omega t, i)\mathrm{dn}(\omega t, i)(1 - (q(0)/C)^2\mathrm{sn}^2(\omega t, i))}{(1 + (q(0)/C)^2\mathrm{sn}^2(\omega t, i))^2}$$

$$- \frac{2m\omega(q(0)/C)\mathrm{sn}(\omega t, i)((q(0)/C)^2 + \mathrm{sn}^2(\omega t, i))}{(1 + (q(0)/C)^2\mathrm{sn}^2(\omega t, i))^2}$$

where sn, cn and dn are the Jacobi elliptic functions. Thus we have a Lie transformation group.

Problem 46. Consider the two-dimensional Euclidean space with the metric tensor field

$$g = dx \otimes dx + dy \otimes dy.$$

(i) Show that g is invariant under the transformation

$$x' = x\cos(\alpha) - y\sin(\alpha), \qquad y' = x\sin(\alpha) + y\cos(\alpha).$$

(ii) Find the vector field

$$V = V_1(x, y)\frac{\partial}{\partial x} + V_2(x, y)\frac{\partial}{\partial y}$$

such that $L_V g = 0$. Such vector fields are called *Killing vector fields*. The Lie derivative $L_V(.)$ is linear and obeys the product rule.

Solution 46. (i) Since $\sin^2(\alpha) + \cos^2(\alpha) = 1$ we have

$$dx' \otimes dx' + dy' \otimes dy' = (\cos(\alpha)dx - \sin(\alpha)dy) \otimes (\cos(\alpha)dx - \sin(\alpha)dy)$$
$$+ (\sin(\alpha)dx + \cos(\alpha)dy) \otimes (\sin(\alpha)dx + \cos(\alpha)dy)$$
$$= dx \otimes dx + dy \otimes dy.$$

(ii) From the condition $L_V g = 0$ we find the Killing vector fields

$$\frac{\partial}{\partial x}, \quad \frac{\partial}{\partial y}, \quad y\frac{\partial}{\partial x} - x\frac{\partial}{\partial y}.$$

Problem 47. Consider the metric tensor field (two-sphere S^2)

$$g = d\theta \otimes d\theta + \sin^2\theta d\phi \otimes d\phi.$$

(i) Find the *Killing vector fields* V of g, i.e. $L_V g = 0$.

(ii) Calculate the commutators of the Killing vector fields and thus show that we have a Lie algebra.

Solution 47. (i) The condition $L_V g = 0$ yields

$$\frac{\partial V_1}{\partial \theta} = 0, \quad \frac{\partial V_1}{\partial \phi} + \sin^2 \theta \frac{\partial V_2}{\partial \theta} = 0, \quad V_1 \cos \theta + \sin \theta \frac{\partial V_2}{\partial \phi} = 0.$$

The solutions of these equations provide the three vector fields

$$W_1 = \frac{\partial}{\partial \phi}, \quad W_2 = \cos \phi \frac{\partial}{\partial \theta} - \cot \theta \sin \phi \frac{\partial}{\partial \phi}, \quad W_3 = -\sin \phi \frac{\partial}{\partial \theta} - \cot \theta \cos \phi \frac{\partial}{\partial \phi}.$$

(ii) The commutators are given by

$$[W_1, W_2] = W_3, \qquad [W_2, W_3] = W_1, \qquad [W_3, W_1] = W_2.$$

Problem 48. The space of constant negative curvature is described by the upper half H of the plane of complex numbers $z = x + iy$ with $y > 0$. The distance ds between neighboring points z and $z + dz$ in H is given by the Riemannian metric

$$ds^2 = \frac{dx^2 + dy^2}{y^2}$$

i.e. the metric tensor field is given by

$$g = \frac{dx \otimes dx + dy \otimes dy}{y^2}.$$

(i) Find the *Killing vector fields* and show they form a basis of a Lie algebra.
(ii) The space H with the metric tensor field g is homogeneous all points and all directions at a given point are equivalent. Let $\zeta = \xi + i\eta$. Consider the transformation

$$\zeta = \frac{az + b}{cz + d}$$

where the real numbers a, b, c, d satisfy $ad - bc = 1$. Show that

$$\frac{d\xi \otimes d\xi + d\eta \otimes d\eta}{\eta^2} = \frac{dx \otimes dx + dy \otimes dy}{y^2}.$$

Solution 48. (i) Let

$$V = V_1(x, y) \frac{\partial}{\partial x} + V_2(x, y) \frac{\partial}{\partial y}$$

be the Killing vector field. Using

$$L_V dx = d(V \rfloor dx) = dV_1 = \frac{\partial V_1}{\partial x} dx + \frac{\partial V_1}{\partial y} dy$$

$$L_V dy = d(V \rfloor dy) = dV_2 = \frac{\partial V_2}{\partial x} dx + \frac{\partial V_2}{\partial y} dy$$

we obtain from $L_V g = 0$ the system of partial differential equations

$$\frac{2}{y^2} \left(-\frac{V_2}{y} + \frac{\partial V_1}{\partial x} \right) = 0$$

$$\frac{1}{y^2} \left(\frac{\partial V_1}{\partial y} + \frac{\partial V_2}{\partial x} \right) = 0$$

$$\frac{1}{y^2} \left(\frac{\partial V_1}{\partial y} + \frac{\partial V_2}{\partial x} \right) = 0$$

$$\frac{2}{y^2} \left(-\frac{V_2}{y} + \frac{\partial V_2}{\partial y} \right) = 0$$

by separating out $dx \otimes dx$, $dx \otimes dy$, $dy \otimes dx$, $dy \otimes dy$. Note that the second and third equation are the same. The solutions of this system of partial differential equations yield the three Killing vector fields

$$W_1 = \frac{\partial}{\partial x}, \quad W_2 = x\frac{\partial}{\partial x} + y\frac{\partial}{\partial y}, \quad W_3 = (x^2 - y^2)\frac{\partial}{\partial x} + 2xy\frac{\partial}{\partial y}.$$

The commutators are

$$[W_1, W_2] = -W_1, \quad [W_2, W_3] = -W_3, \quad [W_3, W_1] = 2W_2.$$

(ii) We have

$$d\zeta = \frac{dz}{|cz + d|^2}, \quad \eta = \frac{y}{|cz + d|^2}.$$

Thus the identity follows. Consider the transformation

$$\zeta = \frac{az + b}{cz + d}$$

where the real numbers a, b, c, d satisfy $ad - bc = 1$. Then

$$\bar\zeta = \frac{a\bar z + b}{c\bar z + d}.$$

Therefore

$$\begin{aligned} \Im(\zeta) &= \frac{1}{2i}(\zeta - \bar\zeta) = \frac{1}{2i} \left(\frac{az + b}{cz + d} - \frac{a\bar z + b}{c\bar z + d} \right) \\ &= \frac{1}{2i} \frac{(az + b)(c\bar z + d) - (a\bar z + b)(cz + d)}{(cz + d)(c\bar z + d)} \\ &= \frac{1}{2i} \frac{(ad - bc)(z - \bar z)}{(cz + d)(c\bar z + d)} = \frac{\Im(z)}{(cz + d)(c\bar z + d)}. \end{aligned}$$

Now

$$d\zeta = \frac{a(cz+d) - c(az+b)}{(cz+d)^2}dz = \frac{ad - bc}{(cz+d)^2}dz = \frac{1}{(cz+d)^2}dz$$

$$d\zeta \wedge \bar{d}\zeta = \frac{dz \wedge \bar{d}z}{(cz+d)^2(c\bar{z}+d)^2}.$$

Thus

$$dz \wedge \bar{d}z = (cz+d)^2(c\bar{z}+d)^2 d\zeta \wedge \bar{d}\zeta$$

$$\frac{dz \wedge \bar{d}z}{(\Im(z))^2} = \frac{(cz+d)^2(c\bar{z}+d)^2 d\zeta \wedge \bar{d}\zeta}{(\Im(z))^2} = \frac{d\zeta \wedge \bar{d}\zeta}{(\Im(\zeta))^2}.$$

Thus the Euclidean metric is invariant under this transformation.

Problem 49. The real hyperbolic space $H^2(\mathbb{R})$ is a noncompact symmetric space of rank one that can be expressed as a quotient of semisimple Lie groups as

$$H^2(\mathbb{R}) \simeq SL(2,\mathbb{R})/SO(2).$$

Various geometric models exist for $H^2(\mathbb{R})$. Consider the unit disk model

$$B_1(0) := \{\, \mathbf{y} \in \mathbb{R}^2 \ : \ |\mathbf{y}| < 1 \,\}$$

where $|\cdot|$ is the usual Riemannian metric in the vector space $\mathbb{R}^2$. Consider the unit disk $B_1(0)$ with the metric tensor field

$$g = \frac{4}{(1-|\mathbf{y}|^2)^2}(dy_1 \otimes dy_1 + dy_2 \otimes dy_2) = \frac{4}{(1-y_1^2-y_2^2)^2}(dy_1 \otimes dy_1 + dy_2 \otimes dy_2).$$

Find the *Killing vector fields* of g.

Solution 49. Let

$$V = V_1(y_1, y_2)\frac{\partial}{\partial y_1} + V_2(y_1, y_2)\frac{\partial}{\partial y_2}.$$

From the condition $L_V g = 0$ we obtain the three Killing vector fields

$$W_1 = (y_1^2 - y_2^2 - 1)\frac{\partial}{\partial y_1} + 2y_1 y_2 \frac{\partial}{\partial y_2}$$

$$W_2 = 2y_1 y_2 \frac{\partial}{\partial y_1} + (-y_1^2 + y_2^2 - 1)\frac{\partial}{\partial y_2}$$

$$R = -y_2\frac{\partial}{\partial y_1} + y_1\frac{\partial}{\partial y_2}.$$

Find the commutators.

Problem 50. Let $A = \mathbb{C}[z, z^{-1}]$ be the algebra of *Laurent polynomials* in z. If

$$p(z) = \sum_{k \in \mathbb{Z}} c_k z^k$$

is a Laurent polynomial, its *residue* of p is defined by

$$\text{Res}(p) := c_{-1}.$$

We have the properties

$$\text{Res}\left(\frac{1}{z}\right) = 1, \qquad \text{Res}\left(\frac{dp}{dz}\right) = 0.$$

The linear functional Res : $A \to \mathbb{C}$ defines a $\mathbb{C}$-bilinear function ϕ on A by

$$\phi(p, q) := \text{Res}\left(\frac{dp}{dz} q\right).$$

(i) Let $p, q \in A$. Show that $\phi(p, q) = -\phi(q, p)$.
(ii) Show that ("Jacobi identity")

$$\phi(pq, r) + \phi(qr, p) + \phi(rp, q) = 0 \quad p, q, r \in A.$$

Solution 50. (i) Since

$$\frac{d}{dz}(pq) = \frac{dp}{dz} q + p \frac{dq}{dz}$$

we have

$$
\begin{aligned}
0 &= \text{Res}(d(pq)/dz) \\
&= \text{Res}((dp/dz)q + p(dq/dz)) \\
&= \text{Res}((dp/dz)q) + \text{Res}(p(dq/dz)).
\end{aligned}
$$

Thus $\phi(p, q) = -\phi(q, p)$.
(ii) We have

$$
\begin{aligned}
\phi(pq, r) &= -\phi(qr, p) - \phi(pr, q) \\
\phi(qr, p) &= -\phi(pr, q) - \phi(pq, r) \\
\phi(rp, q) &= -\phi(pq, r) - \phi(qr, p).
\end{aligned}
$$

When we add these three expressions the sum of the right-hand is equal to the sum on the left-hand side with multiplier (-2). Thus the Jacobi identity holds.

Problem 51.   Consider the system of linear partial differential equations (*Knizhnik-Zamolodchikov equation*)

$$\frac{\partial}{\partial z_i}\Phi = k \sum_{\substack{j \neq i}}^{n} \frac{P^{(ij)}}{z_i - z_j}\Phi$$

where $\Phi = \Phi(z_1, \ldots, z_n)$ is a function of n complex independent variables, which takes values in the tensor product $V \otimes V \otimes \cdots \otimes V$ of N identical vector spaces V. $P^{(ij)}$ is the permutation of the i-th and j-th factors of the tensor product. This is a simple form of the KZ-equation Show that the KZ-equation is consistent. Consider the differential operators

$$O_i := \frac{\partial}{\partial z_i} - k \sum_{\substack{j \neq i}}^{n} \frac{P^{(ij)}}{z_i - z_j}, \qquad i = 1, \ldots, n.$$

Show that O_i and O_j commute.

Solution 51.   The commutator is

$$O_i O_j - O_j O_i = \left(\frac{\partial}{\partial z_i} - k \sum_{\substack{m \neq i}}^{n} \frac{P^{(im)}}{z_i - z_m} \right) \left(\frac{\partial}{\partial z_j} - k \sum_{\substack{m \neq j}}^{n} \frac{P^{(jm)}}{z_j - z_m} \right)$$

$$- \left(\frac{\partial}{\partial z_j} - k \sum_{\substack{m \neq j}}^{n} \frac{P^{(jm)}}{z_j - z_m} \right) \left(\frac{\partial}{\partial z_i} - k \sum_{\substack{m \neq i}}^{n} \frac{P^{(im)}}{z_i - z_m} \right)$$

$$= -\frac{\partial}{\partial z_i} k \sum_{\substack{m \neq j}}^{n} \frac{P^{(jm)}}{z_j - z_m} - k \sum_{\substack{m \neq i}}^{n} \frac{P^{(im)}}{z_i - z_m} \frac{\partial}{\partial z_j}$$

$$+ \frac{\partial}{\partial z_j} k \sum_{\substack{m \neq i}}^{n} \frac{P^{(im)}}{z_i - z_m} + k \sum_{\substack{m \neq j}}^{n} \frac{P^{(jm)}}{z_j - z_m} \frac{\partial}{\partial z_i}.$$

Since

$$\frac{\partial}{\partial z_i} k \sum_{\substack{m \neq j}}^{n} \frac{P^{(jm)}}{z_j - z_m} = \frac{\partial}{\partial z_i} k \frac{P^{(ji)}}{z_j - z_i} + k \sum_{\substack{m \neq j}}^{n} \frac{P^{(jm)}}{z_j - z_m} \frac{\partial}{\partial z_i}$$

$$= k \frac{P^{(ji)}}{(z_j - z_i)^2} + k \sum_{\substack{m \neq j}}^{n} \frac{P^{(jm)}}{z_j - z_m} \frac{\partial}{\partial z_i}$$

we obtain $O_i O_j - O_j O_i = 0$. The permutation operator $P^{(ij)}$ is symmetric $P^{(ij)} = P^{(ji)}$.

Problem 52.   Consider the group of volume preserving diffeomorphisms of $(S^1 \otimes S^1)$. This is an infinite group with composition as the group law.

We denote this group by SDiff. Its Lie algebra is the set of smooth vector fields on $(S^1 \otimes S^1)$ with zero divergence

$$\mathrm{div}(\mathbf{p}) = \frac{\partial p_1}{\partial x_1} + \frac{\partial p_2}{\partial x_2} = 0$$

and equipped with the commutator $[\mathbf{p}, \mathbf{q}] = \mathbf{p}(\mathbf{q}) - \mathbf{q}(\mathbf{p})$. . Show that the commutator of two vector fields with zero divergence is again a vector field with zero divergence.

Solution 52. For the commutator we find

$$\mathbf{p}(\mathbf{q}) - \mathbf{q}(\mathbf{p}) = \left(p_1 \frac{\partial q_1}{\partial x_1} + p_2 \frac{\partial q_1}{\partial x_2} - q_1 \frac{\partial p_1}{\partial x_1} - q_2 \frac{\partial p_1}{\partial x_2} \right) \frac{\partial}{\partial x_1}$$

$$+ \left(p_1 \frac{\partial q_2}{\partial x_1} + p_2 \frac{\partial q_2}{\partial x_2} - q_1 \frac{\partial p_2}{\partial x_1} - q_2 \frac{\partial p_2}{\partial x_2} \right) \frac{\partial}{\partial x_2}.$$

For the divergence we find

$$\frac{\partial}{\partial x_1} \left(p_1 \frac{\partial q_1}{\partial x_1} + p_2 \frac{\partial q_1}{\partial x_2} - q_1 \frac{\partial p_1}{\partial x_1} - q_2 \frac{\partial p_1}{\partial x_2} \right) +$$

$$\frac{\partial}{\partial x_2} \left(p_1 \frac{\partial q_2}{\partial x_1} + p_2 \frac{\partial q_2}{\partial x_2} - q_1 \frac{\partial p_2}{\partial x_1} - q_2 \frac{\partial p_2}{\partial x_2} \right) = 0$$

where we used that $\partial p_1 / p x_1 + \partial p_2 / \partial x_2 = 0$ and $\partial q_1 / \partial x_1 + \partial q_2 / \partial x_2 = 0$.

Problem 53. Consider the differential operators

$$J_+ = -x \left(y \frac{\partial}{\partial y} - x \frac{\partial}{\partial x} + n \right), \quad J_- = -\frac{1}{x} \left((y-1) \frac{\partial}{\partial y} + x \frac{\partial}{\partial x} + n \right)$$

$$J_3 = x \frac{\partial}{\partial x}$$

where n is an integer.
(i) Calculate the commutators $[J_+, J_-]$, $[J_+, J_3]$, $[J_-, J_3]$ and thus show that we have a basis of the Lie algebra $s\ell(2, \mathbb{R})$.
(ii) Find

$$\exp(\gamma J_3) J_+ \exp(-\gamma J_3), \qquad \exp(\gamma J_3) J_- \exp(-\gamma J_3)$$

$$\exp(\alpha J_+) J_- \exp(-\alpha J_+), \qquad \exp(\alpha J_+) J_3 \exp(-\alpha J_+)$$

$$\exp(\beta J_-) J_+ \exp(-\beta J_-), \qquad \exp(\beta J_-) J_3 \exp(-\beta J_-).$$

(iii) Use the result form (ii) to calculate

$$e^{\alpha J_+} e^{\beta J_-} (\lambda J_3 - m) e^{-\beta J_-} e^{-\alpha J_+}.$$

Solution 53. (i) For the commutators we obtain

$$[J_+, J_-] = 2J_3, \quad [J_3, J_+] = J_+, \quad [J_3, J_-] = -J_-.$$

Thus we have a basis of the Lie algebra $s\ell(2, \mathbb{R})$.
(ii) We obtain

$$\exp(\gamma J_3) J_+ \exp(-\gamma J_3) = \exp(\gamma) J_+$$
$$\exp(\gamma J_3) J_- \exp(-\gamma J_3) = \exp(-\gamma) J_-$$
$$\exp(\alpha J_+) J_- \exp(-\alpha J_+) = -\alpha^2 J_+ + 2\alpha J_3 + J_-$$
$$\exp(\alpha J_+) J_3 \exp(-\alpha J_+) = -\alpha J_+ + J_3$$
$$\exp(\beta J_-) J_+ \exp(-\beta J_-) = J_+ - \beta^2 J_- - 2\beta J_3$$
$$\exp(\beta J_-) J_3 \exp(-\beta J_-) = \beta J_- + J_3.$$

(iii) Using the result from (ii) we obtain

$$e^{\alpha J_+} e^{\beta J_-} (\lambda J_3 - m) e^{-\beta J_-} e^{-\alpha J_+} = \lambda(-\beta\alpha^2 - \alpha) J_+ + \lambda\beta J_- + \lambda(2\alpha\beta + 1) J_3 - m.$$

Problem 54. Consider the autonomous system of N first order ordinary differential equations

$$\frac{du_j}{dt} = au_{j+1}u_{j+2} + bu_{j-1}u_{j-2} + cu_{j+1}u_{j-1}$$

with the cyclic condition $u_j = u_{j+N}$ for $j = 1, \ldots, N$ and $a + b + c = 0$. Consider the case $N = 4$. Show that the system has three integrals, the energy

$$E = \frac{1}{2} \sum_{k=1}^{N} u_k^2$$

and the signs of $(u_1 - u_3)$ and $(u_2 - u_4)$.

Solution 54. For $N = 4$ we have

$$\frac{du_1}{dt} = au_2u_3 + bu_4u_3 + cu_2u_4$$
$$\frac{du_2}{dt} = au_3u_4 + bu_1u_4 + cu_3u_1$$
$$\frac{du_3}{dt} = au_4u_1 + bu_2u_1 + cu_2u_4$$
$$\frac{du_4}{dt} = au_1u_2 + bu_3u_2 + cu_3u_1$$

with the vector field

$$V = (au_2u_3 + bu_4u_3 + cu_2u_4)\frac{\partial}{\partial u_1} + (au_3u_4 + bu_1u_4 + cu_3u_1)\frac{\partial}{\partial u_2}$$

$$+ (au_4u_1 + bu_2u_1 + cu_2u_4)\frac{\partial}{\partial u_3} + (au_1u_2 + bu_3u_2 + cu_3u_1)\frac{\partial}{\partial u_4}.$$

Then

$$L_V E(\mathbf{x}) = \frac{d}{dt}E(\mathbf{x}) = 0.$$

We have

$$\frac{d(u_1 - u_3)}{dt} = -(u_2 + u_4)(u_1 - u_3)$$

$$\frac{d(u_2 - u_4)}{dt} = -(u_1 + u_3)(u_2 - u_4).$$

Problem 55. Consider the vector fields

$$J_+ = x^2\frac{\partial}{\partial x} - xy\frac{\partial}{\partial y}, \quad J_- = -\frac{\partial}{\partial x} - \frac{1}{x}(y-1)\frac{\partial}{\partial y}, \quad J_3 = x\frac{\partial}{\partial x}.$$

(i) Find the commutators and thus show that we have a basis of the Lie algebra $s\ell(2, \mathbb{R})$.
(ii) The vector field J_+ corresponds to the autonomous system of differential equations

$$\frac{dx}{d\epsilon} = x^2, \quad \frac{dy}{d\epsilon} = -xy.$$

Solve the initial value problem and thus find the Lie transformation group.
(iii) The vector field J_- corresponds to the autonomous system of differential equations

$$\frac{dx}{d\epsilon} = -1, \quad \frac{dy}{d\epsilon} = -\frac{1}{x}(y-1).$$

Solve the initial value problem and thus find the Lie transformation group.

Solution 55. (i) We obtain

$$[J_3, J_+] = J_+, \quad [J_3, J_-] = -J_-, \quad [J_+, J_-] = 2J_3.$$

(ii) Direct integration yields the solution of the initial value problem

$$x(\epsilon) = \frac{x(0)}{1 - \epsilon x(0)}, \quad y(\epsilon) = (1 - \epsilon x(0))y(0).$$

(iii) Direct integration yields the solution of the initial value problem

$$x(\epsilon) = x(0) - \epsilon, \quad y(\epsilon) = \frac{1}{x(0)}(y(0)(x(0) - \epsilon) + \epsilon).$$

Problem 56. Consider the differential operators

$$L := \frac{d^2}{dx^2} + f(x), \qquad B := \frac{d^3}{dx^3} + \left(\frac{3}{2}f + \frac{1}{2}x\right)\frac{d}{dx} + \frac{3}{4}\frac{df}{dx}.$$

From the condition $[L, B]\phi(x) = L\phi(x)$ find the ordinary differential equation that the function f obeys.

Solution 56. We obtain the ordinary differential equation

$$\frac{d^3 f}{dx^3} + 6f\frac{df}{dx} - 4f + 2x\frac{df}{dx} = 0.$$

This nonlinear differential equation also appears when we insert

$$u(s, t) = \frac{1}{(3t)^{2/3}} f\left(s(3x)^{-1/3}\right)$$

into the *Korteweg-de Vries equation*

$$\frac{\partial u}{\partial t} + 3u\frac{\partial u}{\partial s} + \frac{1}{2}\frac{\partial^3 u}{\partial s^3} = 0.$$

Problem 57. Consider the linear equations

$$\left(\frac{\partial}{\partial u} - U\right)\phi(u, v, \lambda) = 0, \qquad \left(\frac{\partial}{\partial v} - V\right)\phi(u, v, \lambda) = 0$$

where U, V are functions of u, v, λ. The linear operators

$$L = \frac{\partial}{\partial u} - U, \qquad M = \frac{\partial}{\partial v} - V$$

are the Lax pair. The integrability condition for ϕ is the *zero curvature condition* on the Lax pair

$$[L, M]\phi = 0.$$

Let $g(u, v)$ be an invertible 2×2 matrix and let

$$U = g^{-1}\frac{\partial g}{\partial u}, \qquad V = g^{-1}\frac{\partial g}{\partial v}.$$

Find the nonlinear differential equation for $[L, M] = 0$.

Solution 57. We obtain

$$\frac{\partial}{\partial v}\left(g^{-1}\frac{\partial g}{\partial u}\right) + \frac{\partial}{\partial u}\left(g^{-1}\frac{\partial g}{\partial v}\right) = 0.$$

This is the principal *chiral model equation*.

Problem 58. Consider the Lie algebra with the basis

$$J_+ = \begin{pmatrix} 0 & 1 \\ 0 & 0 \end{pmatrix}, \quad J_- = \begin{pmatrix} 0 & 0 \\ 1 & 0 \end{pmatrix}, \quad J_0 = \begin{pmatrix} 1/2 & 0 \\ 0 & 1/2 \end{pmatrix}$$

with the commutation relations

$$[J_+, J_-] = 2J_0, \quad [J_+, J_0] = -J_+, \quad [J_-, J_0] = J_- .$$

(i) Find the disentangled form

$$\exp(a_+ J_+ + a_0 J_0 + a_- J_-) = \exp(b_+ J_+) \exp(\ln(b_0) J_0) \exp(b_- J_-)$$

i.e. find b_+, b_-, b_0 as functions of a_+, a_-, a_0, where $b_0 > 0$.
(ii) Find the disentangled form

$$\exp(a_+ J_+ + a_0 J_0 + a_- J_-) = \exp(c_- J_-) \exp(\ln(c_0) J_0) \exp(c_+ J_+)$$

i.e. find c_-, c_+, c_0 as functions of a_+, a_-, a_0, where $c_0 > 0$.

Solution 58. (i) We have

$$\exp(a_+ J_+ + a_0 J_0 + a_- J_-) = \begin{pmatrix} \cosh(a) + \frac{a_0}{2a}\sinh(a) & \frac{a_+}{a}\sinh(a) \\ \frac{a_-}{a}\sinh(a) & \cosh(a) - \frac{a_0}{2a}\sinh(a) \end{pmatrix}$$

where $a^2 = \frac{1}{4}a_0^2 + a_+ a_-$ and

$$\exp(b_+ J_+)\exp(\ln(b_0)J_0)\exp(b_- J_-) = \frac{1}{\sqrt{b_0}}\begin{pmatrix} b_0 + b_+ b_- & b_+ \\ b_- & 1 \end{pmatrix}.$$

Comparing the matrices we obtain the equations

$$\frac{1}{\sqrt{b_0}}(b_0 + b_+ b_-) = \cosh(a) + \frac{1}{2}\frac{a_0}{a}\sinh(a)$$

$$\frac{1}{\sqrt{b_0}}b_+ = \frac{a_+}{a}\sinh(a)$$

$$\frac{1}{\sqrt{b_0}}b_- = \frac{a_-}{a}\sinh(a)$$

$$\frac{1}{\sqrt{b_0}} = \cosh(a) - \frac{1}{2}\frac{a_0}{a}\sinh(a).$$

With $a_0 = 0$ and $a_\pm = \mp\frac{1}{2}\delta$ we find

$$b_\pm = \mp\tan(\delta/2), \quad b_0 = \frac{1}{\cos^2(\delta/2)}.$$

(ii) We have

$$\exp(a_+J_+ + a_0J_0 + a_-J_-) = \begin{pmatrix} \cosh(a) + \frac{1}{2}\frac{a_0}{a}\sinh(a) & \frac{a_+}{a}\sinh(a) \\ \frac{a_-}{a}\sinh(a) & \cosh(a) - \frac{1}{2}\frac{a_0}{a}\sinh(a) \end{pmatrix}$$

where $a^2 = \frac{1}{4}a_0^2 + a_+a_-$ and

$$\exp(c_-J_-)\exp(\ln(c_0)J_0)\exp(c_+J_+) = \sqrt{c_0}\begin{pmatrix} 1 & c_+ \\ c_- & 1/c_0 + c_+c_- \end{pmatrix}.$$

Comparing the matrices we obtain the equations

$$\sqrt{c_0} = \cosh(a) + \frac{1}{2}\frac{a_0}{a}\sinh(a)$$

$$\sqrt{c_0}c_+ = \frac{a_+}{a}\sinh(a)$$

$$\sqrt{c_0}c_- = \frac{a_-}{a}\sinh(a)$$

$$\sqrt{c_0}(1/c_0 + c_+c_-) = \cosh(a) - \frac{1}{2}\frac{a_0}{a}\sinh(a)$$

With $a_0 = 0$ and $a_\pm = \mp\frac{1}{2}\delta$ we find

$$c_\pm = \mp\tan(\delta/2), \qquad c_0 = \cos^2(\delta/2).$$

Problem 59. Consider the smooth vector fields in $\mathbb{R}^2$

$$V = V_1(\mathbf{x})\frac{\partial}{\partial x_1} + V_2(\mathbf{x})\frac{\partial}{\partial x_2}, \qquad W = W_1(\mathbf{x})\frac{\partial}{\partial x_1} + W_2(\mathbf{x})\frac{\partial}{\partial x_2}.$$

Assume that $[V, W] = 0$. Calculate the Lie derivatives

$$L_V(V_1W_2 - V_2W_1), \qquad L_W(V_1W_2 - V_2W_1).$$

Discuss.

Solution 59. We have

$$L_V(V_1W_2 - V_2W_1) = V_1\frac{\partial}{\partial x_1}(V_1W_2 - V_2W_1) + V_2\frac{\partial}{\partial x_2}(V_1W_2 - V_2W_1)$$

$$= W_2\left(V_1\frac{\partial V_1}{\partial x_1} + V_2\frac{\partial V_1}{\partial x_2}\right) + V_1\left(V_1\frac{\partial W_2}{\partial x_1} + V_2\frac{\partial W_2}{\partial x_2}\right)$$

$$- W_1\left(V_1\frac{\partial V_2}{\partial x_1} + V_2\frac{\partial V_2}{\partial x_2}\right) - V_2\left(V_1\frac{\partial W_1}{\partial x_1} + V_2\frac{\partial W_1}{\partial x_2}\right).$$

and

$$L_W(V_1 W_2 - V_2 W_1) = W_1 \frac{\partial}{\partial x_1}(V_1 W_2 - V_2 W_1) + W_2 \frac{\partial}{\partial x_2}(V_1 W_2 - V_2 W_1)$$

$$= W_2 \left(W_1 \frac{\partial V_1}{\partial x_1} + W_2 \frac{\partial V_1}{\partial x_2} \right) + V_1 \left(W_1 \frac{\partial W_2}{\partial x_1} + W_2 \frac{\partial W_2}{\partial x_2} \right)$$

$$- W_1 \left(W_1 \frac{\partial V_2}{\partial x_1} + W_2 \frac{\partial V_2}{\partial x_2} \right) - V_2 \left(W_1 \frac{\partial W_1}{\partial x_1} + W_2 \frac{\partial W_1}{\partial x_2} \right).$$

Since $[V, W] = 0$ we have

$$V_1 \frac{\partial W_1}{\partial x_1} + V_2 \frac{\partial W_1}{\partial x_2} - W_1 \frac{\partial V_1}{\partial x_1} - W_2 \frac{\partial V_1}{\partial x_2} = 0$$

$$V_1 \frac{\partial W_2}{\partial x_1} + V_2 \frac{\partial W_2}{\partial x_2} - W_1 \frac{\partial V_2}{\partial x_1} - W_2 \frac{\partial V_2}{\partial x_2} = 0.$$

We apply these equalities and obtain

$$L_V(V_1 W_2 - V_2 W_1) = (V_1 W_2 - V_2 W_1) \left(\frac{\partial V_1}{\partial x_1} + \frac{\partial V_2}{\partial x_2} \right).$$

$$L_W(V_1 W_2 - V_2 W_1) = (V_1 W_2 - V_2 W_1) \left(\frac{\partial W_1}{\partial x_1} + \frac{\partial W_2}{\partial x_2} \right).$$

Problem 60. (i) Consider the externally driven *anharmonic oscillator*

$$\frac{d^2 u}{dt^2} + c_1 \frac{du}{dt} + c_2 u + u^3 = f(t)$$

where f is a smooth function of t. Find the conditions on c_1, c_2 and f such that

$$V = -\frac{c_1}{3} \exp\left(\frac{c_1 t}{3} \right) u \frac{\partial}{\partial u} + \exp\left(\frac{c_1 t}{3} \right) \frac{\partial}{\partial t}$$

is a Lie symmetry vector field of the differential equation.
(ii) Insert the *Painlevé expansion*

$$u(t) = \sum_{j=0}^{\infty} u_j (t - t_1)^{j-1}$$

into the differential equation considered in the complex domain. Find the condition on c_1, c_2, f such that the expansion has two arbitrary expansion coefficients.

Solution 60. (i) We obtain

$$c_2 = \frac{2c_1^2}{9}, \qquad \frac{df}{dt} + c_1 f = 0.$$

Thus $f(t) = C \exp(-c_1 t)$.

(ii) At the resonance 4 we obtain the condition

$$-27\sqrt{-2}\frac{df}{dt} - 27\sqrt{-2}c_1 f - 18c_2 c_1^2 + 4c_1^4 = 0.$$

The other free parameter is t_1. From the condition it follows that

$$c_2 = \frac{2c_1^2}{9}, \qquad \frac{df}{dt} + c_1 f = 0.$$

These are the same conditions as we found in (i).

Problem 61. Show that the one-dimensional wave equation

$$\frac{\partial^2 u}{\partial x_2^2} = \frac{\partial^2 u}{\partial x_1^2} \tag{1}$$

is invariant under the transformation

$$\begin{pmatrix} x_1' \\ x_2' \end{pmatrix} = \begin{pmatrix} \cosh \epsilon & \sinh \epsilon \\ \sinh \epsilon & \cosh \epsilon \end{pmatrix} \begin{pmatrix} x_1 \\ x_2 \end{pmatrix} \tag{2a}$$

$$u'(\mathbf{x}'(\mathbf{x})) = u(\mathbf{x}) \tag{2b}$$

where $\mathbf{x} = (x_1, x_2)$ and $\mathbf{x}' = (x_1', x_2')$.

Solution 61. To prove the invariance of the wave equation under the transformation (2) we utilize the *chain rule*

$$\frac{\partial u'}{\partial x_2} = \frac{\partial u'}{\partial x_1'}\frac{\partial x_1'}{\partial x_2} + \frac{\partial u'}{\partial x_2'}\frac{\partial x_2'}{\partial x_2} = \frac{\partial u}{\partial x_2}.$$

It follows that

$$\frac{\partial^2 u'}{\partial x_2^2} = \left(\frac{\partial^2 u'}{\partial x_1'^2}\frac{\partial x_1'}{\partial x_2} + \frac{\partial^2 u'}{\partial x_1' \partial x_2'}\frac{\partial x_2'}{\partial x_2} \right)\frac{\partial x_1'}{\partial x_2} + \left(\frac{\partial^2 u'}{\partial x_1' \partial x_2'}\frac{\partial x_1'}{\partial x_2} + \frac{\partial^2 u'}{\partial x_2'^2}\frac{\partial x_2'}{\partial x_2} \right)\frac{\partial x_2'}{\partial x_2}$$

$$= \frac{\partial^2 u}{\partial x_2^2}.$$

Analogously

$$\frac{\partial^2 u'}{\partial x_1^2} = \left(\frac{\partial^2 u'}{\partial x_1'^2}\frac{\partial x_1'}{\partial x_1} + \frac{\partial^2 u'}{\partial x_1' \partial x_2'}\frac{\partial x_2'}{\partial x_1} \right)\frac{\partial x_1'}{\partial x_1} + \left(\frac{\partial^2 u'}{\partial x_1' \partial x_2'}\frac{\partial x_1'}{\partial x_1} + \frac{\partial^2 u'}{\partial x_2'^2}\frac{\partial x_2'}{\partial x_1} \right)\frac{\partial x_2'}{\partial x_1}$$

$$= \frac{\partial^2 u}{\partial x_1^2}.$$

From (2) we have

$$\frac{\partial x_1'}{\partial x_2} = \sinh \varepsilon, \quad \frac{\partial x_1'}{\partial x_1} = \cosh \varepsilon, \quad \frac{\partial x_2'}{\partial x_2} = \cosh \varepsilon, \quad \frac{\partial x_2'}{\partial x_1} = \sinh \varepsilon.$$

Inserting these equations into the wave equation (1) we find

$$\frac{\partial^2 u'}{\partial x_2'^2} = \frac{\partial^2 u'}{\partial x_1'^2}.$$

Equation (1) is thus invariant under the transformation (2).

Problem 62. Consider the second order ordinary differential equation

$$\frac{d^2 u}{dt^2} = f(du/dt, u, t)$$

where f is an analytic function. Taking the time derivative we obtain

$$\frac{d^3 u}{dt^3} = \frac{\partial f}{\partial \dot{u}} \frac{d^2 u}{dt^2} + \frac{\partial f}{\partial u} \frac{du}{dt} + \frac{\partial f}{\partial t}.$$

Consider the hypersurfaces associated with these two differential equations

$$F_1(u, u_t, u_{tt}, t) = u_{tt} - f(u_t, u, t) = 0$$

$$F_2(u, u_t, u_{tt}, u_{ttt}, t) = u_{ttt} - \frac{\partial f}{\partial u_t} u_{tt} - \frac{\partial f}{\partial u} u_t - \frac{\partial f}{\partial t} = 0.$$

(i) Find the condition that

$$V = g(t, u, u_t) \frac{\partial}{\partial u}$$

is a Lie symmetry vector field (vertical vector field) of the ordinary differential equation. One calculates the *prolongation* of the vector field V

$$\tilde{V} = V + \left(\frac{\partial g}{\partial t} + \frac{\partial g}{\partial u} + \frac{\partial g}{\partial u_t} u_{tt} \right) \frac{\partial}{\partial u_t}$$

$$+ \left(\frac{\partial^2 g}{\partial t^2} + 2\frac{\partial^2 g}{\partial u \partial t} u_t + 2\frac{\partial^2 g}{\partial u_t \partial t} u_{tt} + \frac{\partial^2 g}{\partial u^2} u_t^2 + 2\frac{\partial^2 g}{\partial u \partial u_t} u_t u_{tt} \right.$$

$$\left. + \frac{\partial^2 g}{\partial u_t^2} u_{tt}^2 + \frac{\partial g}{\partial u} u_{tt} + \frac{\partial g}{\partial u_t} u_{ttt} \right) \frac{\partial}{\partial u_{tt}}.$$

The *invariance condition* is then

$$L_{\tilde{V}} F_1 \hat{=} 0$$

where $\hat{=}$ stands for the restriction to solutions of $F_1 = 0$.
(ii) Simplify the result if $f(du/dt, u, t) = 0$.
(iii) Simplify the result if $f(du/dt, u, t) = u$.

Solution 62. (i) From $L_{\widetilde{V}} F_1 \equiv L_{\widetilde{V}}(u_{tt} - f) = 0$ we find the condition

$$-g\frac{\partial f}{\partial u} - \left(\frac{\partial g}{\partial t} + \frac{\partial g}{\partial u}u_t + \frac{\partial g}{\partial u_t}u_{tt}\right)\frac{\partial f}{\partial u_t}$$

$$+ \left(\frac{\partial^2 g}{\partial t^2} + 2\frac{\partial^2 g}{\partial u \partial t}u_t + 2\frac{\partial^2 g}{\partial u_t \partial t}u_{tt} + \frac{\partial^2 g}{\partial u^2}u_t^2 + 2\frac{\partial^2 g}{\partial u \partial u_t}u_t u_{tt}\right.$$

$$\left. + \frac{\partial^2 g}{\partial u_t^2}u_{tt}^2 + \frac{\partial g}{\partial u}u_{tt} + \frac{\partial g}{\partial u_t}u_{ttt}\right) = 0.$$

Inserting

$$u_{tt} = f(u_t, u, t), \qquad u_{ttt} = \frac{\partial f}{\partial u_t}f + \frac{\partial f}{\partial u}u_t + \frac{\partial f}{\partial t}$$

yields

$$-g\frac{\partial f}{\partial u} - \left(\frac{\partial g}{\partial t} + \frac{\partial g}{\partial u}u_t + \frac{\partial g}{\partial u_t}f\right)\frac{\partial f}{\partial u_t}$$

$$+ \left(\frac{\partial^2 g}{\partial t^2} + 2\frac{\partial^2 g}{\partial u \partial t}u_t + 2\frac{\partial^2 g}{\partial u_t \partial t}f + \frac{\partial^2 g}{\partial u^2}u_t^2 + 2\frac{\partial^2 g}{\partial u \partial u_t}u_t f\right.$$

$$\left. + \frac{\partial^2 g}{\partial u_t^2}f^2 + \frac{\partial g}{\partial u}f + \frac{\partial g}{\partial u_t}\frac{\partial f}{\partial u_{tt}} + \frac{\partial g}{\partial u_t}\frac{\partial f}{\partial u}u_t + \frac{\partial g}{\partial u_t}\frac{\partial f}{\partial u_t}f\right) = 0.$$

(ii) For $f(du/dt, u, t) = 0$ the equation simplifies to

$$\frac{\partial^2 g}{\partial t^2} + 2\frac{\partial^2 g}{\partial u \partial t}u_t + \frac{\partial^2 g}{\partial u^2}u_t^2 = 0.$$

Thus the vertical Lie symmetry vector fields are

$$\widetilde{W}_1 = -u_t\frac{\partial}{\partial u}, \quad \widetilde{W}_2 = \frac{\partial}{\partial u}$$

$$\widetilde{W}_3 = -tu_t\frac{\partial}{\partial u}, \quad \widetilde{W}_4 = -uu_t\frac{\partial}{\partial u}$$

$$\widetilde{W}_5 = t\frac{\partial}{\partial u}, \quad \widetilde{W}_6 = u\frac{\partial}{\partial u}$$

$$\widetilde{W}_7 = t(-tu_t + u)\frac{\partial}{\partial u}, \quad \widetilde{W}_8 = u(-tu_t + u)\frac{\partial}{\partial u}.$$

Thus the Lie symmetry vector fields are

$$W_1 = t\frac{\partial}{\partial u}, \quad W_2 = u_t\frac{\partial}{\partial u}, \quad W_3 = ut\frac{\partial}{\partial u} + t^2\frac{\partial}{\partial t}, \quad W_4 = t\frac{\partial}{\partial t}$$

$$W_5 = \frac{\partial}{\partial t}, \quad W_6 = ut\frac{\partial}{\partial t} + u^2\frac{\partial}{\partial u}, \quad W_7 = u\frac{\partial}{\partial u}, \quad W_8 = \frac{\partial}{\partial u}.$$

Calculating the commutators we find that the vector fields form a basis of the Lie algebra $s\ell(3,\mathbb{R})$.

(iii) For $f(du/dt, u, t) = u$ the equation simplifies to

$$-g + \frac{\partial^2 g}{\partial t^2} + 2\frac{\partial^2 g}{\partial u \partial t}u_t + 2\frac{\partial^2 g}{\partial u_t \partial t}u + \frac{\partial^2 g}{\partial u^2}u_t^2 + 2\frac{\partial^2 g}{\partial u \partial u_t}u_t u$$

$$+\frac{\partial^2 g}{\partial u_t^2}u^2 + \frac{\partial g}{\partial u}u + +\frac{\partial g}{\partial u_t} = 0.$$

Thus the second order linear differential equation

$$\frac{d^2 u}{dt^2} + u = 0$$

admits the Lie point symmetry vector fields

$$V_1 = \frac{\partial}{\partial t}, \quad V_2 = u\frac{\partial}{\partial u}, \quad V_3 = \sin t\frac{\partial}{\partial u}, \quad V_4 = \cos t\frac{\partial}{\partial u},$$

$$V_5 = \sin(2t)\frac{\partial}{\partial t} + u\cos(2t)\frac{\partial}{\partial u}, \quad V_6 = \cos(2t)\frac{\partial}{\partial t} + u\sin(2t)\frac{\partial}{\partial u},$$

$$V_7 = u\cos t\frac{\partial}{\partial t} - u^2\sin t\frac{\partial}{\partial u}, \quad V_8 = u\sin t\frac{\partial}{\partial t} + u^2\cos t\frac{\partial}{\partial u}.$$

We have the semisimple Lie algebra $s\ell(3,\mathbb{R})$.

Problem 63. Find the Lie symmetry vector fields for the third order linear differential equation

$$\frac{d^3 u}{dt^3} = 0.$$

Classify the Lie algebra.

Solution 63. We obtain

$$X_1 = \frac{\partial}{\partial t}, \quad X_2 = \frac{\partial}{\partial u}, \quad X_3 = t\frac{\partial}{\partial t},$$

$$X_4 = t\frac{\partial}{\partial u}, \quad X_5 = u\frac{\partial}{\partial u}, \quad X_6 = t^2\frac{\partial}{\partial u}, \quad X_7 = t^2\frac{\partial}{\partial t} + 2ut\frac{\partial}{\partial u}.$$

The Lie algebra is solvable. Its radical is $\{X_2, X_4, X_5, X_6\}$. The radical must contain an abelian ideal. The elements X_2, X_4, X_6 commute. They form an abelian ideal. The elements $\{X_1, X_3, X_7\}$ form a Lie subalgebra which is isomorphic to $s\ell(2,\mathbb{R})$.

Problem 64.   Consider the nonlinear partial differential equation

$$\frac{\partial^3 u}{\partial x^3} + u\left(\frac{\partial u}{\partial t} + c\frac{\partial u}{\partial x}\right) = 0$$

where c is a constant. Show that the partial differential equation admits the Lie symmetry vector fields

$$V_1 = \frac{\partial}{\partial t}, \quad V_2 = \frac{\partial}{\partial x},$$

$$V_3 = 3t\frac{\partial}{\partial t} + (x + 2ct)\frac{\partial}{\partial x}, \quad V_4 = t\frac{\partial}{\partial t} + ct\frac{\partial}{\partial x} + u\frac{\partial}{\partial u}.$$

Solution 64.   Consider the vector field V_4. The autonomous system of first order differential equation corresponding to the vector field is

$$\frac{dt'}{d\epsilon} = t', \quad \frac{dx'}{d\epsilon} = ct', \quad \frac{du'}{d\epsilon} = u'.$$

The solution of the initial value problem is

$$t'(\epsilon) = e^\epsilon t'(0)$$
$$x'(\epsilon) = (e^\epsilon - 1)ct'(0) + x'(0)$$
$$u'(\epsilon) = e^\epsilon u'(0).$$

This leads to the transformation

$$t'(x,t) = e^\epsilon t$$
$$x'(x,t) = (e^\epsilon - 1)ct + x$$
$$u'(x'(x,t), t'(x,t)) = e^\epsilon u(x,t).$$

Then since $\partial t'/\partial x = 0$, $\partial x'/\partial x = 1$ we have

$$\frac{\partial u'}{\partial x} = \frac{\partial u'}{\partial x'}\frac{\partial x'}{\partial x} + \frac{\partial u'}{\partial t'}\frac{\partial t'}{\partial x} = e^\epsilon\frac{\partial u}{\partial x} = \frac{\partial u'}{\partial x'} = e^\epsilon\frac{\partial u}{\partial x}$$

$$\frac{\partial u'}{\partial t} = \frac{\partial u'}{\partial t'}\frac{\partial t'}{\partial t} + \frac{\partial u'}{\partial x'}\frac{\partial x'}{\partial t} = e^\epsilon\frac{\partial u}{\partial t} = \frac{\partial u'}{\partial t'}e^\epsilon + \frac{\partial u'}{\partial x'}(e^\epsilon - 1)c = e^\epsilon\frac{\partial u}{\partial t}$$

$$\frac{\partial^2 u'}{\partial x^2} = \frac{\partial^2 u'}{\partial x'^2}\frac{\partial x'}{\partial x} + \frac{\partial^2 u'}{\partial t'\partial x'}\frac{\partial t'}{\partial x} = e^\epsilon\frac{\partial^2 u}{\partial x^2} = \frac{\partial^2 u'}{\partial x'^2} = e^\epsilon\frac{\partial^2 u}{\partial x^2}$$

$$\frac{\partial^3 u'}{\partial x^3} = \frac{\partial^3 u'}{\partial x'^3}\frac{\partial x'}{\partial x} = e^\epsilon\frac{\partial^3 u}{\partial x^3} = \frac{\partial^3 u'}{\partial x'^3} = e^\epsilon\frac{\partial^3 u}{\partial x^3}.$$

Inserting these expression into the partial differential equation yields

$$e^{-\epsilon}\frac{\partial^3 u'}{\partial x'^3} + e^{-\epsilon}u'\left(\frac{\partial u'}{\partial t'} + e^{-\epsilon}\frac{\partial u'}{\partial x'}(e^\epsilon - 1)c + ce^{-\epsilon}\frac{\partial u'}{\partial x'}\right) = 0.$$

Thus

$$e^{-\epsilon}\frac{\partial^3 u'}{\partial x'^3} + e^{-\epsilon}u'\left(\frac{\partial u'}{\partial t'} + c\frac{\partial u'}{\partial x'}\right) = 0$$

with $e^{-\epsilon} \neq 0$.

Problem 65. Consider the stationary incompressible *Prandtl boundary layer equation*

$$\frac{\partial^3 u}{\partial \eta^3} = \frac{\partial u}{\partial \eta}\frac{\partial^2 u}{\partial \eta\partial \xi} - \frac{\partial u}{\partial \xi}\frac{\partial^2 u}{\partial \eta\partial \xi}.$$

Using the classical Lie method we obtain the *similarity reduction*

$$u(\xi, \eta) = \xi^\beta y(x(\xi, \eta)), \qquad x(\xi, \eta) = \eta\xi^{\beta-1} + f(\xi)$$

where f is an arbitrary differentiable function of ξ. Find the ordinary differential equation for y.

Solution 65. Differentiation and applying the chain rule provides

$$\frac{d^3 y}{dx^3} = \beta y\frac{d^2 y}{dx^2} - (2\beta - 1)\left(\frac{dy}{dx}\right)^2$$

which, in the special case $\beta = 2$, is the *Chazy equation*.

Problem 66. Given the partial differential equation

$$\frac{\partial^2 u}{\partial x\partial t} = f(u)$$

where $f : \mathbb{R} \to \mathbb{R}$ is a smooth function. Find the condition that

$$V = a(x, t, u)\frac{\partial}{\partial x} + b(x, t, u)\frac{\partial}{\partial t} + c(x, t, u)\frac{\partial}{\partial u}$$

is a symmetry vector field of the partial differential equation. Start with the corresponding vertical vector field

$$V_v = (-a(x, t, u)u_x - b(x, t, u)u_t + c(x, t, u))\frac{\partial}{\partial u}$$

and calculate first the prolongation. Utilize the differential consequencies which follow from the partial differential equations

$$u_{xt} - f(u) = 0, \qquad u_{xxt} - \frac{df}{du}u_x = 0, \qquad u_{xtt} - \frac{df}{du}u_t = 0.$$

Solution 66.   The prolongation $\widetilde{V}_v$ of V_v is given by

$$
\widetilde{V}_v = (-au_x - bu_t + c)\frac{\partial}{\partial u}
$$

$$
+\left(-\frac{\partial^2 a}{\partial x\partial t}u_x - \frac{\partial^2 a}{\partial x\partial u}u_tu_x - \frac{\partial a}{\partial x}u_{xt} - \frac{\partial^2 a}{\partial u\partial t}u_x^2 - \frac{\partial^2 a}{\partial u^2}u_tu_x^2 - 2\frac{\partial a}{\partial u}u_xu_{xt}\right.
$$

$$
-\frac{\partial a}{\partial t}u_{xx} - \frac{\partial a}{\partial u}u_tu_{xx} - au_{xxt} - \frac{\partial^2 b}{\partial x\partial t}u_t - \frac{\partial^2 b}{\partial x\partial u}u_t^2 - \frac{\partial b}{\partial x}u_{tt}
$$

$$
-\frac{\partial^2 b}{\partial u\partial t}u_xu_t - \frac{\partial^2 b}{\partial u^2}u_xu_t^2 - \frac{\partial b}{\partial u}u_{xt}u_t - \frac{\partial b}{\partial u}u_xu_{tt}
$$

$$
\left.+\frac{\partial^2 c}{\partial x\partial t} + \frac{\partial^2 c}{\partial x\partial u}u_t + \frac{\partial^2 c}{\partial u\partial t}u_x + \frac{\partial^2 c}{\partial u^2}u_tu_x + \frac{\partial c}{\partial u}u_{xt}\right)\frac{\partial}{\partial u_{xt}}\,.
$$

Using the Lie derivative the invariance condition is

$$
L_{\widetilde{V}_v}(u_{xt} - f(u)) = 0\,.
$$

It follows that

$$
-c\frac{\partial f}{\partial u} - \frac{\partial^2 a}{\partial x\partial t}u_x - \frac{\partial^2 a}{\partial x\partial u}u_tu_x - \frac{\partial a}{\partial x}f
$$

$$
-\frac{\partial^2 a}{\partial u\partial t}u_x^2 - \frac{\partial^2 a}{\partial u^2}u_tu_x^2 - 2\frac{\partial a}{\partial u}u_xf
$$

$$
-\frac{\partial a}{\partial t}u_{xx} - \frac{\partial a}{\partial u}u_tu_{xx} - \frac{\partial^2 b}{\partial x\partial t}u_t - \frac{\partial^2 b}{\partial x\partial u}u_t^2 - \frac{\partial b}{\partial x}u_{tt}
$$

$$
-\frac{\partial^2 b}{\partial t\partial u}u_xu_t - \frac{\partial^2 b}{\partial u^2}u_t^2u_x - \frac{\partial b}{\partial u}u_tf - \frac{\partial b}{\partial u}u_xu_{tt}
$$

$$
-\frac{\partial b}{\partial t}f - \frac{\partial b}{\partial u}u_tf + \frac{\partial^2 c}{\partial x\partial t} + \frac{\partial^2 c}{\partial u\partial x}u_t + \frac{\partial^2 c}{\partial t\partial u}u_x + \frac{\partial^2 c}{\partial u^2}u_tu_x + \frac{\partial c}{\partial u}f = 0
$$

where we have used the differential consequencies. We obtain 12 conditions for the coefficients 1, u_x, u_t, u_xu_t, u_x^2, u_t^2, u_{xx}, u_{tt}, $u_x^2u_t$, $u_xu_t^2$, $u_{xx}u_t$, u_xu_{tt}

$$
-c\frac{\partial f}{\partial u} - \frac{\partial a}{\partial x}f - \frac{\partial b}{\partial t}f + \frac{\partial^2 c}{\partial x\partial t} + \frac{\partial c}{\partial u}f = 0
$$

$$
-\frac{\partial^2 a}{\partial x\partial t} - 2\frac{\partial a}{\partial u}f + \frac{\partial^2 c}{\partial u\partial t} = 0
$$

$$
-\frac{\partial^2 b}{\partial x\partial t} - 2\frac{\partial b}{\partial u}f + \frac{\partial^2 c}{\partial u\partial x} = 0
$$

$$
-\frac{\partial^2 a}{\partial x\partial u} - \frac{\partial^2 b}{\partial t\partial u} + \frac{\partial^2 c}{\partial u^2} = 0
$$

$$
\frac{\partial^2 a}{\partial u\partial t} = 0
$$

$$
\frac{\partial^2 b}{\partial u\partial x} = 0
$$

$$\frac{\partial a}{\partial t} = 0$$

$$\frac{\partial b}{\partial x} = 0$$

$$\frac{\partial^2 a}{\partial u^2} = 0$$

$$\frac{\partial^2 b}{\partial u^2} = 0$$

$$\frac{\partial a}{\partial u} = 0$$

$$\frac{\partial b}{\partial u} = 0.$$

From these conditions we conclude that a depends only on x and b depends only on t. It follows that

$$\frac{\partial^2 c}{\partial u^2} = 0, \qquad \frac{\partial^2 c}{\partial u \partial x} = 0, \qquad \frac{\partial^2 c}{\partial u \partial t} = 0.$$

Thus we find

$$\frac{\partial c}{\partial u} = g(x, t)$$

but from the other two equations we have

$$\frac{\partial g}{\partial x} = \frac{\partial g}{\partial t} = 0.$$

Then from $\partial c / \partial u = K$ we find $c(x, t, u) = Ku + h(x, t)$ for some smooth function h and K is a constant. Thus from the first equation of the conditions we obtain

$$-Ku \frac{df(u)}{du} + h(x, t) \frac{df(u)}{du} - \left(\frac{da(x)}{dx} + \frac{db(t)}{dt} - K \right) f(u) + \frac{\partial^2 h(x, t)}{\partial x \partial t} = 0.$$

In general the functions udf/du, df/du, f, 1 will be linearly independent. Hence $K = 0$, $h(x, t) = 0$ and

$$\frac{da(x)}{dx} + \frac{db(t)}{dt} = 0.$$

From this equation we obtain $a(x) = Ax + B$, $b(t) = -At + D$ for suitable constants A, B, D. Thus we find the general form for the vector field V

$$V = (Ax + B) \frac{\partial}{\partial x} + (-At + D) \frac{\partial}{\partial t}.$$

This leads to the symmetry vector fields

$$V_1 = x \frac{\partial}{\partial x} - t \frac{\partial}{\partial t}, \quad V_2 = \frac{\partial}{\partial x}, \quad V_3 = \frac{\partial}{\partial t}$$

which form a basis of a Lie algebra. The form of the smooth function f may admit other symmetry vector fields. This will be the case when the functions

$$u\frac{df}{du}, \quad \frac{df}{du}, \quad f, \quad 1$$

are not linearly independent.

If $f(u) = u$ then we obtain the vector fields

$$V_1 = x\frac{\partial}{\partial x} - t\frac{\partial}{\partial t}, \quad V_2 = \frac{\partial}{\partial x}, \quad V_3 = \frac{\partial}{\partial t}, \quad V_4 = \frac{\partial}{\partial u}$$

and

$$V_5 = h(x,t)\frac{\partial}{\partial u}$$

where the smooth function $h(x,t)$ satisfies the partial differential equation $\partial^2 h/\partial x\partial t + h = 0$.

If $f(u) = u^n$ with $n = 2, 3, \ldots$ we find the vector fields

$$V_1 = x\frac{\partial}{\partial x} - t\frac{\partial}{\partial t}, \quad V_2 = \frac{\partial}{\partial x}, \quad V_3 = \frac{\partial}{\partial t}, \quad V_4 = -(n-1)t\frac{\partial}{\partial t} + u\frac{\partial}{\partial u}.$$

Problem 67. The *Lie derivative* of a differential form ω with respect to a vector field W can be expressed as

$$L_W\omega \equiv W\rfloor d\omega + d(W\rfloor\omega)$$

where d is the exterior derivative and $\rfloor$ denotes the contraction. Let α be a smooth differential one-form and V be smooth vector field. Assume that

$$L_V\alpha = f\alpha$$

where f is a smooth function and $L_V(.)$ denotes the Lie derivative. Define the function F as

$$F := V\rfloor\alpha$$

Show that

$$dF = f\alpha - V\rfloor d\alpha.$$

Solution 67. Taking the exterior derivative of the function F yields

$$dF = d(V\rfloor\alpha) = L_V\alpha - V\rfloor d\alpha.$$

Using $L_V\alpha = f\alpha$ we have $dF = f\alpha - V\rfloor d\alpha$.

Problem 68. Consider the n-dimensional smooth manifold $M = \mathbb{R}^n$ with coordinates $(x_1, \ldots, x_n)$ and an arbitrary smooth first order partial differential equation on M

$$F(x_1, \ldots, x_n, \partial u/\partial x_1, \ldots, \partial u/\partial x_n, u) = 0.$$

Find the symmetry vector fields (sometimes called the infinitesimal symmetries) of this first order partial differential equation. Consider the cotangent bundle $T^*(M)$ over the manifold M with coordinates

$$(x_1, \ldots, x_n, p_1, \ldots, p_n)$$

and construct the product manifold $T^*(M) \times \mathbb{R}$. Then $T^*(M)$ has a canonical differential one-form

$$\sum_{j=1}^{n} p_j \, dx_j$$

which provides the contact differential one-form

$$\alpha = du - \sum_{j=1}^{n} p_j \, dx_j$$

on $T^*(M) \times \mathbb{R}$. The solutions of the partial differential equation are surfaces in $T^*(M) \times \mathbb{R}$

$$F(x_1, \ldots, x_n, p_1, \ldots, p_n, u) = 0$$

which annul the differential one-form α. One constructs the closed ideal I defined by

$$F(x_1, \ldots, x_n, p_1, \ldots, p_n, u)$$

$$\alpha = du - \sum_{j=1}^{n} p_j \, dx_j$$

$$dF = \sum_{j=1}^{n} \left(\frac{\partial F}{\partial x_j} dx_j + \frac{\partial F}{\partial p_j} dp_j \right) + \frac{\partial F}{\partial u} du$$

$$d\alpha = \sum_{j=1}^{n} dx_j \wedge dp_j$$

where $\wedge$ is the exterior product and d the exterior derivative. The surfaces in $T^*(M) \times \mathbb{R}$ which annul I will be the solutions of the first order partial differential equation. Let

$$V(x_1, \ldots, x_n, p_1, \ldots, p_n, u) = \sum_{j=1}^{n} V_{x_j} \frac{\partial}{\partial x_j} + \sum_{j=1}^{n} V_{p_j} \frac{\partial}{\partial p_j} + V_u \frac{\partial}{\partial u}$$

be a smooth vector field. Let L_V denote the Lie derivative. The Lie derivative is linear and satisfies the product rule with respect to the exterior
product $\wedge$. Then the conditions for V to be a symmetry vector field are

$$L_V F = gF$$

$$L_V \alpha = \lambda \alpha + \eta dF + \left(\sum_{j=1}^{n} (A_j dx_j + B_j dp_j) \right) F .$$

Here λ, η, A_j, B_j are smooth functions of $x_1, \ldots, x_n, p_1, \ldots, p_n$ and u on
$T^*(\mathbb{R}^n) \times \mathbb{R}$, where g, A_j, B_j must be nonsingular in a neighborhood of
$F = 0$. Find V.

Solution 68.   The Lie derivative and the exterior derivative commute,
i.e.

$$d(L_V(.)) = L_V(d(.)) .$$

Thus the one-form dF and the two-form $d\alpha$ are in I when a symmetry
vector field is applied. Thus the two equations are sufficient to define the
symmetry condition for all of the ideal I. Since we have the identities

$$L_V f \equiv V \rfloor df, \qquad L_V \omega \equiv d(V \rfloor \omega) + V \rfloor d\omega$$

we obtain from the second condition

$$V_{x_j} = -\frac{\partial (V \rfloor \alpha)}{\partial p_j} + \eta \frac{\partial F}{\partial p_j} + B_j F$$

$$V_{p_j} = \frac{\partial (V \rfloor \alpha)}{\partial x_j} + p_j \frac{\partial (V \rfloor \alpha)}{\partial u} - \eta \left(\frac{\partial F}{\partial x_j} + p_j \frac{\partial F}{\partial u} \right) - A_j F$$

$$V_u = (V \rfloor \alpha) - \sum_{j=1}^{n} p_j \frac{(V \rfloor \alpha)}{\partial p_j} + \eta \sum_{j=1}^{n} p_j \frac{(V \rfloor \alpha)}{\partial p_j} + \sum_{j=1}^{n} B_j p_j F .$$

From the first condition $L_V F = gF$ we find

$$V \rfloor dF = \sum_{j=1}^{n} \frac{\partial F}{\partial x_j} V_{x_j} + \sum_{j=1}^{n} \frac{\partial F}{\partial p_j} V_{p_j} + \frac{\partial F}{\partial u} V_u = gF .$$

Inserting this expression into the second condition we find the autonomous
system of first order differential equations

$$\frac{dx_j}{d\epsilon} = \frac{\partial F}{\partial p_j}$$

$$\frac{du}{d\epsilon} = \sum_{j=1}^{n} p_j \frac{\partial F}{\partial p_j}$$

$$\frac{dp_j}{d\epsilon} = -\left(\frac{\partial F}{\partial x_j} + p_j \frac{\partial F}{\partial u}\right)$$

$$\frac{d(V \lrcorner \alpha)}{d\epsilon} = \left(g - \sum_{j=1}^{n} \frac{\partial F}{\partial x_j} B_j + \sum_{j=1}^{n} \frac{\partial F}{\partial p_j} A_j - \sum_{j=1}^{n} p_j B_j \frac{\partial F}{\partial u}\right) F + \frac{\partial F}{\partial u}(V \lrcorner \alpha).$$

This is the characteristic system of the partial differential equation.

Problem 69. Consider the nonlinear one-dimensional diffusion equation

$$\frac{\partial u}{\partial t} - \frac{\partial}{\partial x}\left(u^n \frac{\partial u}{\partial x}\right) = 0$$

where $n = 1, 2, \ldots$. An equivalent set of differential forms is given by

$$\alpha = du - u_t dt - u_x dx$$
$$\beta = (u_t - nu^{n-1}u_x^2)dx \wedge dt - u^n du_x \wedge dt$$

with the coordinates t, x, u, u_t, u_x. Since $ddu = 0$ the exterior derivative of α is given

$$d\alpha = -du_t \wedge dt - du_x \wedge dx.$$

Consider the vector field

$$V = V_t \frac{\partial}{\partial t} + V_x \frac{\partial}{\partial x} + V_u \frac{\partial}{\partial u} + V_{u_t} \frac{\partial}{\partial u_t} + V_{u_x} \frac{\partial}{\partial u_x}.$$

Then the symmetry vector fields of the partial differential equation are determined by

$$L_V \alpha = g\alpha, \qquad L_V \beta = h\beta + w\alpha + rd\alpha$$

where $L_V(.)$ denotes the Lie derivative, g, h, r are smooth functions depending on t, x, u, u_t, u_x and w is a differential one-form also depending on t, x, u, u_t, u_x. Find the symmetry vector fields from these two conditions. Note that we have

$$L_V(d\alpha) = d(L_V \alpha) = d(g\alpha) = (dg) \wedge \alpha + gd\alpha.$$

Solution 69. The first condition $L_V \alpha = g\alpha$ provides 5 equations. The second condition provides 10 equations. Solving these 15 equations yields

$$V_t = c_1 + c_3 t, \quad V_x = c_2 + c_4 x, \quad V_u = \frac{1}{n}(2c_4 - c_3)u$$

$$V_{u_t} = \frac{1}{n}(2c_4 - (n+1)c_3)u_t, \quad V_{u_x} = \frac{1}{n}((2-n)c_4 - c_3)u_x$$

where c_j $(j = 1, 2, 3, 4)$ are constants. Thus we obtain the symmetry vector fields

$$V_1 = \frac{\partial}{\partial t}, \qquad V_2 = \frac{\partial}{\partial x}$$

$$V_3 = -\frac{1}{n}\left(u\frac{\partial}{\partial u} + (n+1)u_t\frac{\partial}{\partial u_t} + u_x\frac{\partial}{\partial u_x}\right) + t\frac{\partial}{\partial t}$$

$$V_4 = \frac{1}{n}\left(2u\frac{\partial}{\partial u} + 2u_t\frac{\partial}{\partial u_t} + (2-n)u_x\frac{\partial}{\partial u_x}\right) + x\frac{\partial}{\partial x}.$$

We have the commutators

$$[V_1, V_2] = 0, \quad [V_1, V_3] = V_1, \quad [V_2, V_4] = V_2, \quad [V_3, V_4] = 0.$$

Problem 70. Consider the *Hilbert space* of all $n \times n$ matrices over $\mathbb{R}$ with *scalar product*

$$\langle X, Y \rangle := \operatorname{tr}(XY^T).$$

Consider the Lie group $G = SO(n, \mathbb{R})$.
(i) Let $X \in G$. Find $\langle X, X \rangle$.
(ii) The infinitesimal generators of the Lie group $SO(3, \mathbb{R})$ are given by

$$A_1 = \begin{pmatrix} 0 & 0 & 1 \\ 0 & 0 & 0 \\ -1 & 0 & 0 \end{pmatrix}, \quad A_2 = \begin{pmatrix} 0 & 1 & 0 \\ -1 & 0 & 0 \\ 0 & 0 & 0 \end{pmatrix}, \quad A_3 = \begin{pmatrix} 0 & 0 & 0 \\ 0 & 0 & 1 \\ 0 & -1 & 0 \end{pmatrix}.$$

Find the scalar products

$$\langle A_1, A_2 \rangle, \quad \langle A_2, A_3 \rangle, \quad \langle A_3, A_1 \rangle$$

and $(t \in \mathbb{R})$

$$\langle e^{tA_1}, e^{tA_2} \rangle, \quad \langle e^{tA_2}, e^{tA_3} \rangle, \quad \langle e^{tA_3}, e^{tA_1} \rangle.$$

Solution 70. (i) Since $XX^T = I_n$ we find

$$\langle X, X \rangle = \operatorname{tr}(XX^T) = \operatorname{tr}(I_n) = n.$$

(ii) Note that for $A, B \in so(n, \mathbb{R})$ we have $A^T = -A$ and $B^T = -B$. From direct calculations follows

$$\langle A_1, A_2 \rangle = \langle A_2, A_3 \rangle = \langle A_3, A_1 \rangle = 0.$$

We find

$$\langle e^{tA_1}, e^{tA_2} \rangle = 2\cos(t) + \cos^2(t)$$
$$\langle e^{tA_2}, e^{tA_3} \rangle = 2\cos(t) + \cos^2(t)$$
$$\langle e^{tA_2}, e^{tA_1} \rangle = 2\cos(t) + \cos^2(t).$$

Problem 71. Consider the nonlinear Schrödinger equation

$$i\frac{\partial\psi}{\partial t} + \Delta\psi = c_0\psi + c_1|\psi|^2\psi + c_2|\psi|^4\psi$$

in $3 + 1$ dimensions, where

$$\psi(x_1, x_2, x_3, t) = u_1(x_1, x_2, x_3, t) + iu_2(x_1, x_2, x_3, t)$$

and u_1, u_2 are real fields. Furthermore

$$\Delta := \frac{\partial^2}{\partial x_1^2} + \frac{\partial^2}{\partial x_2^2} + \frac{\partial^2}{\partial x_3^2}.$$

Find the Lie symmetry vector fields of this partial differential equation, i.e. consider the Lie symmetry vector field

$$V = \eta_1\frac{\partial}{\partial x_1} + \eta_2\frac{\partial}{\partial x_2} + \eta_3\frac{\partial}{\partial x_3} + \eta_4\frac{\partial}{\partial t} + \xi_1\frac{\partial}{\partial u_1} + \xi_2\frac{\partial}{\partial u_2}$$

where η_j $(j = 1, 2, 3, 4)$ and ξ_k $(k = 1, 2)$ are smooth functions of x_1, x_2, x_3, t, u_1, u_2. Assume that c_0, c_1, c_2 are nonzero.

Solution 71. The conditions on the functions η_j and ξ_k are determined from the requirement that the second prolongation of the Lie symmetry vector field annihilates the equation on the solution set of the equation. We find the following vector fields:
-three space translations

$$\frac{\partial}{\partial x_1}, \quad \frac{\partial}{\partial x_2}, \quad \frac{\partial}{\partial x_3}$$

-one time translations

$$\frac{\partial}{\partial t} + c_0\left(u_2\frac{\partial}{\partial u_1} - u_1\frac{\partial}{\partial u_2}\right)$$

-three rotations

$$x_3\frac{\partial}{\partial x_2} - x_2\frac{\partial}{\partial x_3}, \quad x_1\frac{\partial}{\partial x_3} - x_3\frac{\partial}{\partial x_1}, \quad x_2\frac{\partial}{\partial x_1} - x_1\frac{\partial}{\partial x_2}$$

-three proper Galilei transformations

$$t\frac{\partial}{\partial x_1} - \frac{1}{2}x_1\left(u_2\frac{\partial}{\partial u_1} - u_1\frac{\partial}{\partial u_2}\right)$$

$$t\frac{\partial}{\partial x_2} - \frac{1}{2}x_2\left(u_2\frac{\partial}{\partial u_1} - u_1\frac{\partial}{\partial u_2}\right)$$

$$t\frac{\partial}{\partial x_3} - \frac{1}{2}x_3\left(u_2\frac{\partial}{\partial u_1} - u_1\frac{\partial}{\partial u_2}\right)$$

-change of phase generator

$$u_2\frac{\partial}{\partial u_1} - u_1\frac{\partial}{\partial u_2}.$$

This is the extended *Galilei group* which contains three space translations, three rotations, three proper Galilei transformations, one time translation and change of phase generator.

Problem 72. The Euler-Poisson-Darboux equation is given by

$$\left(\frac{\partial^2}{\partial t^2} - \frac{\partial^2}{\partial r^2} - \frac{1}{r}\frac{\partial}{\partial r} + \frac{m^2}{r^2}\right)u(t,r) = 0$$

where $r \geq 0$ and $-\infty < t < \infty$. The symmetry Lie algebra of the Euler-Poisson-Darboux equation is the set of all linear differential operators

$$S = a_1(t,r)\frac{\partial}{\partial t} + a_2(t,r)\frac{\partial}{\partial r} + b(t,r)$$

such that Su is a local solution of the Euler-Poisson-Darboux equation whenever u is a local solution.
(i) Find the set of all S.
(ii) Show the partial differential equation can be expressed using the differential operators from (i).

Solution 72. (i) We obtain

$$S_1 = \frac{1}{2}\left((1 - t^2 - r^2)\frac{\partial}{\partial t} - 2tr\frac{\partial}{\partial r} - t\right)$$

$$S_2 = -\left(\frac{1}{2} + t\frac{\partial}{\partial t} + r\frac{\partial}{\partial r}\right)$$

$$S_3 = \frac{1}{2}\left((1 + t^2 + r^2)\frac{\partial}{\partial t} + 2tr\frac{\partial}{\partial r} + t\right)$$

where $S_1 + S_3 = \partial/\partial t$. The commutators are

$$[S_1, S_2] = -S_3, \quad [S_1, S_3] = -S_2, \quad [S_2, S_3] = S_1.$$

Thus the Lie algebra is isomorphic to $s\ell(2,\mathbb{R})$.
(ii) We have

$$(S_3^2 - S_1^2 - S_2^2)u = \left(\frac{1}{4} - m^2\right)u$$

where $S_3^1 - S_1^2 - S_2^2$ is the Casimir operator of the Lie algebra.

Problem 73. Consider the three differential operators

$$J_3 = t\frac{\partial}{\partial t}$$

$$J_- = -\frac{1}{t}\left((w-1)\frac{\partial}{\partial w} + t\frac{\partial}{\partial t} + \sigma\right)$$

$$J_+ = -t\left(w\frac{\partial}{\partial w} - t\frac{\partial}{\partial t} + \sigma\right).$$

(i) Find the commutators $[J_3, J_+]$, $[J_3, J_-]$, $[J_+, J_-]$ and thus show that we have a basis of the Lie algebra $s\ell(2, \mathbb{C})$.

(ii) Let $\epsilon_1, \epsilon_2, \epsilon_3 \in \mathbb{R}$. Let f be an analytic function. Calculate

$$\exp(\epsilon_1 J_+)\exp(\epsilon_2 J_-)\exp(\epsilon_3 J_3)f(w, t)$$

by solving the initial value problem of the autonomous system of first order differential equation for $\exp(\epsilon_3 J_3)$

$$\frac{dt}{d\epsilon_3} = t, \quad t(0) = t,$$

and by solving the initial value problem of the autonomous system of first order differential equations for $\exp(\epsilon_2 J_-)$

$$\frac{dw}{d\epsilon_2} = \frac{1-w}{t}, \quad \frac{dt}{d\epsilon_2} = -1, \quad \frac{dv}{d\epsilon_2} = -\frac{\sigma v}{t}, \quad w(0) = w, \quad t(0) = t, \quad v(0) = 1$$

and by solving the initial value problem of the autonomous system of first order differential equations for $\exp(\epsilon_1 J_+)$

$$\frac{dw}{d\epsilon_1} = -tw, \quad \frac{dt}{d\epsilon_1} = t^2, \quad \frac{dv}{d\epsilon_1} = -\sigma tv, \quad w(0) = w, \quad t(0) = t, \quad v(0) = 1.$$

(iii) Calculate

$$\exp(\epsilon_3 J_3)J_+ \exp(-\epsilon_3 J_3), \quad \exp(\epsilon_3 J_3)J_- \exp(-\epsilon_3 J_3),$$

$$\exp(\epsilon_1 J_+)J_- \exp(-\epsilon_1 J_+), \quad \exp(\epsilon_1 J_+)J_3 \exp(-\epsilon_1 J_+),$$

$$\exp(\epsilon_2 J_-)J_+ \exp(-\epsilon_2 J_-), \quad \exp(\epsilon_2 J_-)J_3 \exp(-\epsilon_2 J_-).$$

Solution 73. (i) For the commutators we obtain

$$[J_3, J_+] = J_+, \quad [J_3, J_-] = -J_-, \quad [J_+, J_-] = 2J_3.$$

(ii) For $\exp(\epsilon_3 J_3)$ we find

$$t(\epsilon_3) = e^{\epsilon_3}t.$$

For $\exp(\epsilon_2 J_-)$ we find

$$t(\epsilon_2) = t - \epsilon_2, \quad w(\epsilon_2) = w - \epsilon_2 \frac{w}{t} + \frac{\epsilon_2}{t}, \quad v(\epsilon_2) = \left(1 - \frac{\epsilon_2}{t}\right)^\sigma v.$$

For $\exp(\epsilon_1 J_+)$ we find

$$t(\epsilon_1) = \frac{t}{1 - \epsilon_1 t}, \quad w(\epsilon_1) = (1 - \epsilon_1 t)w, \quad v(\epsilon_1) = (1 - \epsilon_1 t)^\sigma v.$$

(iii) We find

$$\exp(\epsilon_3 J_3) J_+ \exp(-\epsilon_3 J_3) = e^{\epsilon_3} J_+$$
$$\exp(\epsilon_3 J_3) J_- \exp(-\epsilon_3 J_3) = e^{-\epsilon_3} J_-$$
$$\exp(\epsilon_1 J_+) J_- \exp(-\epsilon_1 J_+) = -\epsilon_1^2 J_+ + 2\epsilon_1 J_3 + J_-$$
$$\exp(\epsilon_1 J_+) J_3 \exp(-\epsilon_1 J_+) = -\epsilon_1 J_+ + J_3$$
$$\exp(\epsilon_2 J_-) J_+ \exp(-\epsilon_2 J_-) = J_+ - \epsilon_2^2 J_- - 2\epsilon_2 J_3$$
$$\exp(\epsilon_2 J_-) J_3 \exp(-\epsilon_2 J_-) = \epsilon_2 J_- + J_3.$$

Problem 74.   Consider the $n \times n$ tridiagonal matrices

$$A = \begin{pmatrix} a_1 & b_1 & 0 & \cdots & 0 \\ b_1 & a_2 & b_2 & \cdots & 0 \\ \vdots & \vdots & \vdots & \vdots & b_{n-1} \\ 0 & \cdots & 0 & b_{n-1} & a_n \end{pmatrix}, \quad B = \begin{pmatrix} 0 & b_1 & 0 & \cdots & 0 \\ -b_1 & 0 & b_2 & \cdots & 0 \\ \vdots & \vdots & \vdots & \vdots & b_{n-1} \\ 0 & \cdots & 0 & -b_{n-1} & 0 \end{pmatrix}$$

where the entries a_j and b_j are real analytic functions of t. Thus A is symmetric over $\mathbb{R}$ and B is skew-symmetric over $\mathbb{R}$.
(i) Let $n = 2$. Find the equations of motion (*Lax representation*)

$$\frac{dA}{dt} = [A, B](t).$$

Find a first integral.
(ii) Let $n = 3$. Find the equations of motion

$$\frac{dA}{dt} = [A, B](t).$$

Find a first integral.

Solution 74.   (i) For $n = 2$ the equations of motion are

$$\frac{da_1}{dt} = -2b_1^2, \quad \frac{db_1}{dt} = b_1(a_1 - a_2), \quad \frac{da_2}{dt} = 2b_1^2.$$

From these equations we see

$$\frac{da_1}{dt} + \frac{da_2}{dt} = 0$$

thus $I_1 = a_1 + a_2$ is a first integral. Similarly

$$\frac{da_1}{dt} - \frac{da_2}{dt} = -4b_1^2.$$

Thus

$$\frac{db_1}{b_1(a_1 - a_2)} = \frac{d(a_1 - a_2)}{-4b_1^2}$$

and

$$I_2 = 4b_1^2 + (a_1 - a_2)^2$$

is another first integral. From the general theory of the Lax pair we know that the spectrum of the A matrix is constant. Thus traces of the powers of matrix A are constant. The first coefficient is equal to the first integral I_1. The second coefficient is

$$\text{tr}(A^2) = a_1^2 + a_2^2 + 2b_1^2 = \frac{1}{2}(I_1^2 + I_2).$$

(ii) For $n = 3$ the equations of motion are

$$\frac{da_1}{dt} = -2b_1^2, \quad \frac{da_2}{dt} = 2(b_1^2 - b_2^2), \quad \frac{da_3}{dt} = 2b_2^2,$$

$$\frac{db_1}{dt} = b_1(a_1 - a_2), \quad \frac{db_2}{dt} = b_2(a_2 - a_3).$$

From these equations we see

$$\frac{da_1}{dt} + \frac{da_2}{dt} + \frac{da_3}{dt} = 0.$$

Thus $I_1 = a_1 + a_2 + a_3$ is a first integral. The spectrum of the matrix A is constant. Thus traces of the powers of matrix A are constant. This gives two new first integrals

$$I_2 = \text{tr}(A^2) = a_1^2 + a_2^2 + a_3^2 + 2b_1^2 + 2b_2^2$$

$$I_3 = \text{tr}(A^3) = a_1^3 + 3a_1b_1^2 + a_2^3 + 3a_2b_1^2 + 3a_2b_2^2 + a_3^3 + 3a_3b_2^2.$$

Problem 75. (i) Show that the nonlinear autonomous system of first order differential equations

$$\frac{du_1}{dt} = u_2, \quad \frac{du_2}{dt} = u_1u_3, \quad \frac{du_3}{dt} = -u_1u_2$$

admits the Lax representation $dL/dt = [B, L](t)$ with

$$L = \begin{pmatrix} u_3 & (u_2 - \frac{1}{2}u_1^2)/\sqrt{2} \\ (u_2 + \frac{1}{2}u_1^2)/\sqrt{2} & \frac{1}{2}u_1^2 \end{pmatrix}, \quad B = \begin{pmatrix} -u_1 & -u_1/\sqrt{2} \\ u_1/\sqrt{2} & 0 \end{pmatrix}.$$

(ii) Find the first integrals from $\operatorname{tr}(L)$ and $\frac{1}{2}\operatorname{tr}(L^2)$.

Solution 75. (i) For the commutator we obtain

$$[B, L] = \begin{pmatrix} -u_1 u_2 & -u_1(u_2 - u_3)/\sqrt{2} \\ u_1(u_2 + u_3)/\sqrt{2} & u_1 u_2 \end{pmatrix}$$

and for the time derivative we find

$$\frac{dL}{dt} = \begin{pmatrix} du_3/dt & (-u_1 du_1/dt + du_2/dt)/\sqrt{2} \\ (u_1 du_1/dt + du_2/dt)/\sqrt{2} & u_1 du_1/dt \end{pmatrix}.$$

(ii) We have

$$I_1 = \operatorname{tr}(L) = \frac{1}{2}(u_1^2 + 2u_3), \qquad I_2 = \operatorname{tr}(L^2) = \frac{1}{2}(u_2^2 + u_3^2).$$

Problem 76. Consider the vector fields

$$V_1 = \frac{d}{dz}, \quad V_2 = z\frac{d}{dz}, \quad V_3 = z^2\frac{d}{dz}.$$

They satisfy the commutation relations

$$[V_2, V_1] = -V_1, \qquad [V_2, V_3] = V_3, \qquad [V_3, V_1] = -2V_2.$$

Thus we have a realization of the Lie algebra $s\ell(2, \mathbb{C})$. Extend the vector fields to the operators

$$W_1 = \frac{d}{dz} + f_1(z), \quad W_2 = z\frac{d}{dz} + f_2(z), \quad W_3 = z^2\frac{d}{dz} + f_3(z).$$

Find the condition on the analytic functions f_1, f_2, f_3 such that W_1, W_2, W_3 also satisfy the commutation relations

$$[W_2, W_1] = -W_1, \quad [W_2, W_3] = W_3, \quad [W_3, W_1] = -2W_2.$$

Solution 76. From the commutation relations we find the conditions

$$\frac{df_2}{dz} - z\frac{df_1}{dz} = f_1, \qquad \frac{df_3}{dz} - z^2\frac{df_1}{dz} = 2f_2, \qquad z\frac{df_3}{dz} - z^2\frac{df_2}{dz} = f_3.$$

The solution of this system of linear differential equations is

$$f_2(z) = zf_1(z) + c, \quad f_3(z) = z^2 f_1(z) + 2cz$$

where f_1 is an arbitrary smooth function and c is a constant.

Problem 77. Consider the operators

$$H_1 = 1, \quad H_2 = x, \quad H_3 = \frac{\partial^2}{\partial x^2}, \quad H_4 = i\frac{\partial}{\partial x}.$$

(i) Show that we have a *nilpotent Lie algebra* under the commutator.
(ii) Let

$$\alpha_1(t) = cf(t), \quad \alpha_2(t) = c, \quad \alpha_3(t) = -\frac{1}{2}, \quad \alpha_4(t) = \frac{df}{dt}.$$

Consider the Hamilton operator

$$K = \sum_{j=1}^4 \alpha_j(t)H_j$$

and the Schrödinger equation

$$i\frac{\partial \psi}{\partial t} = K\psi.$$

We write the solution of the Schrödinger equation in the form

$$\psi(x,t) = U(t,0)\psi(x,0)$$

where the unitary time evolution operator is given by

$$U(t,0) = \exp(\beta_1(t)H_1)\exp(\beta_2(t)H_2)\exp(\beta_3(t)H_3)\exp(\beta_4(t)H_4).$$

Find the system of ordinary differential equations for $\beta_j(t)$ ($j = 1,2,3,4$) and solve them.

Solution 77. (i) From the commutators

$$[H_j, H_k] = \sum_{l=1}^4 c_{jk}^l H_l$$

we obtain the nonzero structure constants $c_{23}^4 = 2i$, $c_{24}^1 = -i$. Since all other structure constants are zero we have a nilpotent Lie algebra.

(ii) Differentiating the ansatz $U(t, 0)$ with respect to t we obtain

$$i\frac{d\beta_1}{dt} + \beta_2\frac{d\beta_4}{dt} + i\frac{d\beta_3}{dt}\beta_2^2 = cf(t)$$

$$i\frac{d\beta_2}{dt} = c$$

$$i\frac{d\beta_3}{dt} = -\frac{1}{2}$$

$$i\frac{d\beta_4}{dt} - 2\frac{d\beta_3}{dt}\beta_2 = \frac{df}{dt}$$

with the initial conditions $\beta_j(0) = 0$, $j = 1, 2, 3, 4$. The solution of this nonlinear system of ordinary differential equations is

$$\beta_1(t) = -i\left(c\int_0^t f(s)ds + c\int_0^t s\frac{df}{ds}ds + c^2\frac{t^3}{6}\right)$$

$$\beta_2(t) = -ict$$

$$\beta_3(t) = \frac{i}{2}t$$

$$\beta_4(t) = -i\left(f(t) - f(0) + \frac{c}{2}t^2\right).$$

Problem 78. Many soliton equations in $1 + 1$ dimensions can be written as an integrability condition for a set of linear equations called the linear scattering problem

$$d\mathbf{v} = \Omega\mathbf{v}$$

where Ω is an $N \times N$ matrix consisting of a one-parameter ζ (spectral parameter) family of differential one-forms. The integrability condition is $\Theta = 0$ (*zero curvature equation*), where

$$\Theta := d\Omega - \Omega \wedge \Omega$$

d is the exterior derivative and $\wedge$ is the exterior product. Assume that

$$\Omega = (\zeta D + P(x, t))dx + Q(\zeta, x, t)dt, \qquad \mathrm{tr}(\Omega) = 0$$

where ζ is the complex spectral parameter, D is a diagonal $N \times N$ matrix with d_{jj} complex and pairwise distinct. The $N \times N$ matrix $P(x, t)$ is off-diagonal and it is assumed to satisfy the boundary conditions

$$P_{jk}(x, t), \frac{\partial P_{jk}(x, t)}{\partial x}, \ldots, \left(\frac{\partial}{\partial x}\right)^m P_{jk}(x, t), \cdots \to 0 \quad \text{as} \quad |x| \to \infty.$$

(i) Write down the linear problem $d\mathbf{v} = \Omega\mathbf{v}$ for this case.

(ii) Give the zero-curvature equation $d\Omega - \Omega \wedge \Omega = 0$ for this case.

Solution 78. (i) We obtain

$$\frac{\partial \mathbf{v}}{\partial x} = (\zeta D + P)\mathbf{v}, \qquad \frac{\partial \mathbf{v}}{\partial t} = Q\mathbf{v}.$$

(ii) The zero-curvature equation yields

$$\frac{\partial P}{\partial t} - \frac{\partial Q}{\partial x} + [\zeta D + P, Q] = 0$$

where $[\,,\,]$ denotes the commutator.

Problem 79. Consider the Lie group $SL(2, \mathbb{R})$. Let $X \in SL(2, \mathbb{R})$, i.e.

$$X = \begin{pmatrix} x_{11} & x_{12} \\ x_{21} & x_{22} \end{pmatrix}, \qquad x_{11}x_{22} - x_{12}x_{21} = 1, \quad x_{jk} \in \mathbb{R}.$$

Consider the matrix differential one-forms

$$\Omega = X^{-1}dX.$$

(i) Find Ω.
(ii) Find $\mathrm{tr}(\Omega)$.
(iii) Find the left-invariant one-forms.

Solution 79. (i) The inverse of X is given by

$$X^{-1} = \begin{pmatrix} x_{22} & -x_{12} \\ -x_{21} & x_{11} \end{pmatrix}.$$

Now

$$dX = \begin{pmatrix} dx_{11} & dx_{12} \\ dx_{21} & dx_{22} \end{pmatrix}.$$

Thus

$$\Omega = X^{-1}dX = \begin{pmatrix} x_{22} & -x_{12} \\ -x_{21} & x_{11} \end{pmatrix} \begin{pmatrix} dx_{11} & dx_{12} \\ dx_{21} & dx_{22} \end{pmatrix}$$

$$= \begin{pmatrix} x_{22}dx_{11} - x_{12}dx_{21} & x_{22}dx_{12} - x_{12}dx_{22} \\ -x_{21}dx_{11} + x_{11}dx_{21} & -x_{21}dx_{12} + x_{11}dx_{22} \end{pmatrix}$$

with $\det(X) = x_{11}x_{22} - x_{12}x_{21} = 1$. Taking the exterior derivative of $\det(X) = 1$ yields

$$x_{22}dx_{11} + x_{11}dx_{22} - x_{12}dx_{21} - x_{21}dx_{12} = 0$$

or

$$x_{11}dx_{22} - x_{21}dx_{12} = -x_{22}dx_{11} + x_{12}dx_{21}.$$

Inserting this expression into $X^{-1}dX$ gives

$$\Omega = \begin{pmatrix} x_{22}dx_{11} - x_{12}dx_{21} & x_{22}dx_{12} - x_{12}dx_{22} \\ -x_{21}dx_{11} + x_{11}dx_{21} & -x_{22}dx_{11} + x_{12}dx_{21} \end{pmatrix}.$$

(ii) From Ω given in (i) we obtain $\text{tr}(\Omega) = 0$.
(iii) We have (A a fixed element of $SL(2, \mathbb{R})$)

$$L_A X = AX = \begin{pmatrix} a_{11} & a_{12} \\ a_{21} & a_{22} \end{pmatrix} \begin{pmatrix} x_{11} & x_{12} \\ x_{21} & x_{22} \end{pmatrix}$$

$$= \begin{pmatrix} a_{11}x_{11} + a_{12}x_{21} & a_{11}x_{12} + a_{12}x_{22} \\ a_{21}x_{11} + a_{22}x_{21} & a_{21}x_{12} + a_{22}x_{22} \end{pmatrix}$$

with $\det(A) = a_{11}a_{22} - a_{12}a_{21} = 1$. Since $A^{-1}A = I_2$ we have

$$(AX)^{-1}d(AX) = (X^{-1}A^{-1})AdX = X^{-1}dX.$$

Thus Ω is a left-invariant matrix form. The entries of the matrix Ω are left-invariant forms.

Problem 80. Let $i, j \in \{1, 2, \ldots, n\}$. Consider the Lie algebra spanned by the operators $E_{ij} = (E_{ij})^\dagger$, $B_i^\dagger$, $B_i = (B_i^\dagger)^\dagger$, I (identity operator) and the non-zero commutators

$$[E_{ij}, E_{k\ell}] = \delta_{jk}E_{i\ell} - \delta_{i\ell}E_{kj}$$
$$[B_i, B_j^\dagger] = \delta_{ij}I$$
$$[E_{ij}, B_k^\dagger] = \delta_{jk}B_i^\dagger$$
$$[E_{ij}, B_k] = -\delta_{ik}B_j.$$

We define the operators

$$\widetilde{E}_{ij} := E_{ij} - B_i^\dagger B_j.$$

Find the commutators $[\widetilde{E}_{ij}, B_k^\dagger]$, $[\widetilde{E}_{ij}, B_k]$, $[\widetilde{E}_{ij}, \widetilde{E}_{k\ell}]$.

Solution 80. We obtain

$$[\widetilde{E}_{ij}, B_k^\dagger] = 0, \quad [\widetilde{E}_{ij}, B_k] = 0, \quad [\widetilde{E}_{ij}, \widetilde{E}_{k\ell}] = \delta_{jk}\widetilde{E}_{i\ell} - \delta_{i\ell}\widetilde{E}_{kj}.$$

Problem 81. The *parity operator* Π is defined by

$$\Pi \mathbf{r} := -\mathbf{r}, \quad \Pi(r, \theta, \phi) := (r, \pi - \theta, \pi + \phi).$$

Let $Y_{\ell m}$ be the spherical harmonics, where $\ell = 0, 1, \ldots$ and $m = -\ell, -\ell + 1, \ldots, +\ell$. Find $\Pi Y_{\ell m}$.

Solution 81. Since

$$Y_{\ell m}(\pi - \theta, \pi + \phi) = (-1)^{\ell} Y_{\ell m}(\theta, \phi)$$

we find

$$\Pi Y_{\ell m} = (-1)^{\ell} Y_{\ell m}.$$

This is an eigenvalue equation with eigenvalue $(-1)^{\ell}$.

Problem 82. Consider the Lie algebra $u(n)$ and $A_j \in u(n)$. Thus $A_j^* = -A_j$. The curvature

$$F_A = \sum_{j,k=1}^{4} F_{jk} dx_j \wedge dx_k$$

of a $u(n)$-valued connection differential one-form

$$A = \sum_{j=1}^{4} A_j dx_j$$

on $\mathbb{R}^4$ is

$$F_{jk} = [\nabla_j, \nabla_k] = -\frac{\partial}{\partial x_j} A_k + \frac{\partial}{\partial x_k} A_j + [A_j, A_k]$$

where

$$\nabla_j := \frac{\partial}{\partial x_j} - A_j$$

The connection (matrix-valued differential one-form) A is anti-self-dual Yang-Mills on $\mathbb{R}^4$ if

$$*F_A = -F_A$$

where $*$ is the Hodge star operator with respect to the Euclidean metric tensor field

$$g = dx_1 \otimes dx_1 + dx_2 \otimes dx_2 + dx_3 \otimes dx_3 + dx_4 \otimes dx_4.$$

The Hodge star operator is f-linear. We have $*(dx_1 \wedge dx_2) = dx_3 \wedge dx_4$, $*(dx_1 \wedge dx_3) = -dx_2 \wedge dx_4$.
(i) Write down the anti-self-dual Yang-Mills in coordinates.
(ii) Show that anti-self-dual Yang-Mills equation has a Lax pair representation. Set

$$z = x_1 + ix_2, \qquad w = x_3 + ix_4.$$

Thus

$$\nabla_z = \frac{1}{2}(\nabla_1 - i\nabla_2) = \frac{\partial}{\partial z} - A_z, \quad \nabla_{\bar{z}} = \frac{1}{2}(\nabla_1 + i\nabla_2) = \frac{\partial}{\partial \bar{z}} - A_{\bar{z}}$$

and ∇_w, $\nabla_{\bar{w}}$, analogously.

Solution 82.   (i) We have

$$F_{12} = -F_{34}, \quad F_{13} = -F_{42}, \quad F_{14} = -F_{23}.$$

(ii) Since $A_j \in u(n)$ we have $A_{\bar{z}} = -A_z^*$ and $A_{\bar{w}} = -A_w^*$, where $*$ denotes the transpose and complex conjugate. Thus we have the Lax representation

$$[\nabla_{\bar{w}} + \mu \nabla_z, \nabla_w - \mu^{-1}\nabla_{\bar{z}}] = 0$$

where $\mu \in \mathbb{C} \setminus \{0\}$. This system of partial differential equation is equivalent to the anti-self-dual Yang-Mills equation on $\mathbb{R}^4$. The equation holds for all $\mu \in \mathbb{C} \setminus \{0\}$ if and only if the coefficients of μ, 1, and μ^{-1} are zero, which is the anti-self-dual Yang-Mills equation.

Problem 83.   The equation of motion for the generalized anisotropic model is given by the autonomous system of first order differential equations

$$\frac{dr}{dt} = p_r$$
$$\frac{d\theta}{dt} = \frac{1}{r^2}p_\theta$$
$$\frac{dp_r}{dt} = \frac{1}{r^3}p_\theta^2 - \frac{dV}{dr}f(\theta)$$
$$\frac{dp_\theta}{dt} = -V(r)\frac{df}{d\theta}$$

where $f(\theta) = (\mu \cos^2 \theta + \sin^2 \theta)^{-x/2}$, μ is the mass ration and $x \in (0, 2)$. The conditions on the potential V are

$$\lim_{r \to +\infty} V(r) = 0, \quad \lim_{r \to 0^+} V(r) = -\infty$$

and there exist $c \in \mathbb{R}^+$, $x \in (0, 2)$ such that

$$\lim_{r \to 0^+} r^x V(r) = -c, \quad \lim_{r \to 0^+} r^{x+1}\frac{dV}{dr} = xc$$

$$\lim_{r \to 0^+} \frac{d}{dr}\left(r^{x+1}\frac{dV}{dr}\right) = c_1 \in \mathbb{R}, \quad \lim_{r \to 0^+} \frac{d}{dr}(r^x V(r)) = c_2 \in \mathbb{R}$$

and $dV/dr \geq -xr^{-1}V(r)$. The equation of motion can be derived from the Hamilton function

$$H(r, \theta, p_r, p_\theta) = \frac{p_r^2}{2} + \frac{p_\theta^2}{2r^2} + V(r)f(\theta).$$

Show that the singularity at $r = 0$ can be removed by a non-canonical change of variables of *McGehee's type*

$$(r, \theta, p_r, p_\theta, t) \mapsto (r, \theta, v = r^{x/2}p_r, u = r^{(x-2)/2}p_\theta, \tau)$$

where $dt/d\tau = r^{x/2+1}$.

Solution 83. We obtain the nonlinear autonomous system of first order differential equations

$$\frac{dr}{d\tau} = rv$$

$$\frac{d\theta}{d\tau} = u$$

$$\frac{dv}{d\tau} = \frac{1}{2}xv^2 + u^2 - r^{x+1}\frac{dV}{dr}f(\theta)$$

$$\frac{du}{d\tau} = \frac{1}{2}(x-2)uv - r^x V(r)\frac{df}{d\theta}.$$

These equations are regular at $r = 0$. The transformation maps the singularity $r = 0$ onto the two-dimensional torus

$$T = \{(r, \theta, v, u) \ : \ r = 0, \ \theta \in S^1, \ u^2 + v^2 = 2cf(\theta)\}.$$

Problem 84. (i) Let $b^\dagger$, b be *Bose creation* and *annihilation operators* with the commutation relations

$$[b, b^\dagger] = I, \qquad [b, b] = 0, \qquad [b^\dagger, b^\dagger] = 0$$

where I is the identity operator. Consider the operators

$$b^\dagger b, \quad b^\dagger, \quad b, \quad I.$$

Find the commutators.

(ii) Show that there is a non-hermitian faithful representation by 3×3 matrices

$$b^\dagger b \to M_{22} = \begin{pmatrix} 0 & 0 & 0 \\ 0 & 1 & 0 \\ 0 & 0 & 0 \end{pmatrix}$$

$$b^\dagger \to M_{23} = \begin{pmatrix} 0 & 0 & 0 \\ 0 & 0 & 1 \\ 0 & 0 & 0 \end{pmatrix}, \quad b \to M_{12} = \begin{pmatrix} 0 & 1 & 0 \\ 0 & 0 & 0 \\ 0 & 0 & 0 \end{pmatrix}$$

and

$$I \to M_{13} = \begin{pmatrix} 0 & 0 & 1 \\ 0 & 0 & 0 \\ 0 & 0 & 0 \end{pmatrix}.$$

Note that the identity operator I is not mapped into the 3×3 identity matrix.

Solution 84.   (i) For the commutators we find

$$[b^\dagger b, b^\dagger] = b^\dagger, \quad [b^\dagger b, b] = -b, \quad [b^\dagger, b] = -I.$$

All the other commutators are 0.
(ii) For the commutators of the matrices M_{22}, M_{23}, M_{12} and M_{13} we find

$$[M_{22}, M_{23}] = M_{23}, \quad [M_{22}, M_{12}] = -M_{12}, \quad [M_{22}, M_{13}] = 0$$

$$[M_{23}, M_{12}] = M_{23}, \quad [M_{23}, M_{13}] = 0, \quad [M_{12}, M_{13}] = 0.$$

Thus we have a faithful representation of the Lie algebra given by the operators $b^\dagger b$, $b^\dagger$, b, I.

Problem 85.   Consider the Lie algebra $su(1,1)$ of the noncompact Lie group $SU(1,1)$ with a basis given by K_+, K_-, K_3 satisfying the commutation relations

$$[K_3, K_+] = K_+, \quad [K_3, K_-] = -K_-, \quad [K_+, K_-] = -2K_3.$$

A faithful matrix representation is given by

$$K_+ = \begin{pmatrix} 0 & 1 \\ 0 & 0 \end{pmatrix}, \quad K_- = \begin{pmatrix} 0 & 0 \\ -1 & 0 \end{pmatrix}, \quad K_3 = \frac{1}{2}\begin{pmatrix} 1 & 0 \\ 0 & -1 \end{pmatrix}.$$

(i) Let $x \in \mathbb{R}$. Calculate

$$\exp(x(K_- - K_+)).$$

(ii) Let $\alpha, \beta, \gamma \in \mathbb{R}$. Calculate

$$\exp(\alpha K_+) \exp(\beta K_3) \exp(\gamma K_-).$$

(iii) Solve

$$\exp(x(K_- - K_+)) = \exp(\alpha K_+) \exp(\beta K_3) \exp(\gamma K_-)$$

for α, β, γ.

(iv) The result derived in (iii) is valid for all faithful representations. Let b, $b^\dagger$ be Bose annihilation and creation operators. Consider the faithful representation

$$K_+ = \frac{1}{2}b^\dagger b^\dagger, \quad K_- = \frac{1}{2}bb, \quad K_3 = \frac{1}{4}(b^\dagger b + bb^\dagger).$$

Find

$$\exp(\tanh(x)K_+)b\exp(-\tanh(x)K_+),$$

$$\exp(\tanh(x)K_-)b^\dagger \exp(-\tanh(x)K_-),$$

$$\exp(-2\ln(\cosh(x))K_3)(b^\dagger + \tanh(x)b)\exp(2\ln(\cosh(x))K_3).$$

(v) Let $|0\rangle$ be the vacuum state, i.e. $b|0\rangle = 0$. Calculate the state

$$\exp(-2\ln(\cosh(x))K_3)|0\rangle.$$

Solution 85. (i) We obtain

$$\exp(x(K_+ - K_-)) = \begin{pmatrix} \cosh(x) & -\sinh(x) \\ -\sinh(x) & \cosh(x) \end{pmatrix}.$$

(ii) We obtain

$$\exp(\alpha K_+)\exp(\beta K_3)\exp(\gamma K_-) = \begin{pmatrix} \exp(\beta/2) - \alpha\gamma\exp(-\beta/2) & \alpha\exp(-\beta/2) \\ -\gamma\exp(-\beta/2) & \exp(-\beta/2) \end{pmatrix}.$$

(iii) Solving the equation

$$\begin{pmatrix} \cosh(x) & -\sinh(x) \\ -\sinh(x) & \cosh(x) \end{pmatrix} = \begin{pmatrix} \exp(\beta/2) - \alpha\gamma\exp(-\beta/2) & \alpha\exp(-\beta/2) \\ -\gamma\exp(-\beta/2) & \exp(-\beta/2) \end{pmatrix}$$

for α, β, γ yields

$$\alpha = -\tanh(x), \quad \beta = -2\ln(\cosh(x)), \quad \gamma = \tanh(x).$$

(iv) We obtain

$$\exp(\tanh(x)K_+)b\exp(-\tanh(x)K_+) = b - \tanh(x)b^\dagger$$
$$\exp(\tanh(x)K_-)b^\dagger \exp(-\tanh(x)K_-) = b^\dagger + \tanh(x)b$$

and

$$\exp(-2\ln(\cosh(x))K_3)(b^\dagger + \tanh(x)b)\exp(2\ln(\cosh(x))K_3) = \mathrm{sech}(x)b^\dagger + \sinh(x)b$$

(v) Since $b|0\rangle = 0$ we obtain the state

$$\exp(-2\ln(\cosh(x))K_3)|0\rangle = (\operatorname{sech}(x))^{1/2}|0\rangle \,.$$

Problem 86. Let $b^\dagger$, b be Bose creation and annihilation operators with $[b, b^\dagger] = I$. Here I denotes the identity operator. Show that $b^\dagger b^\dagger$, bb, $b^\dagger b$, I form a Lie algebra under the commutator.

Solution 86. We find the non-zero commutators

$$[b^\dagger b^\dagger, bb] = -2I - 4b^\dagger b, \quad [b^\dagger b^\dagger, b^\dagger b] = -2b^\dagger b^\dagger, \quad [b^\dagger b, bb] = -2bb\,.$$

We see that the right-hand sides of the commutators are linear combinations of the basis elements.

Problem 87. In the Bose oscillator representation the generators of the Lie algebra $u(2)$ take the form

$$K_1 = b_1^\dagger b_1, \qquad K_2 = b_2^\dagger b_2, \qquad K_+ = b_2^\dagger b_1, \qquad K_- = b_1^\dagger b_2\,.$$

(i) Find the Casimir operator.
(ii) Write $u(2)$ as $u(2) = su(2) \oplus u(1)$.

Solution 87. (i) The *number operator*

$$\hat{N} = b_2^\dagger b_2 + b_1^\dagger b_1$$

is the Casimir operator, i.e. $[K_1, \hat{N}] = [K_2, \hat{N}] = [K_+, \hat{N}] = [K_-, \hat{N}] = 0$.
(ii) We can write

$$u(2) = \{K_+, K_-, K_0 = \frac{1}{2}(b_2^\dagger b_2 - b_1^\dagger b_1)\} \oplus \{\hat{N}\}$$

where the generators K_+, K_-, K_0 span the Lie subalgebra $su(2)$

$$[K_+, K_-] = 2K_0, \qquad [K_0, K_\pm] = \pm K_\pm\,.$$

Problem 88. Consider the Bose creation and annihilation operators $b^\dagger$ and b, respectively. Do the operators $b^\dagger b$ and $b^\dagger + b$ form a Lie algebra under the commutator? If not extend it so that one has a Lie algebra.

Solution 88. We find for the commutators

$$[b^\dagger b, b^\dagger + b] = b^\dagger - b, \quad [b^\dagger b, b^\dagger - b] = b^\dagger + b, \quad [b^\dagger + b, b^\dagger - b] = 2I\,.$$

The Jacobi identity is satisfied. Thus the operators $b^\dagger b$, $b^\dagger + b$, $b^\dagger - b$, I form a Lie algebra.

Problem 89. Let $b_j^\dagger$, b_j ($j = 1, 2$) be Bose creation and annihilation operators. Consider the Hamilton operator

$$\hat{H} = \epsilon(b_1^\dagger b_1 + b_2^\dagger b_2) + \gamma_1(b_1^\dagger b_1^\dagger b_1 b_1 + b_2^\dagger b_2^\dagger b_2 b_2) + \gamma_2 b_1^\dagger b_2^\dagger b_1 b_2 \,.$$

The Hamilton operator describes a quantum system of two nonlinear interacting oscillators. The two-mode representation of the Lie algebra $su(1, 1)$ is given by

$$J_- = b_1 b_2, \quad J_+ = b_1^\dagger b_2^\dagger, \quad J_0 = \frac{1}{2}(b_1^\dagger b_1 + b_2^\dagger b_2 + I) \,.$$

Express the Hamilton operator $\hat{H}$ using J_-, J_+, J_0.

Solution 89. We obtain

$$\hat{H} = (2\gamma_1 - \epsilon)I + (2\epsilon - 6\gamma_1)J_0 + 4\gamma_1 J_0^2 + (\gamma_2 - 2\gamma_1)J_+ J_- \,.$$

Problem 90. Consider the Bose creation $b_j^\dagger$ and Bose annihilation operators b_j, respectively where $j = 1, 2, \ldots, n$.
(i) Show that the vector space of operators

$$\{\, b_j^\dagger b_k \ : \ j, k = 1, 2, \ldots, n \,\}$$

form a Lie algebra under the commutator. There are n^2 such operators.
(ii) Consider the operator (so-called *number operator*)

$$\hat{N} = \sum_{j=1}^{n} b_j^\dagger b_j \,.$$

Calculate the commutator $[\hat{N}, b_k^\dagger b_\ell]$. Discuss.

Solution 90. (i) Applying the commutators $[b_j, b_k^\dagger] = \delta_{jk} I$ we obtain for the commutators

$$[b_j^\dagger b_k, b_\ell^\dagger b_m] = -\delta_{mj} b_\ell^\dagger b_k + \delta_{\ell k} b_j^\dagger b_m \,.$$

Thus the set is closed under the commutator. The Jacobi identity is also satisfied. Thus we have a non-abelian Lie algebra.

(ii) We obtain

$$[\hat{N}, b_k^\dagger b_\ell] = 0$$

for all $k, \ell = 1, 2, \ldots, n$. Thus the operator is an element of the center of the Lie algebra. Are there other nonzero elements in the center?

Problem 91.   Consider the Lie algebra $su(1,1)$ with the basis J_+, J_-, J_0 and the commutation relation

$$[J_+, J_-] = -2J_0, \qquad [J_0, J_\pm] = \pm J_\pm .$$

A two-mode representation of $su(1,1)$ is given by

$$J_- = b_1 b_2, \quad J_+ = b_1^\dagger b_2^\dagger, \quad J_0 = \frac{1}{2}(b_1^\dagger b_1 + b_2^\dagger b_2 + I)$$

with the *Casimir operator*

$$C = -\frac{1}{4}I + \frac{1}{4}(b_1^\dagger b_1 - b_2^\dagger b_2)^2 .$$

A single bosonic realization is

$$J_- = (2kI + b^\dagger b)^{1/2} b, \quad J_+ = b^\dagger (2kI + b^\dagger b)^{1/2}, \quad J_0 = kI + b^\dagger b$$

where $k = 0, 1, 2, \ldots$ and I is the identity operator. The Casimir operator is $C = k(k-1)I$. Consider the Hamilton operator

$$\hat{H}(t) = a_1(t) J_+ + a_2(t) J_0 + a_3(t) J_-$$

where a_1, a_2, a_3 are smooth functions of t. The *Schrödinger equation* is

$$\hat{H}(t)|\psi(t)\rangle = i\hbar \frac{\partial}{\partial t}|\psi(t)\rangle .$$

We define the evolution operator $U(t, 0)$ such that $|\phi(t)\rangle = U(t)|\phi(0)\rangle$, where $|\phi(0)\rangle$ is the initial state at time $t = 0$. Thus

$$\hat{H}(t)U(t) = i\hbar \frac{\partial U(t)}{\partial t}, \qquad U(0) = I .$$

Since J_-, J_0, J_+ form a basis of the finite dimensional Lie algebra $su(1,1)$, the evolution operator can be expressed in the form

$$U(t) = \exp(c_1(t)J_+) \exp(c_2(t)J_0) \exp(c_3(t)J_-) .$$

Find the system of ordinary differential equations for c_1, c_2, c_3 depending on $a_1(t)$, $a_2(t)$, $a_3(t)$.

Solution 91. By differentiation of $U(t)$ with respect to t we obtain

$$\frac{\partial U(t)}{\partial t} = \left(\left(\frac{dc_1}{dt} - c_1 \frac{dc_2}{dt} + c_1^2 \exp(-c_2) \frac{dc_3}{dt} \right) J_+ \right.$$

$$\left. + \left(\frac{dc_2}{dt} - 2c_1 \exp(-c_2) \frac{dc_3}{dt} \right) J_0 + \exp(-c_2) \frac{dc_3}{dt} J_- \right) U(t).$$

Inserting the Hamilton operator $\hat{H}$, the ansatz for $U(t)$, and $\partial U(t)/\partial t$ into the evolution equation for $U(t)$ and comparing the coefficients we obtain the system of first ordinary differential equations

$$\frac{dc_1}{dt} - c_1 \frac{dc_2}{dt} + c_1^2 \exp(-c_2) \frac{dc_3}{dt} = \frac{1}{i\hbar} a_1(t)$$

$$\frac{dc_2}{dt} - 2c_1 \exp(-c_2) \frac{dc_3}{dt} = \frac{1}{i\hbar} a_2(t)$$

$$\exp(-c_2) \frac{dc_3}{dt} = \frac{1}{i\hbar} a_3(t).$$

It follows that

$$\frac{dc_1}{dt} = \frac{1}{i\hbar} (a_1(t) + a_2(t)c_1 + a_3(t)c_1^2)$$

$$\frac{dc_2}{dt} = \frac{1}{i\hbar} (a_2(t) + 2a_3(t)c_1)$$

$$\frac{dc_3}{dt} = \frac{1}{i\hbar} a_3(t) \exp(c_2)$$

with the initial conditions $c_1(0) = c_2(0) = c_3(0) = 0$. The first equation is a *Riccati equation* for c_1. After solving this equation we obtain c_2 and c_3 as

$$c_2(t) = \frac{1}{i\hbar} \int_0^t (a_2(s) + 2a_3(s)c_1(s))ds, \quad c_3(t) = \frac{1}{i\hbar} \int_0^t a_3(s) \exp(c_2(s))ds.$$

Problem 92. Let b, $b^\dagger$ be Bose annihilation and creation operators. Consider the operators

$$T_1 = \frac{1}{4}((b^\dagger)^2 + b^2), \quad T_2 = \frac{i}{4}(b^2 - (b^\dagger)^2), \quad T_3 = \frac{1}{2}b^\dagger b + \frac{1}{4}I$$

where I is the identity operator. Find the commutators $[T_1, T_2]$, $[T_2, T_3]$, $[T_3, T_1]$. Discuss.

Solution 92. The commutators are

$$[T_1, T_2] = -iT_3, \quad [T_1, T_3] = -iT_2, \quad [T_2, T_3] = iT_1.$$

The Jacobi identity is satisfied. Thus the operators form a basis of a simple Lie algebra.

Problem 93.   Consider the differential operators

$$b := \frac{1}{\sqrt{2}} \left(x + \frac{d}{dx} \right), \qquad b^\dagger := \frac{1}{\sqrt{2}} \left(x - \frac{d}{dx} \right)$$

acting on the vector space $S(\mathbb{R})$. Find the commutator $[b, b^\dagger]$. Find the operator $\hat{N} = b^\dagger b$.

Solution 93.   Let $f \in S(\mathbb{R})$. We have

$$[b, b^\dagger]f = bb^\dagger f - b^\dagger b f$$

$$= \frac{1}{2} \left(x^2 f - \frac{d}{dx}(xf) - x\frac{df}{dx} - \frac{d^2 f}{dx^2} \right)$$

$$- \frac{1}{2} \left(x^2 f - \frac{d}{dx}(xf) + x\frac{df}{dx} - \frac{d^2 f}{dx^2} \right)$$

$$= \frac{d}{dx}(xf) - x\frac{df}{dx}$$

$$= f.$$

We obtain

$$\hat{N} = \frac{1}{2} \left(x^2 - 1 - \frac{d^2}{dx^2} \right).$$

Problem 94.   Let $c^\dagger$, c be *Fermi creation* and *Fermi annihilation operators*, i.e.

$$[c, c^\dagger]_+ = I$$

where $[\,,\,]_+$ denotes the anti-commutator and I is the identity operator. We also have $c^2 = 0$ and $(c^\dagger)^2 = 0$. Calculate the commutator

$$[c^\dagger c, c^\dagger + c].$$

Do the operators $c^\dagger c$, $(c^\dagger + c)$ form a basis of a Lie algebra? Can the set of operators be extended so that we have a basis of a Lie algebra?

Solution 94.   Since $c^\dagger c + cc^\dagger = I$, where I is the identity operator we find

$$[c^\dagger c, c^\dagger + c] = c^\dagger - c.$$

We see that we do not have a basis of a Lie algebra. Now we add the operator $(c^\dagger - c)$ to the set and calculate all the commutators. We obtain

$$[c^\dagger c, c^\dagger - c] = c^\dagger + c, \qquad [c^\dagger + c, c^\dagger - c] = 2(I - 2c^\dagger c).$$

From this result we see that we have a basis

$$\{ c^\dagger c, \ c^\dagger + c, \ c^\dagger - c, \ I \}$$

of a four dimensional Lie algebra. Thus we could also use the basis

$$\{ c^\dagger c, \ c^\dagger, \ c, \ I \}.$$

Is the Lie algebra semisimple? Extend the exercise to the operators

$$c_1^\dagger c_1, \quad c_2^\dagger c_2, \quad c_1^\dagger + c_1, \quad c_2^\dagger + c_2.$$

Problem 95. Let $c_\uparrow^\dagger$, $c_\downarrow^\dagger$ be Fermi creation operators with spin up and down, respectively. Let $c_\uparrow$, $c_\downarrow$ be Fermi annihilation operators with spin up and down, respectively. Let

$$\mathbf{c}^\dagger = \begin{pmatrix} c_\uparrow^\dagger & c_\downarrow^\dagger \end{pmatrix}, \qquad \mathbf{c} = \begin{pmatrix} c_\uparrow \\ c_\downarrow \end{pmatrix}.$$

The Fermi operators satisfy the anti-commutation relations

$$c_\uparrow^\dagger c_\uparrow + c_\uparrow c_\uparrow^\dagger = I, \quad c_\downarrow^\dagger c_\downarrow + c_\downarrow c_\downarrow^\dagger = I, \quad c_\uparrow^\dagger c_\downarrow + c_\downarrow c_\uparrow^\dagger = 0,$$

$$c_\uparrow^\dagger c_\uparrow^\dagger + c_\uparrow^\dagger c_\uparrow^\dagger = 0, \qquad c_\downarrow^\dagger c_\downarrow^\dagger + c_\downarrow^\dagger c_\downarrow^\dagger = 0,$$

$$c_\uparrow c_\uparrow + c_\uparrow c_\uparrow = 0, \qquad c_\downarrow c_\downarrow + c_\downarrow c_\downarrow = 0,$$

$$c_\uparrow^\dagger c_\downarrow^\dagger + c_\downarrow^\dagger c_\uparrow^\dagger = 0, \qquad c_\uparrow c_\downarrow + c_\downarrow c_\uparrow = 0.$$

(i) Let M, N be 2×2 matrices over $\mathbb{R}$. Show that

$$[\mathbf{c}^\dagger M \mathbf{c}, \mathbf{c}^\dagger N \mathbf{c}] = \mathbf{c}^\dagger [M, N] \mathbf{c}.$$

(ii) Thus we have a mapping $M \mapsto \mathbf{c}^\dagger M \mathbf{c}$. Consider the case that these matrices M and N are given by

$$X_+ = \begin{pmatrix} 0 & 1 \\ 0 & 0 \end{pmatrix}, \quad X_- = \begin{pmatrix} 0 & 0 \\ 1 & 0 \end{pmatrix}, \quad H = \begin{pmatrix} 1 & 0 \\ 0 & -1 \end{pmatrix}.$$

Find $\mathbf{c}^\dagger X_+ \mathbf{c}$, $\mathbf{c}^\dagger X_+ \mathbf{c}$, $\mathbf{c}^\dagger H \mathbf{c}$. Show that these operators form a Lie algebra under the commutator.

Solution 95. (i) We have

$$\mathbf{c}^\dagger M \mathbf{c} = \begin{pmatrix} c_\uparrow^\dagger c_\downarrow^\dagger \end{pmatrix} \begin{pmatrix} m_{11} & m_{12} \\ m_{21} & m_{22} \end{pmatrix} \begin{pmatrix} c_\uparrow \\ c_\downarrow \end{pmatrix} = m_{11} c_\uparrow^\dagger c_\uparrow + m_{12} c_\uparrow^\dagger c_\downarrow + m_{21} c_\downarrow^\dagger c_\uparrow + m_{22} c_\downarrow^\dagger c_\downarrow.$$

Analogously we obtain

$$\mathbf{c}^\dagger N \mathbf{c} = n_{11} c_\uparrow^\dagger c_\uparrow + n_{12} c_\uparrow^\dagger c_\downarrow + n_{21} c_\downarrow^\dagger c_\uparrow + n_{22} c_\downarrow^\dagger c_\downarrow .$$

Thus for the left-hand side we find

$$
\begin{aligned}
[\mathbf{c}^\dagger M \mathbf{c}, \mathbf{c}^\dagger N \mathbf{c}] = &-c_\downarrow^\dagger c_\downarrow m_{12} n_{21} + c_\downarrow^\dagger c_\downarrow m_{21} n_{12} - c_\downarrow^\dagger c_\uparrow m_{11} n_{21} + c_\downarrow^\dagger c_\uparrow m_{21} n_{11} \\
&- c_\downarrow^\dagger c_\uparrow m_{21} n_{22} + c_\downarrow^\dagger c_\uparrow m_{22} n_{21} + c_\uparrow^\dagger c_\downarrow m_{11} n_{12} - c_\uparrow^\dagger c_\downarrow m_{12} n_{11} \\
&+ c_\uparrow^\dagger c_\downarrow m_{12} n_{22} - c_\uparrow^\dagger c_\downarrow m_{22} n_{12} + c_\uparrow^\dagger c_\uparrow m_{12} n_{21} - c_\uparrow^\dagger c_\uparrow m_{21} n_{12} .
\end{aligned}
$$

From the commutator $[M, N]$ we obtain

$$
\begin{aligned}
\mathbf{c}^\dagger [M, N] \mathbf{c} = &-c_\downarrow^\dagger c_\downarrow m_{12} n_{21} + c_\downarrow^\dagger c_\downarrow m_{21} n_{12} - c_\downarrow^\dagger c_\uparrow m_{11} n_{21} + c_\downarrow^\dagger c_\uparrow m_{21} n_{11} \\
&- c_\downarrow^\dagger c_\uparrow m_{21} n_{22} + c_\downarrow^\dagger c_\uparrow m_{22} n_{21} + c_\uparrow^\dagger c_\downarrow m_{11} n_{12} - c_\uparrow^\dagger c_\downarrow m_{12} n_{11} \\
&+ c_\uparrow^\dagger c_\downarrow m_{12} n_{22} - c_\uparrow^\dagger c_\downarrow m_{22} n_{12} + c_\uparrow^\dagger c_\uparrow m_{12} n_{21} - c_\uparrow^\dagger c_\uparrow m_{21} n_{12}
\end{aligned}
$$

and the identity follows.

(ii) We obtain

$$\mathbf{c}^\dagger X_+ \mathbf{c} = c_\uparrow^\dagger c_\downarrow, \quad \mathbf{c}^\dagger X_- \mathbf{c} = c_\downarrow^\dagger c_\uparrow, \quad \mathbf{c}^\dagger H \mathbf{c} = c_\uparrow^\dagger c_\uparrow - c_\downarrow^\dagger c_\downarrow .$$

Thus for the commutators we find

$$
\begin{aligned}
[\mathbf{c}^\dagger X_+ \mathbf{c}, \mathbf{c}^\dagger X_- \mathbf{c}] &= \mathbf{c}^\dagger H \mathbf{c}, \\
[\mathbf{c}^\dagger X_+ \mathbf{c}, \mathbf{c}^\dagger H \mathbf{c}] &= -2 \mathbf{c}^\dagger X_+ \mathbf{c}, \\
[\mathbf{c}^\dagger X_- \mathbf{c}, \mathbf{c}^\dagger H \mathbf{c}] &= 2 \mathbf{c}^\dagger X_- \mathbf{c} .
\end{aligned}
$$

Problem 96.   *Fermi creation* $c_j^\dagger$ *and Fermi annihilation* c_j *operators* $(j = 1, \ldots, n)$ satisfy the anti-commutation relations

$$[c_j^\dagger, c_k]_+ = \delta_{jk} I$$

and $c_j^2 = 0$, $(c_j)^2 = 0$.

(i) Show that the vector space of operators

$$\{ c_j^\dagger c_k \; : \; j, k = 1, 2, \ldots, n \}$$

form a Lie algebra under the commutator. There are n^2 such operators.

(ii) Consider the operator (so-called *number operator*)

$$\hat{N} = \sum_{j=1}^{n} c_j^\dagger c_j .$$

Calculate the commutator $[\hat{N}, c_k^\dagger c_\ell]$. Discuss.

Solution 96. (i) Applying that $c_j^\dagger c_j^\dagger = 0$ and $c_j c_j = 0$ we obtain for the commutators

$$[c_j^\dagger c_k, c_\ell^\dagger c_m] = -\delta_{mj} c_\ell^\dagger c_k + \delta_{\ell k} c_j^\dagger c_m$$

Thus the set is closed under the commutator. The Jacobi identity is also satisfied. Thus we have a non-abelian Lie algebra.
(ii) We obtain

$$[\hat{N}, c_k^\dagger c_\ell] = 0$$

for all $k, \ell = 1, 2, \ldots, n$. Thus the operator is an element of the center of the Lie algebra. Are there other nonzero elements in the center?

Problem 97. Consider the *Jaynes-Cummings model*. It describes a two-level atom coupled linearly with a single bosonic mode. Let $c^\dagger$, $b^\dagger$ be Fermi and Bose creation operators, respectively. The Hamilton operator $\hat{H}$ is given by

$$\hat{H} = 2\omega_1 b^\dagger b \otimes I_F + 2\omega_2 I_B \otimes c^\dagger c + \lambda b \otimes c^\dagger + b^\dagger \otimes c\bar{\lambda}$$

where I_B is the identity operator (infinite dimensional unit matrix) for the bosons and I_F is the identity operator (2×2 unit matrix) for fermions. Thus the atom has the eigenstates with eigenvalues $E = 2\omega_2$ and $E = 0$. The interaction constant λ may be considered as a Grassmann or an ordinary c-valued number.
(i) Find the anticommutators

$$[b \otimes c^\dagger, b^\dagger \otimes c]_+, \qquad [b \otimes c^\dagger, b \otimes c^\dagger]_+, \qquad [b \otimes c, b \otimes c]_+ .$$

(ii) Find the commutators

$$[b \otimes I_F, b^\dagger \otimes c], \qquad [I_B \otimes c, b \otimes c^\dagger] .$$

(iii) Find the anticommutator

$$[I_B \otimes c, b \otimes c^\dagger]_+ .$$

Solution 97. (i) We obtain

$$\begin{aligned}
[b \otimes c^\dagger, b^\dagger \otimes c]_+ &= bb^\dagger \otimes c^\dagger c + b^\dagger b \otimes cc^\dagger \\
&= (I_B + b^\dagger b) \otimes c^\dagger c + b^\dagger b \otimes (I_F - c^\dagger c) \\
&= I_B \otimes c^\dagger c + b^\dagger b \otimes I_F .
\end{aligned}$$

Analogously we find since $(c^\dagger)^2 = 0_F$ and $c^2 = 0_F$

$$[bc^\dagger, b \otimes c^\dagger]_+ = 0_B \otimes 0_F, \qquad [b \otimes c, b \otimes c]_+ = 0_B \otimes 0_F .$$

(ii) For the commutators we obtain

$$[b \otimes I_F, b^\dagger \otimes c] = bb^\dagger \otimes c - b^\dagger b \otimes c$$
$$= (I_B + b^\dagger b) \otimes c - b^\dagger b \otimes c$$
$$= I_B \otimes c$$

and

$$[I_B \otimes c, b \otimes c^\dagger] = b \otimes I_F - 2b \otimes c^\dagger c.$$

(iii) We obtain

$$[I_B \otimes c, b \otimes c^\dagger]_+ = b \otimes I_F.$$

Problem 98.   Consider the Lie algebra $s\ell(2, \mathbb{R})$ with the basis H, X, Y and the commutators

$$[H, X] = 2X, \qquad [H, Y] = -2Y, \qquad [X, Y] = H.$$

Give faithful representations of $s\ell(2, \mathbb{R})$ using (i) 2×2 matrices, (ii) 3×3 matrices, (iii) vector fields, (iv) second order differential operators, (v) Bose operators, (vi) Fermi operators.

Solution 98.   (i) For the 2×2 matrices we have

$$x = \begin{pmatrix} 0 & 1 \\ 0 & 0 \end{pmatrix}, \quad y = \begin{pmatrix} 0 & 0 \\ 1 & 0 \end{pmatrix}, \quad h = \begin{pmatrix} 1 & 0 \\ 0 & -1 \end{pmatrix}.$$

(ii) For the 3×3 matrices we have

$$x = \begin{pmatrix} 0 & 0 & -2 \\ 0 & 0 & 0 \\ 0 & 1 & 0 \end{pmatrix}, \quad y = \begin{pmatrix} 0 & 0 & 0 \\ 0 & 0 & 2 \\ -1 & 0 & 0 \end{pmatrix}, \quad h = \begin{pmatrix} 2 & 0 & 0 \\ 0 & -2 & 0 \\ 0 & 0 & 0 \end{pmatrix}.$$

This is the adjoint representation of $s\ell(2, \mathbb{R})$.
(iii) A vector field representation is

$$V_x = z_2 \frac{\partial}{\partial z_1}, \quad V_y = z_1 \frac{\partial}{\partial z_2}, \quad V_h = z_1 \frac{\partial}{\partial z_1} - z_2 \frac{\partial}{\partial z_2}.$$

(iv) Second order differential operators are

$$D_x = z_2 \frac{\partial^2}{\partial z_1 \partial z_3}, \quad D_y = z_1 \frac{\partial^2}{\partial z_2 \partial z_3}, \quad D_h = z_1 \frac{\partial^2}{\partial z_1 \partial z_3} - z_2 \frac{\partial^2}{\partial z_2 \partial z_3}.$$

(v) Bose operators satisfy the relation $bb^\dagger = b^\dagger b + I$. A representation with Bose operators is

$$x = -\frac{i}{2} b^\dagger b^\dagger, \quad y = -\frac{i}{2} bb, \quad h = \frac{1}{2}(bb^\dagger + b^\dagger b).$$

(vi) Fermi operators satisfy the anti-commutation relation $cc^\dagger + c^\dagger c = I$, where c is the Fermi annihilation operator and $c^\dagger$ the Fermi creation operator. A representation with Fermi operators is

$$x = \sqrt{2}c, \quad y = \sqrt{2}c^\dagger, \quad h = \frac{1}{2}(cc^\dagger - c^\dagger c).$$

Problem 99. The $g\ell(1|1)$ superalgebra has two even and two odd generators we denote by h, z and e, f, respectively. The following commutation relations hold

$$[z, e] = [z, f] = [z, h] = 0, \quad [h, e] = e, \quad [h, f] = -f$$

and

$$[e, f]_+ = z, \qquad e^2 = f^2 = 0$$

where $[\,,\,]_+$ denotes the anticommutator. Find a representation by 2×2 matrices.

Solution 99. Let $\alpha, \beta \in \mathbb{R}$. Then

$$h = \begin{pmatrix} \alpha & 0 \\ 0 & \alpha - 1 \end{pmatrix}, \quad z = \begin{pmatrix} \beta & 0 \\ 0 & \beta \end{pmatrix}, \quad e = \begin{pmatrix} 0 & \beta \\ 0 & 0 \end{pmatrix}, \quad f = \begin{pmatrix} 0 & 0 \\ 1 & 0 \end{pmatrix}.$$

Problem 100. The Lie superalgebra $osp(1/2)$ has five generators K_0, K_+, K_-, F_+, F_-. The commutation relations are given by

$$[K_0, K_\pm] = \pm K_\pm, \qquad [K_+, K_-] = -2K_0.$$

$$[K_0, F_\pm] = \pm\frac{1}{2}F_\pm, \quad [K_\pm, F_\pm] = 0, \quad [K_\pm, F_\mp] = \mp F_\pm.$$

The anticommutation relations are given by

$$[F_\pm, F_\pm]_+ = K_\pm, \quad [F_+, F_-]_+ = K_0.$$

Thus the Lie superalgebra contains the sub Lie algebra $su(1, 1)$ spanned by K_0, K_+, K_-.
(i) Find the Casimir operator of the Lie superalgebra $osp(1/2)$.
(ii) Find a faithful representation with Bose operators $b^\dagger$, b and the identity operator I.
(iii) Introduce the number states (Fock states) $|n\rangle$ $(n = 0, 1, \ldots,)$ and apply the operators from (ii) to the number states.
(iv) Let $\beta, \gamma \in \mathbb{C}$. Introduce the operators

$$D(\beta) = \exp(\beta F_+ - \beta^* F_-), \qquad S(\gamma) = \exp(\gamma K_+ - \gamma^* K_-).$$

Calculate

$$D^\dagger(\beta)\begin{pmatrix} F_- \\ F_+ \end{pmatrix}D(\beta), \qquad S^\dagger(\gamma)\begin{pmatrix} F_- \\ F_+ \end{pmatrix}S(\gamma),$$

$$D^\dagger(\beta)\begin{pmatrix} K_- \\ K_+ \end{pmatrix}D(\beta), \qquad S^\dagger(\gamma)\begin{pmatrix} K_- \\ K_+ \end{pmatrix}S(\gamma).$$

(v) Consider the *coherent-squeezed state*

$$|\beta\gamma\rangle := S(\gamma)D(\beta)|0\rangle.$$

Calculate the state $|\beta\gamma\rangle$ using the number states $|n\rangle$ and the identity

$$\sum_{n=0}^\infty \frac{(t/2)^n}{n!} H_n(x)H_n(y) \equiv \frac{1}{\sqrt{1-t^2}}\exp((1-t)^{-1}(2xyt-(x^2+y^2)t))$$

where H_n $(n = 0, 1, \dots)$ are the Hermite polynomials.

Solution 100. (i) We obtain

$$C_2 = K_0^2 - \frac{1}{2}(K_+K_- + K_-K_+) + \frac{1}{2}(F_+F_- - F_-F_+).$$

(ii) Using the $[b, b^\dagger] = I$ we have

$$K_+ = \frac{1}{2}(b^\dagger)^2, \qquad K_- = \frac{1}{2}b^2, \qquad K_0 = \frac{1}{2}\left(b^\dagger b + \frac{1}{2}I\right)$$

$$F_+ = \frac{1}{2}b^\dagger, \qquad F_- = \frac{1}{2}b.$$

(iii) Applying the operators to the number state yields

$$K_+|n\rangle = \frac{1}{2}\sqrt{(n+1)(n+2)}|n+2\rangle,$$

$$K_-|n\rangle = \frac{1}{2}\sqrt{n(n-1)}|n-2\rangle,$$

$$K_0|n\rangle = \frac{1}{2}\left(n+\frac{1}{2}\right)|n\rangle,$$

$$F_+|n\rangle = \frac{1}{2}\sqrt{n+1}|n+1\rangle,$$

$$F_-|n\rangle = \frac{1}{2}\sqrt{n}|n-1\rangle.$$

(iv) Using the Baker-Campbell-Hausdorff formula yields

$$D^\dagger(\beta)\begin{pmatrix} F_- \\ F_+ \end{pmatrix}D(\beta) = \begin{pmatrix} F_- + \frac{1}{4}\beta I \\ F_+ + \frac{1}{4}\beta^* I \end{pmatrix}$$

$$S^\dagger(\gamma) \begin{pmatrix} F_- \\ F_+ \end{pmatrix} S(\gamma) = \begin{pmatrix} \cosh(r) & e^{i\theta}\sinh(r) \\ e^{-i\theta}\sinh(r) & \cosh(r) \end{pmatrix} \begin{pmatrix} F_- \\ F_+ \end{pmatrix}$$

$$D^\dagger(\beta) \begin{pmatrix} K_- \\ K_+ \end{pmatrix} D(\beta) = \begin{pmatrix} K_- + \beta F_- + \frac{1}{8}\beta^2 I \\ K_+ + \beta^* F_+ + \frac{1}{8}(\beta^*)^2 I \end{pmatrix}$$

$$S^\dagger(\gamma) \begin{pmatrix} K_- \\ K_+ \end{pmatrix} S(\gamma) = \begin{pmatrix} \cosh^2(r) & e^{2i\theta}\sinh^2(r) \\ e^{-2i\theta}\sinh^2(r) & \cosh^2(r) \end{pmatrix} \begin{pmatrix} K_- \\ K_+ \end{pmatrix}$$

$$+ \cosh(r)\sinh(r) \begin{pmatrix} e^{i\theta} I \\ e^{-i\theta} I \end{pmatrix}.$$

(v) Let $\gamma = re^{i\theta}$. We obtain

$$|\beta\gamma\rangle = \sum_{n=0}^{\infty} (n!\cosh(r))^{-1/2} \left(\frac{1}{2}e^{i\theta}\tanh(r) \right)^{n/2}$$

$$\times \exp\left(-\frac{1}{8}(|\beta|^2 - \beta^2 e^{i\theta}\tanh(r)) \right) H_n\left(\frac{\beta}{2}(e^{i\theta}\sinh(r)^{-1/2}) \right) |n\rangle.$$

Programming Problems

Problem 101.   Let $P = \{\, P_0,\, P_2,\, \ldots,\, P_{k-1} \,\}$ be a *semi-group* of permutations on the set $\Omega = \{\, 0,\, 1,\, \ldots,\, n-1 \,\}$. A semi-group is closed under the group operation, but the inverse and identity need not be a member of the set. The transitivity set (or orbit) containing $i \in \Omega$ is the set of images under the action of products of elements of P. Write a C++ program which finds the orbits given P and Ω.

Solution 101.   Since P is closed under products of elements of P, we need only consider the actions of elements of P. The image $P_j(i)$ of $i \in \Omega$ under the action of $P_j \in P$ is written as $\mathtt{im[j][i]} = P_j(i)$ in the program. In the program we consider the group of permutations on 3 elements and a subgroup of the group of permutations on 4 elements.

```
// transitivity.cpp

#include <iostream>
#include <map>
#include <vector>
using namespace std;

void orbits(map<int,map<int,int> > im,map<int,int> &ind,
            map<int,vector<int> > &orb)
{
  int k = im.size();
  int n = im.begin()->second.size();
  int i, j, orbits = 0, orbit;
  vector<bool> hasorbit(n,0);

  for(i=0;i<n;i++)
    if(!hasorbit[i])               // this is a new orbit
    {
      orbit = orbits;
      orbits++;
      orb[orbit] = vector<int>(1);
      orb[orbit][0] = i;           // the orbit starts with i
      ind[i] = orbit;
      hasorbit[i] = 1;
      for(j=0;j<k;j++)
      {
        int q = im[j][i];          // find each image of i
        if(!hasorbit[q])
        {
          hasorbit[q] = 1;
          ind[q] = orbit;          // each image is in the same orbit
          orb[orbit].push_back(q);
```

```cpp
        }
      }
    }
}

int main(void)
{
  int j, k;
  map<int,map<int,int> > im;
  map<int,int> ind;
  map<int,vector<int> > orb;
  cout << "Full permutation group." << endl;
  im[0][0] = 0; im[0][1] = 1; im[0][2] = 2; // identity
  im[1][0] = 0; im[1][1] = 2; im[1][2] = 1;
  im[2][0] = 1; im[2][1] = 0; im[2][2] = 2;
  im[3][0] = 1; im[3][1] = 2; im[3][2] = 0;
  im[4][0] = 2; im[4][1] = 0; im[4][2] = 1;
  im[5][0] = 2; im[5][1] = 1; im[5][2] = 0;
  orbits(im,ind,orb);
  cout << "Group element orbit indices:" << endl;
  for(j=0;j<int(ind.size());j++)
   cout << "ind[" << j << "] = " << ind[j] << endl;
  cout << "Orbits:" << endl;
  for(j=0;j<int(orb.size());j++)
  {
   cout << "orb[" << j << "] = ";
   for(k=0;k<int(orb[j].size());k++) cout << orb[j][k] << " ";
   cout << endl;
  }
  im.clear();
  cout << endl;
  cout << "Permutation subgroup." << endl;
  im[0][0] = 0; im[0][1] = 1; im[0][2] = 2; im[0][3] = 3;
  im[1][0] = 1; im[1][1] = 0; im[1][2] = 3; im[1][3] = 2;
  orbits(im,ind,orb);
  cout << "Group element orbit indices:" << endl;
  for(j=0;j<int(ind.size());j++)
   cout << "ind[" << j << "] = " << ind[j] << endl;
  cout << "Orbits:" << endl;
  for(j=0;j<int(orb.size());j++)
  {
   cout << "orb[" << j << "] = ";
   for(k=0;k<int(orb[j].size());k++) cout << orb[j][k] << " ";
   cout << endl;
  }
  return 0;
}
```

The program output is as follows. For the permutation group on 3 elements there is only one orbit. For the subgroup of the permutation group of 4 elements there are two orbits.

```
Full permutation group.
Group element orbit indices:
ind[0] = 0
ind[1] = 0
ind[2] = 0
Orbits:
orb[0] = 0 1 2

Permutation subgroup.
Group element orbit indices:
ind[0] = 0
ind[1] = 0
ind[2] = 1
ind[3] = 1
Orbits:
orb[0] = 0 1
orb[1] = 2 3
```

Problem 102.   Let $\mathcal{B}_n$ denote the *braid group* on $n-1$ strands. $\mathcal{B}_n$ is generated by the elementary braids (generators) $\{\, b_1, b_2, \ldots, b_{n-1}\,\}$ with the *braid relations*

$$b_j b_{j+1} b_j = b_{j+1} b_j b_{j+1}, \qquad 1 \le j < n-1,$$

$$b_j b_k = b_k b_j, \qquad |j-k| \ge 2.$$

Actually one should better write b_{12}, b_{23}, $\ldots$, $b_{n-1\,n}$ instead of b_1, b_2, $\ldots$, b_{n-1}. Let $\{\, e_1, e_2, \ldots, e_n \,\}$ denote the standard basis in $\mathbb{R}^n$. Then $\mathbf{u} \in \mathbb{R}^n$ can be written as

$$\mathbf{u} = \sum_{k=1}^{n} c_k e_k, \qquad c_1, c_2, \ldots, c_n \in \mathbb{R}.$$

Consider the operators B_j ($\alpha, \beta, \gamma, \delta \in \mathbb{R}$ and $\alpha, \gamma \ne 0$) defined by

$$B_j \mathbf{u} := c_1 e_1 + \ldots + (\alpha c_{j+1} + \beta) e_j + (\gamma c_{j+1} + \delta) e_{j+1} + \ldots + c_n e_n$$

and the corresponding inverse operation

$$B_j^{-1} \mathbf{u} := c_1 e_1 + \ldots + \frac{1}{\gamma}(c_{j+1} - \delta) e_j + \frac{1}{\alpha}(\gamma c_j - \beta) e_{j+1} + \ldots + c_n e_n.$$

Use computer algebra to show that B_1, B_2, $\ldots$, B_{n-1} satisfy the braid condition

$$B_j B_{j+1} B_j \mathbf{u} = B_{j+1} B_j B_{j+1} \mathbf{u}$$

if

$$\gamma\beta + \delta = \alpha\delta + \beta.$$

Solution 102. We use SymbolicC++. The function B(k,u) implements B_k acting on $u \in \mathbb{R}^n$ expressed in terms of the standard basis e[1], ..., e[n].

```cpp
// braid.cpp

#include <iostream>
#include "symbolicc++.h"
using namespace std;

Symbolic e = ~Symbolic("e");

Symbolic B(int j,const Symbolic &u)
{
 UniqueSymbol x;
 Symbolic r,alpha("alpha"),beta("beta"),gamma("gamma"),delta("delta");
 r = u[e[j]==gamma*x];
 r = r[e[j+1]==alpha*e[j]];
 r = r[x==e[j+1]];
 r += beta*e[j]+delta*e[j+1];
 return r;
}

int main(void)
{
 int n = 5, j, k, m;
 Symbolic c("c"), u;

 for(k=1;k<=n;k++) u += c[k]*e[k];

 for(k=1;k<n-1;k++)
 {
  Symbolic r = B(k,B(k+1,B(k,u)))-B(k+1,B(k,B(k+1,u)));
  cout << r << endl;
  for(j=1;j<=n;j++)
   cout << e[j] << ": " << (r.coeff(e[j])==0) << endl;
 }
 return 0;
}
```

The program output is

```
beta*gamma*e[2]+delta*e[2]-delta*alpha*e[2]-beta*e[2]
e[1]: 0 == 0
```

```
e[2]: beta*gamma+delta-delta*alpha-beta == 0
e[3]: 0 == 0
e[4]: 0 == 0
e[5]: 0 == 0
beta*gamma*e[3]+delta*e[3]-delta*alpha*e[3]-beta*e[3]
e[1]: 0 == 0
e[2]: 0 == 0
e[3]: beta*gamma+delta-delta*alpha-beta == 0
e[4]: 0 == 0
e[5]: 0 == 0
beta*gamma*e[4]+delta*e[4]-delta*alpha*e[4]-beta*e[4]
e[1]: 0 == 0
e[2]: 0 == 0
e[3]: 0 == 0
e[4]: beta*gamma+delta-delta*alpha-beta == 0
e[5]: 0 == 0
```

Problem 103.   Write a C++ program that generates all $n \times n$ permutation matrices. In main we use these matrices to find the permutation matrix that satisfies

$$P \begin{pmatrix} a_{00} & 0 & 0 & a_{01} \\ 0 & b_{00} & b_{01} & 0 \\ 0 & b_{10} & b_{11} & 0 \\ a_{10} & 0 & 0 & a_{11} \end{pmatrix} P^T = \begin{pmatrix} a_{00} & a_{01} \\ a_{10} & a_{11} \end{pmatrix} \oplus \begin{pmatrix} b_{00} & b_{01} \\ b_{10} & b_{11} \end{pmatrix}$$

where $\oplus$ is the direct sum.

Solution 103.

```
// permute.cpp

#include <iostream>
#include <vector>
#include "symbolicc++.h"
using namespace std;

vector<Symbolic> permutations(int n)
{
 Symbolic In = Symbolic("",n,n).identity();
 int lvar = 0;              // which loop variable
 int *lvars = new int[n];   // all the loop variables
 Symbolic permutation = In; // current permutation
 vector<Symbolic> permutations;
 int i, j;

 for(i=0;i<n;i++) lvars[i] = -1;
```

```
do
{
 // increment this loop's variable until we find an unused index
 // i.e. lvars[lvar]++;
 do
   for(lvars[lvar]++,i=lvar-1;i>=0;i--)
     if(lvars[i]==lvars[lvar]) break; // repetition of index
 while(i >= 0);                       // repeated, increment and try again

 if(lvars[lvar] < n)       // if the loop has not completed
 {
 for(j=0;j<n;j++) permutation(lvar,j) = In(lvars[lvar],j);
 if(lvar<n-1) lvar++;                 // descend into the subloop
 else permutations.push_back(permutation); // no more loops
 }                                    // store permutation
 else lvars[lvar--] = -1;  // reset this loop and return to
 }                         // the outer loop
 while(lvars[0] != -1);
 delete[] lvars;
 return permutations;
}

int main(void)
{
 int i;
 Symbolic a("a",2,2), b("b",2,2);
 Symbolic M1 = ((     a(0,0),        0,       0, a(0,1)),
               (Symbolic(0), b(0,0), b(0,1),       0),
               (Symbolic(0), b(1,0), b(1,1),       0),
               (     a(1,0),        0,       0, a(1,1)) );
 Symbolic M2 = dsum(a,b);

 vector<Symbolic> p = permutations(4);
 for(i=0;i<(int)p.size();i++)
   if(p[i]*M1*p[i].transpose()==M2) cout << p[i] << endl;
 return 0;
}
```

Problem 104. Let n be a positive integer. There are exactly as many irreducible representations of the *permutation group* S_n as there are *partitions* λ_j of n

$$n = \lambda_1 + \lambda_2 + \cdots + \lambda_n, \quad \lambda_1 \geq \lambda_2 \geq \cdots \geq \lambda_n \geq 0.$$

For example for $n = 4$ we have the 5 partitions

4000   3100   2200   2110   1111

Let k, n be positive integers. The number of partitions $p(1, n)$ can be found from the recursion

$$p(k,n) = \begin{cases} 0 & \text{if} \quad k > n \\ 1 & \text{if} \quad k = n \\ p(k+1,n) + p(k,n-k) \text{ otherwise} \end{cases}$$

(i) Give a C++ implementation of this recursion.
(ii) Give a recursive implementation to find the partitions.

Solution 104.   (i) The C++ implementation to find $p(1, k)$ is

```
// NoofPartitions.cpp

#include <iostream>
using namespace std;

int p(int k,int n)
{
 if(k > n) return 0;
 if(k==n) return 1;
 else return p(k+1,n) + p(k,n-k);
}

int main(void)
{
  int r1 = p(2,7);
  cout << "r1 = " << r1 << endl;
  int r2 = p(1,10);
  cout << "r2 = " << r2 << endl;
  return 0;
}
```

(ii) The find to actual partitions the C++ implementation is

```
// partitionsYorick.cpp

#include <iostream>
#include <vector>
using namespace std;

void partitions(int n,vector<int> p=vector<int>(0),int var=0,int sum=0)
{
 int i, max;

 if(p.size()==0) { p.resize(n,0); max = n; }
```

```
if(var==0) max = n; else max = p[var-1];

if(sum==n)
{
  for(i=0;i<var;i++) cout << p[i] << " ";
  for(i=var;i<n;i++) cout << 0 << " ";
  cout << endl;
}

if(var==n || sum >= n) return;

for(i=0;i<=max;i++) { p[var] = i; partitions(n,p,var+1,sum+i); }
}

int main(void)
{
  cout << "Partitions of 75" << endl;
  partitions(12);
  return 0;
}
```

Problem 105. A number of Hamilton systems can be written in the form

$$\frac{dL}{dt} = [A, L](t)$$

where L and A are time-dependent $n \times n$ matrices. This is called the *Lax representation* of the Hamilton system. We find that

$$\frac{dL^k}{dt} = [A, L^k](t)$$

and the that $\text{tr}(L^k)$ $(k = 1, 2, \ldots)$ are first integrals. Consider the Hamilton function (*Toda lattice*)

$$H(\mathbf{p}, \mathbf{q}) = \frac{1}{2}(p_1^2 + p_2^2 + p_3^2) + \exp(q_1 - q_2) + \exp(q_2 - q_3) + \exp(q_3 - q_1).$$

Introducing the quantities

$$a_j := \frac{1}{2}\exp\left(\frac{1}{2}(q_j - q_{j+1})\right), \qquad b_j := \frac{1}{2}p_j$$

and cyclic boundary conditions (i.e., $q_4 \equiv q_1$) we find that the Hamilton equations of motion take the form (with $b_3 = 0$)

$$\frac{da_j}{dt} = a_j(b_j - b_{j+1}), \qquad \frac{db_1}{dt} = -2a_1^2, \qquad \frac{db_2}{dt} = 2(a_1^2 - a_2^2), \qquad \frac{db_3}{dt} = 2a_2^2$$

where $j = 1, 2$. Introducing the matrices (*Lax pair*)

$$L := \begin{pmatrix} b_1 & a_1 & 0 \\ a_1 & b_2 & a_2 \\ 0 & a_2 & b_3 \end{pmatrix}, \qquad A := \begin{pmatrix} 0 & -a_1 & 0 \\ a_1 & 0 & -a_2 \\ 0 & a_2 & 0 \end{pmatrix}$$

the equations of motion can be written as Lax representation. From L we find the first integral as $\mathrm{tr}(L^n)$, where $n = 1, 2, \ldots$, where tr denotes the trace. We obtain

$$\mathrm{tr}L = b_1 + b_2 + b_3, \qquad \mathrm{tr}(L^2) = b_1^2 + b_2^2 + b_3^2 + 2a_1^2 + 2a_2^2.$$

Write a SymbolicC++ program `lax.cpp` that finds $[L, A]$ and shows that $\mathrm{tr}(L)$, $\mathrm{tr}(L^2)$ and the determinant of L are first integrals.

Solution 105.

```
// lax.cpp

#include <iostream>
#include "symbolicc++.h"
using namespace std;

int main(void)
{
 Symbolic L("L",3,3), A("A",3,3), Lt("Lt",3,3); // Lt=dL/dt
 Symbolic a1("a1"), a2("a2"), b1("b1"), b2("b2"), b3("b3"),
          a1t, a2t, b1t, b2t, b3t;
 L(0,0) = b1; L(0,1) = a1;  L(0,2) = 0;
 L(1,0) = a1; L(1,1) = b2;  L(1,2) = a2;
 L(2,0) = 0;  L(2,1) = a2;  L(2,2) = b3;
 A(0,0) = 0;  A(0,1) = -a1; A(0,2) = 0;
 A(1,0) = a1; A(1,1) = 0;   A(1,2) = -a2;
 A(2,0) = 0;  A(2,1) = a2;  A(2,2) = 0;
 Lt = A*L-L*A;
 cout << "Lt = " << Lt << endl;
 b1t = Lt(0,0); b2t = Lt(1,1); b3t = Lt(2,2);
 a1t = Lt(0,1); a2t = Lt(1,2);
 cout << "b1t = " << b1t << ", b2t = " << b2t
      << ", b3t = " << b3t << endl;
 cout << "a1t = " << a1t << ", a2t = " << a2t << endl;
 cout << endl;

 // I(0),I(1),I(2) are first integrals
 int n = 3;
 Symbolic result;
 Symbolic I("I",n);
```

```
I(0) = L.trace();      cout << "I(0) = " << I(0) << endl;
I(1) = (L*L).trace();  cout << "I(1) = " << I(1) << endl;
I(2) = L.determinant(); cout << "I(2) = " << I(2) << endl;
cout << endl;
for(int i=0;i<n;i++)
{
result = b1t*df(I(i),b1)+b2t*df(I(i),b2) + b3t*df(I(i),b3)
         +a1t*df(I(i),a1)+a2t*df(I(i),a2);
cout << "result" << i+1 << " = " << result << endl;
}
return 0;
}
```

The output is

```
Lt = [-2*a1^(2) -a1*b2+b1*a1 0]
[a1*b1-b2*a1 2*a1^(2)-2*a2^(2) -a2*b3+b2*a2]
[0 a2*b2-b3*a2 2*a2^(2)]
b1t = -2*a1^(2), b2t = 2*a1^(2)-2*a2^(2), b3t = 2*a2^(2)
a1t = -a1*b2+b1*a1, a2t = -a2*b3+b2*a2
I[0] = b1+b2+b3
I[1] = b1^(2)+2*a1^(2)+b2^(2)+2*a2^(2)+b3^(2)
I[2] = b1*b2*b3-b1*a2^(2)-a1^(2)*b3
result1 = 0
result2 = 0
result3 = 0
```

Problem 106. The n-qubit *Pauli group* is defined by

$$\mathcal{P}_n := \{ I_2, \sigma_x, \sigma_y, \sigma_z \}^{\otimes n} \otimes \{ \pm 1, \pm i \} \qquad (1)$$

where σ_x, σ_y, σ_z are the 2×2 Pauli matrices and I_2 is the 2×2 identity matrix. The dimension of the Hilbert space under consideration is dim $\mathcal{H} = 2^n$. Thus each element of the Pauli group $\mathcal{P}_n$ is (up to an overall phase ± 1, $\pm i$) a Kronecker product of Pauli matrices and 2×2 identity matrices acting on n qubits. The order of the Pauli group is 2^{2n+2}. Thus for $n = 1$ we have the order 16. For $n = 2$ we have order 64. Write a SymbolicC++ program that implements the Pauli group for $n = 2$.

Solution 106. All 64 elements of $\mathcal{P}_2$ are generated. Then we calculate the product

$$(\sigma_x \otimes \sigma_x)(\sigma_z \otimes \sigma_z) = -\sigma_y \otimes \sigma_y .$$

The group element $\sigma_x \otimes \sigma_x$ is given by **g(5)** since

$$5 = 0 \cdot 16 + 3 \cdot 4 + 1$$

i.e. $j = 0$, $k = 1$, $\ell = 1$. The group element **g**(15) is given by $\sigma_z \otimes \sigma_z$ since

$$15 = 0 \cdot 16 + 3 \cdot 4 + 3$$

i.e. $j = 0$, $k = 3$, $\ell = 3$. For the element **g**(42) we have

$$42 = 2 \cdot 16 + 2 \cdot 4 + 2$$

i.e. $j = 2$, $k = 2$, $\ell = 2$.

```cpp
// Pauli_group.cpp

#include <iostream>
#include <map>
#include <sstream>
#include <string>
#include "symbolicc++.h"
using namespace std;

int operator < (const Symbolic &s1,const Symbolic &s2)
{
ostringstream os1, os2;
os1 << s1; os2 << s2;
return (os1.str() < os2.str());
}

int main(void)
{
int j,k,l;
using SymbolicConstant::i;
map<Symbolic,Symbolic> rep;
Symbolic I2 = Symbolic("",2,2).identity();
Symbolic I4 = Symbolic("",4,4).identity();
Symbolic sx = ((Symbolic(0),Symbolic(1)),
               (Symbolic(1),Symbolic(0)));
Symbolic sy = ((Symbolic(0),-i),
               (i,Symbolic(0)));
Symbolic sz = ((Symbolic(1),Symbolic(0)),
               (Symbolic(0),-Symbolic(1)));
Symbolic g("g",64), gr("gr",64), s("s",4);
s(0) = I2; s(1) = sx; s(2) = sy; s(3) = sz;
for(j=0;j<4;j++)
for(k=0;k<4;k++)
for(l=0;l<4;l++)
{
gr(j*16+k*4+l) = ((i^j)*kron(s(k),s(l)))[(i^3)==-i];
rep[gr(j*16+k*4+l)] = g(j*16+k*4+l);
cout << gr(j*16+k*4+l) << " => " << rep[gr(j*16+k*4+l)] << endl;
```

```
}
cout << gr(5)*gr(15) << " => " << rep[gr(5)*gr(15)] << endl;
return 0;
}
```

Problem 107. Let $b_1^\dagger$, $b_2^\dagger$ be Bose creation operator and let I be the identity operator. The semi-simple Lie algebra $su(1,1)$ is generated by

$$K_+ := b_1^\dagger b_2^\dagger, \quad K_- := b_1 b_2, \quad K_0 := \frac{1}{2}(b_1^\dagger b_1 + b_2^\dagger b_2 + I)$$

with the commutation relations

$$[K_0, K_+] = K_+, \quad [K_0, K_-] = -K_-, \quad [K_-, K_+] = 2K_0.$$

We use the ordering K_+, K_-, K_0 for the basis. Write a computer algebra program that finds the *ajoint representation* for the Lie algebra.

Solution 107. Using the ordering given above for the basis we have

$$(\mathrm{ad}K_+)K_+ = [K_+, K_+] = 0$$
$$(\mathrm{ad}K_+)K_- = [K_+, K_-] = -2K_0$$
$$(\mathrm{ad}K_+)K_0 = [K_+, K_0] = -K_+.$$

Thus

$$(K_+ \quad K_- \quad K_0) \begin{pmatrix} 0 & 0 & -1 \\ 0 & 0 & 0 \\ 0 & -2 & 0 \end{pmatrix} = (0 \quad -2K_0 \quad -K_+).$$

Since K_+, K_-, K_0 is a basis of the Lie algebra it follows that

$$\mathrm{ad}K_+ = \begin{pmatrix} 0 & 0 & -1 \\ 0 & 0 & 0 \\ 0 & -2 & 0 \end{pmatrix}.$$

Analogously we find

$$\mathrm{ad}K_- = \begin{pmatrix} 0 & 0 & 0 \\ 0 & 0 & 1 \\ 2 & 0 & 0 \end{pmatrix}, \quad \mathrm{ad}K_0 = \begin{pmatrix} 1 & 0 & 0 \\ 0 & -1 & 0 \\ 0 & 0 & 0 \end{pmatrix}.$$

With $X(0) = K_+$, $X(1) = K_-$, $X(2) = K_0$ the SymbolicC++ implementation is:

```
// adjointrep.cpp
// g++ -I . adjoint.cpp -o adjoint

#include <iostream>
#include "symbolicc++.h"
using namespace std;

int main(void)
{
 int i, j, k;
 const int n=3;
 Symbolic ad[n];
 for(j=0;j<n;j++) ad[j] = Symbolic("",n,n);

 Symbolic X("X",n);
 Symbolic Y("Y",n,n);
 Y(0,0) = 0; Y(0,1) = -2*X(2); Y(0,2) = -X(0);
 Y(1,0) = 2*X(2); Y(1,1) = 0; Y(1,2) = X(1);
 Y(2,0) = X(0); Y(2,1) = -X(1); Y(2,2) = 0;

 for(i=0;i<n;i++)
  for(j=0;j<n;j++)
   for(k=0;k<n;k++) ad[i](j,k) = Y(i,k).coeff(X(j));

 for(j=0;j<n;j++)
   cout << "ad[" << j << "]" << "=" << ad[j] << endl;
 return 0;
}
```

Problem 108.   The exceptional Lie algebra g_2 has rank 2 and dimension 14. A basis is given by

$$H_1, H_2, X_1, X_2, X_3, X_4, X_5, X_6, Y_1, Y_2, Y_3, Y_4, Y_5, Y_6$$

with the commutation relations $[H_1, H_2] = 0$ and

$$H_1 = [X_1, Y_1], \qquad H_2 = [X_2, Y_2], \qquad [H_1, X_1] = 2X_1, \qquad [H_2, X_2] = 2X_2 .$$

Thus

$$[H_1, Y_1] = -2Y_1, \qquad [H_2, Y_2] = -2Y_2 .$$

Thus H_1, X_1, Y_1 and H_2, X_2, Y_2 each span the sub Lie algebra $s\ell(2, \mathbb{C})$ of g_2. The commutator table for g_2 is given in Table 4.1. Note that $s\ell(3, \mathbb{C})$ is a sub Lie algebra of g_2. Write a SymbolicC++ program that finds the adjoint representation.

Solution 108.   The two-dimensional array c(i,j) stores the commutators.

Table 4.1: Commutator table for g_2

	H_2	X_1	Y_1	X_2	Y_2	X_3	Y_3	X_4	Y_4	X_5	Y_5	X_6	Y_6
H_1	0	$2X_1$	$-2Y_1$	$-3X_2$	$3Y_2$	$-X_3$	Y_3	X_4	$-Y_4$	$3X_5$	$-3Y_5$	0	0
H_2		$-X_1$	Y_1	$2X_2$	$-2Y_2$	X_3	$-Y_3$	0	0	$-X_5$	Y_5	X_6	$-Y_6$
X_1			H_1	X_3	0	$2X_4$	$-3Y_2$	$-3X_5$	$-2Y_3$	0	Y_4	0	0
Y_1				0	$-Y_3$	$3X_2$	$-2Y_4$	$2X_3$	$3Y_5$	$-X_4$	0	0	0
X_2					H_2	0	Y_1	0	0	$-X_6$	0	0	Y_5
Y_2						$-X_1$	0	0	0	0	Y_6	$-X_5$	0
X_3							H_1+3H_2	$-3X_6$	$2Y_1$	0	0	0	Y_4
Y_3								$-2X_1$	$3Y_6$	0	0	$-X_4$	0
X_4									$2H_1+3H_2$	0	$-Y_1$	0	Y_3
Y_4										X_1	0	X_3	0
X_5											H_1+H_2	0	$-Y_2$
Y_5												X_2	0
X_6													H_1+2H_2

```
// adjoint2.cpp

#include <iostream>
#include "symbolicc++.h"
using namespace std;

int main(void)
{
  int n=14, i, j ,k;
  Symbolic V("V",n), H("H",2), X("X",6), Y("Y",6);
  Symbolic c("",n,n);
  Symbolic adV("adV",n);

  V(0) = H(0); V(1) = H(1);
  V(2) = X(0); V(3) = X(1); V(4) = X(2); V(5) = X(3);
  V(6) = X(4); V(7) = X(5); V(8) = Y(0); V(9) = Y(1);
  V(10) = Y(2); V(11) = Y(3); V(12) = Y(4); V(13) = Y(5);

  c(0,0) = Symbolic(0); c(0,1) = Symbolic(0);
  c(0,2) = -2*X(0); c(0,3) = -3*X(1); c(0,4) = -X(2);
  c(0,5) = X(3); c(0,6) = 3*X(4); c(0,7) = Symbolic(0);
  c(0,8) = -2*Y(0); c(0,9) = -3*Y(1); c(0,10) = Y(2);
  c(0,11) = -Y(3); c(0,12) = -3*Y(4); c(0,13) = Symbolic(0);
  c(1,1) = Symbolic(0); c(1,2) = -X(0); c(1,3) = 2*X(1);
  c(1,4) = X(2); c(1,5) = Symbolic(0); c(1,6) = -X(4); c(1,7) = X(5);
  c(1,8) = Y(0); c(1,9) = -2*Y(1); c(1,10) = -Y(2);
  c(1,11) = Symbolic(0); c(1,12) = Y(4); c(1,13) = -Y(5);
  c(2,2) = Symbolic(0); c(2,3) = X(2); c(2,4) = 2*X(3);
  c(2,5) = -3*X(4); c(2,6) = Symbolic(0); c(2,7) = Symbolic(0);
  c(2,8) = H(0); c(2,9) = Symbolic(0); c(2,10) = -3*Y(1);
  c(2,11) = -2*Y(2); c(2,12) = Y(3); c(2,13) = Symbolic(0);
  c(3,3) = Symbolic(0); c(3,4) = Symbolic(0); c(3,5) = Symbolic(0);
  c(3,6) = -X(5); c(3,7) = Symbolic(0);
  c(3,8) = Symbolic(0); c(3,9) = H(1); c(3,10) = Y(0);
  c(3,11) = Symbolic(0); c(3,12) = Symbolic(0); c(3,13) = Y(4);
  c(4,4) = Symbolic(0); c(4,5) = -3*X(5); c(4,6) = Symbolic(0);
  c(4,7) = Symbolic(0); c(4,8) = -3*X(1); c(4,9) = X(0);
  c(4,10) = H(0)+3*H(1); c(4,11) = 2*Y(0); c(4,12) = Symbolic(0);
  c(4,13) = Y(3);
  c(5,5) = Symbolic(0); c(5,6) = Symbolic(0); c(5,7) = Symbolic(0);
  c(5,8) = -2*X(2); c(5,9) = Symbolic(0); c(5,10) = 2*X(0);
  c(5,11) = 2*H(0)+3*H(1); c(5,12) = -Y(0); c(5,13) = Y(2);
  c(6,6) = Symbolic(0); c(6,7) = Symbolic(0); c(6,8) = X(3);
  c(6,9) = Symbolic(0); c(6,10) = Symbolic(0); c(6,11) = -X(0);
  c(6,12) = H(0)+H(1); c(6,13) = -Y(1);
  c(7,7) = Symbolic(0); c(7,8) = Symbolic(0); c(7,9) = X(4);
  c(7,10) = X(3); c(7,11) = -X(2); c(7,12) = -X(1); c(7,13) = H(0)+H(1);
```

```
c(8,8) = Symbolic(0); c(8,9) = -Y(2); c(8,10) = -2*Y(3);
c(8,11) = 3*Y(4); c(8,12) = Symbolic(0); c(8,13) = Symbolic(0);
c(9,9) = Symbolic(0); c(9,10) = Symbolic(0); c(9,11) = Symbolic(0);
c(9,12) = Y(5); c(9,13) = Symbolic(0);
c(10,10) = Symbolic(0); c(10,11) = 3*Y(5); c(10,12) = Symbolic(0);
c(10,13) = Symbolic(0);
c(11,11) = Symbolic(0); c(11,12) = Symbolic(0); c(11,13) = Symbolic(0);
c(12,12) = Symbolic(0); c(12,13) = Symbolic(0);
c(13,13) = Symbolic(0);

for(i=0; i<n; i++)
 for(j=0; j<i; j++) c(i,j) = -c(j,i);

for(i=0; i<n; i++)
{
 Symbolic rep("",n,n);
 for(j=0;j<n;j++)
  for(k=0;k<n;k++) rep(k,j) = c(i,j).coeff(V(k));
 adV(i) = rep;
 cout << "ad" << V(i) << " = " << adV(i) << endl;
}
 return 0;
}
```

Problem 109. Let b_1, b_2, b_3 be Bose annihilation operators. Show that

$$H_1 = b_1^\dagger b_1 - b_2^\dagger b_2, \qquad H_2 = b_2^\dagger b_2 - b_3^\dagger b_3$$

$$E_{12} = b_1^\dagger b_2, \quad E_{23} = b_2^\dagger b_3, \quad E_{13} = b_1^\dagger b_3$$

$$E_{21} = b_2^\dagger b_1, \quad E_{32} = b_3^\dagger b_2, \quad E_{31} = b_3^\dagger b_1$$

are a representation of the Lie algebra $su(3)$. Write a SymbolicC++ program that implements this representation.

Solution 109. The commutators are $[H_1, H_2] = 0$ and

$$[H_1, E_{12}] = 2E_{12}, \quad [H_1, E_{23}] = -E_{23}, \quad [H_1, E_{13}] = E_{13}$$

$$[H_2, E_{12}] = -E_{12}, \quad [H_2, E_{23}] = 2E_{23}, \quad [H_2, E_{13}] = E_{13}$$

$$[E_{12}, E_{21}] = H_1, \quad [E_{23}, E_{32}] = H_2, \quad [E_{12}, E_{23}] = E_{13}\,.$$

The program is

```
// comm.cpp

#include <iostream>
```

```
#include "symbolicc++.h"
using namespace std;

int main(void)
{
 int j, k;
 Symbolic b = ~Symbolic("b",3), bd = ~Symbolic("bd",3);
 Symbolic H1 = bd(0)*b(0)-bd(1)*b(1), H2 = bd(1)*b(1)-bd(2)*b(2);
 Symbolic E12 = bd(0)*b(1), E23 = bd(1)*b(2), E13 = bd(0)*b(2);
 Symbolic E21 = bd(1)*b(0), E32 = bd(2)*b(1), E31 = bd(2)*b(0);
 Equations rules;

 for(j=0;j<3;j++)
  for(k=0;k<3;k++)
  {
   if(k!=j) rules = (rules,b(k)*bd(j)==bd(j)*b(k));
   else     rules = (rules,b(j)*bd(j)==1+bd(j)*b(j));
   if(j > k)
    rules = (rules,b(j)*b(k)==b(k)*b(j),bd(j)*bd(k)==bd(k)*bd(j));
  }

 rules = (rules,H1==Symbolic("H1"),H2==Symbolic("H2"),
         E12==Symbolic("E12"),E23==Symbolic("E23"),
         E13==Symbolic("E13"),E21==Symbolic("E21"),
         E32==Symbolic("E32"),E31==Symbolic("E31"));

 cout << "[H1,H2] = " << (H1*H2 - H2*H1).subst_all(rules) << endl;
 cout << "[H1,E12] = " << (H1*E12-E12*H1).subst_all(rules) << endl;
 cout << "[H1,E23] = " << (H1*E23-E23*H1).subst_all(rules) << endl;
 cout << "[H1,E13] = " << (H1*E13-E13*H1).subst_all(rules) << endl;
 cout << "[H2,E12] = " << (H2*E12-E12*H2).subst_all(rules) << endl;
 cout << "[H2,E23] = " << (H2*E23-E23*H2).subst_all(rules) << endl;
 cout << "[H2,E13] = " << (H2*E13-E13*H2).subst_all(rules) << endl;
 cout << "[E12,E21] = " << (E12*E21-E21*E12).subst_all(rules) << endl;
 cout << "[E23,E32] = " << (E23*E32-E32*E23).subst_all(rules) << endl;
 cout << "[E12,E23] = " << (E12*E23-E23*E12).subst_all(rules) << endl;
 return 0;
}
```

Problem 110. Consider the fundamental representation of the superalgebra $su(2|1)$. Let σ_1, σ_2, σ_3 be the Pauli spin matrices. Let $\oplus$ denote the direct sum. Its generators are given by the 3×3 matrices

$$L_1 = \frac{1}{2}\sigma_1 \oplus (0), \quad L_2 = \frac{1}{2}\sigma_2 \oplus (0), \quad L_3 = \frac{1}{2}\sigma_3 \oplus (0), \quad L_4 = \frac{1}{2}\begin{pmatrix} 1 & 0 & 0 \\ 0 & 1 & 0 \\ 0 & 0 & 2 \end{pmatrix}$$

$$V_1 = \frac{1}{2} \begin{pmatrix} 0 & 0 & 1 \\ 0 & 0 & 0 \\ 1 & 0 & 0 \end{pmatrix}, \quad V_2 = \frac{1}{2} \begin{pmatrix} 0 & 0 & -i \\ 0 & 0 & 0 \\ i & 0 & 0 \end{pmatrix}$$

$$W_1 = \frac{1}{2} \begin{pmatrix} 0 & 0 & 0 \\ 0 & 0 & 1 \\ 0 & 1 & 0 \end{pmatrix}, \quad W_2 = \frac{1}{2} \begin{pmatrix} 0 & 0 & 0 \\ 0 & 0 & -i \\ 0 & i & 0 \end{pmatrix}$$

Let

$$L_\pm = L_1 \pm iL_2, \quad V_\pm = V_1 \pm iV_2, \quad W_\pm = W_1 \pm iW_2\,.$$

Here L_1, L_2, L_3, L_4 are the generators forming the Lie subalgebra $su(2) \otimes u(1)$ of $su(2|1)$ and V_1, V_2, W_1, W_2 are the supergenerators. Let $c^\dagger$, c be Fermi creation and annihilation operators. Let $b_1^\dagger$, $b_2^\dagger$ be Bose creation operators. A Boson-Fermion realization is given by

$$L_+ = b_1^\dagger b_2 \otimes I_F, \quad L_- = b_2^\dagger b_1 \otimes I_F$$

$$L_3 = \frac{1}{2}(b_1^\dagger b_1 \otimes I_F - b_2^\dagger b_2 \otimes I_F), \quad L_4 = \frac{1}{2}(b_1^\dagger b_1 \otimes I_F + b_2^\dagger b_2 \otimes I_F) + I_B \otimes c^\dagger c$$

$$V_+ = b_1^\dagger \otimes c, \quad V_- = b_1 \otimes c^\dagger, \quad W_+ = b_2^\dagger \otimes c, \quad W_- = b_2 \otimes c^\dagger\,.$$

Show that the representations are isomorphic. Give a SymbolicC++ implementation for the Bose-Fermi realization.

Solution 110. We have

$$V_+ = V_1 + iV_2 = \begin{pmatrix} 0 & 0 & 1 \\ 0 & 0 & 0 \\ 0 & 0 & 0 \end{pmatrix}, \quad V_- = V_1 - iV_2 = \begin{pmatrix} 0 & 0 & 0 \\ 0 & 0 & 0 \\ 1 & 0 & 0 \end{pmatrix}\,.$$

Thus for the anticommutator $[V_+, V_-]_+$ we find

$$[V_+, V_-]_+ = \begin{pmatrix} 1 & 0 & 0 \\ 0 & 0 & 0 \\ 0 & 0 & 1 \end{pmatrix} = L_3 + L_4\,.$$

On the other hand for the Bose-Fermi realization we have

$$[b_1^\dagger \otimes c, b_1 \otimes c^\dagger]_+ = b_1^\dagger b_1 \otimes I_F + I_B \otimes c^\dagger c = L_3 + L_4\,.$$

We have

$$W_+ = \begin{pmatrix} 0 & 0 & 0 \\ 0 & 0 & 1 \\ 0 & 0 & 0 \end{pmatrix}, \quad W_- = \begin{pmatrix} 0 & 0 & 0 \\ 0 & 0 & 0 \\ 0 & 1 & 0 \end{pmatrix}$$

with

$$[W_+, W_-]_+ = \begin{pmatrix} 0 & 0 & 0 \\ 0 & 1 & 0 \\ 0 & 0 & 1 \end{pmatrix} = -L_3 + L_4\,.$$

For the Bose-Fermi realization we have

$$[W_+, W_-]_+ = b_2^\dagger b_2 \otimes I_F + I_B \otimes c^\dagger c = -L_3 + L_4 .$$

The SymbolicC++ program is

```
// super.cpp

#include <iostream>
#include "symbolicc++.h"
using namespace std;

int main(void)
{
 Symbolic A("A"), B("B"), C("C"), D("D"), I("I");
 Symbolic b = ~Symbolic("b"), bd = ~Symbolic("bd");
 Symbolic c = ~Symbolic("c"), cd = ~Symbolic("cd");
 Equations simplify;
 Equations algrules = ((A,B,C,D,kron(A,B)*kron(C,D)==kron(A*C,B*D)),
                       (A,B,C,kron(A,B+C)==kron(A,B)+kron(A,C)),
                       (A,B,C,kron(A+B,C)==kron(A,C)+kron(B,C)),
                       (A,B,kron(A,-B)==-kron(A,B)),
                       (A,B,kron(-A,B)==-kron(A, B)),
                       (A,kron(A,0)==0),(A,kron(0,A)==0),
                       (A,I*A==A),(A,A*I==A),(A,(I^A)==I),
                       b*bd==I+bd*b,c*cd==I-cd*c,
                       c*c==0,cd*cd==0);

 Symbolic Lp = kron(bd,kron(b,I));
 Symbolic Lm = kron(b,kron(bd,I));
 Symbolic L3 = (kron(bd*b,kron(I,I))-kron(I,kron(bd*b,I)))/2;
 Symbolic L4 = (kron(bd*b,kron(I,I))+kron(I,kron(bd*b,I)))/2
             + kron(I,kron(I,cd*c));
 simplify = (algrules,
             Lp==Symbolic("Lp"),Lm==Symbolic("Lm"),
             L3==Symbolic("L3"),2*L3==2*Symbolic("L3"),
             L4==Symbolic("L4"));
 cout << "[Lp,Lm]=" << (Lp*Lm-Lm*Lp).subst_all(simplify) << endl;
 cout << "[Lp,L3]=" << (Lp*L3-L3*Lp).subst_all(simplify) << endl;
 cout << "[Lp,L4]=" << (Lp*L4-L4*Lp).subst_all(simplify) << endl;
 cout << "[Lm,L3]=" << (Lm*L3-L3*Lm).subst_all(simplify) << endl;
 cout << "[Lm,L4]=" << (Lm*L4-L4*Lm).subst_all(simplify) << endl;
 cout << "[L3,L4]=" << (L3*L4-L4*L3).subst_all(simplify) << endl;
 cout << endl;

 Symbolic Vp = kron(bd,kron(I,c));
 Symbolic Vm = kron(b,kron(I,cd));
 Symbolic Wp = kron(I,kron(bd,c));
```

```
Symbolic Wm = kron(I,kron(b,cd));
simplify = (algrules,
          Lp==Symbolic("Lp"),Lm==Symbolic("Lm"),
          L3==Symbolic("L3"),L4==Symbolic("L4"),
          Vp==Symbolic("Vp"),Vm==Symbolic("Vm"),
          Wp==Symbolic("Wp"),Wm==Symbolic("Wm"),
          kron(bd*b,kron(I,I))==Symbolic("L4")+Symbolic("L3")
                          -kron(I,kron(I,cd*c)),
          kron(I,kron(bd*b,I))==Symbolic("L4")-Symbolic("L3")
                          -kron(I,kron(I,cd*c)));
cout << "[Lp,Vp]=" << (Lp*Vp-Vp*Lp).subst_all(simplify) << endl;
cout << "[Lp,Vm]=" << (Lp*Vm-Vm*Lp).subst_all(simplify) << endl;
cout << "[Lp,Wp]=" << (Lp*Wp-Wp*Lp).subst_all(simplify) << endl;
cout << "[Lp,Wm]=" << (Lp*Wm-Wm*Lp).subst_all(simplify) << endl;
cout << "[Lm,Vp]=" << (Lm*Vp-Vp*Lm).subst_all(simplify) << endl;
cout << "[Lm,Vm]=" << (Lm*Vm-Vm*Lm).subst_all(simplify) << endl;
cout << "[Lm,Wp]=" << (Lm*Wp-Wp*Lm).subst_all(simplify) << endl;
cout << "[Lm,Wm]=" << (Lm*Wm-Wm*Lm).subst_all(simplify) << endl;
cout << "[L3,Vp]=" << (L3*Vp-Vp*L3).subst_all(simplify) << endl;
cout << "[L3,Vm]=" << (L3*Vm-Vm*L3).subst_all(simplify) << endl;
cout << "[L3,Wp]=" << (L3*Wp-Wp*L3).subst_all(simplify) << endl;
cout << "[L3,Wm]=" << (L3*Wm-Wm*L3).subst_all(simplify) << endl;
cout << "[L4,Vp]=" << (L4*Vp-Vp*L4).subst_all(simplify) << endl;
cout << "[L4,Vm]=" << (L4*Vm-Vm*L4).subst_all(simplify) << endl;
cout << "[L4,Wp]=" << (L4*Wp-Wp*L4).subst_all(simplify) << endl;
cout << "[L4,Wm]=" << (L4*Wm-Wm*L4).subst_all(simplify) << endl;
cout << endl;
cout << "[Vp,Vm]+=" << (Vp*Vm+Vm*Vp).subst_all(simplify) << endl;
cout << "[Vp,Wp]+=" << (Vp*Wp+Wp*Vp).subst_all(simplify) << endl;
cout << "[Vp,Wm]+=" << (Vp*Wm+Wm*Vp).subst_all(simplify) << endl;
cout << "[Vm,Wp]+=" << (Vm*Wp+Wp*Vm).subst_all(simplify) << endl;
cout << "[Vm,Wm]+=" << (Vm*Wm+Wm*Vm).subst_all(simplify) << endl;
cout << "[Wp,Wm]+=" << (Wp*Wm+Wm*Wp).subst_all(simplify) << endl;
return 0;
}
```

Supplementary Problems

Problem 111. (i) For $n = 4$ the transform matrix for the *Daubechies wavelet* is given by

$$D_4 = \begin{pmatrix} c_0 & c_1 & c_2 & c_3 \\ c_3 & -c_2 & c_1 & -c_0 \\ c_2 & c_3 & c_0 & c_1 \\ c_1 & -c_0 & c_3 & -c_2 \end{pmatrix}, \qquad \begin{pmatrix} c_0 \\ c_1 \\ c_2 \\ c_3 \end{pmatrix} = \frac{1}{4\sqrt{2}} \begin{pmatrix} 1 + \sqrt{3} \\ 3 + \sqrt{3} \\ 3 - \sqrt{3} \\ 1 - \sqrt{3} \end{pmatrix}.$$

Is D_4 orthogonal? Prove or disprove.

(ii) For $n = 8$ the transform matrix for the Daubechies wavelet is given by

$$D_8 = \begin{pmatrix} c_0 & c_1 & c_2 & c_3 & 0 & 0 & 0 & 0 \\ c_3 & -c_2 & c_1 & -c_0 & 0 & 0 & 0 & 0 \\ 0 & 0 & c_0 & c_1 & c_2 & c_3 & 0 & 0 \\ 0 & 0 & c_3 & -c_2 & c_1 & -c_0 & 0 & 0 \\ 0 & 0 & 0 & 0 & c_0 & c_1 & c_2 & c_3 \\ 0 & 0 & 0 & 0 & c_3 & -c_2 & c_1 & -c_0 \\ c_2 & c_3 & 0 & 0 & 0 & 0 & c_0 & c_1 \\ c_1 & -c_0 & 0 & 0 & 0 & 0 & c_3 & -c_2 \end{pmatrix}.$$

Is D_8 orthogonal? Prove or disprove.

Problem 112. Consider the Lie algebra $s\ell(2, \mathbb{R})$ with the basis x, y, h and the commutators

$$[x, h] = -2x, \qquad [x, y] = h, \qquad [h, y] = -2y.$$

Let $t \in \mathbb{R}$.
(i) Find $a(t)$, $b(t)$, $c(t)$ such that

$$\exp(tx)y\exp(-tx) = a(t)x + b(t)y + c(t)h.$$

(ii) Find $a(t)$, $b(t)$, $c(t)$ such that

$$\exp(tx)h\exp(-tx) = a(t)x + b(t)y + c(t)h.$$

(iii) Find $a(t)$, $b(t)$, $c(t)$ such that

$$\exp(ty)h\exp(-ty) = a(t)x + b(t)y + c(t)h.$$

We consider the representation

$$h = \begin{pmatrix} 1 & 0 \\ 0 & -1 \end{pmatrix}, \qquad x = \begin{pmatrix} 0 & 1 \\ 0 & 0 \end{pmatrix}, \qquad y = \begin{pmatrix} 0 & 0 \\ 1 & 0 \end{pmatrix}.$$

Problem 113. A realization of the Lie algebra $s\ell(2, \mathbb{R})$ using vector fields is

$$V_1 = \frac{d}{dx}, \quad V_2 = u\frac{d}{du}, \quad V_3 = u^2\frac{d}{du}.$$

Consider the *Riccati equation*

$$\frac{du}{dt} = 1 + u + u^2.$$

(i) Find the solution of the initial value problem using the *Lie series*

$$u(t) = \exp(tV)u|_{u=u_0}$$

where

$$V = V_1 + V_2 + V_3 = (1 + u + u^2)\frac{d}{du}.$$

(ii) Find the functions $f(t)$, $g(t)$, $h(t)$ such that

$$e^{tV} = e^{f(t)d/du}e^{g(t)ud/du}e^{h(t)u^2d/du}.$$

Problem 114. Let $\mathbb{F}$ be a field. In the following we have $\mathbb{F} = \mathbb{R}$ or $\mathbb{F} = \mathbb{C}$. A *Hopf algebra* is a vector space $\mathcal{A}$ endowed with five linear operations

$$\mu : \mathcal{A} \otimes \mathcal{A} \to \mathcal{A}$$
$$\eta : \mathbb{F} \to \mathcal{A}$$
$$\Delta : \mathcal{A} \to \mathcal{A} \otimes \mathcal{A}$$
$$\epsilon : \mathcal{A} \to \mathbb{F}$$
$$\gamma : \mathcal{A} \to \mathcal{A}$$

which possess the following properties

$$\mu \cdot (id \otimes \mu) = \mu \otimes (\mu \otimes id)$$
$$\mu \cdot (id \otimes \eta) = id = \mu \cdot (\eta \otimes id)$$
$$(id \otimes \Delta) \cdot \Delta = (\Delta \otimes id) \cdot \Delta$$
$$(\epsilon \otimes id) \cdot \Delta = id = (id \otimes \epsilon) \cdot \Delta$$
$$\mu \cdot (id \otimes \gamma) \cdot \Delta = \eta \cdot \epsilon = \mu \cdot (\gamma \otimes id) \cdot \Delta$$

and the co-multiplication Δ and co-unit ϵ are $\mathbb{F}$-algebra morphisms, that is, preserves multiplication. Give an example of a Hopf algebra using a finite group G.

Problem 115. Consider the *nonlinear wave equation*

$$\frac{\partial^2 u}{\partial t^2} - \frac{\partial}{\partial x}\left(u\frac{\partial u}{\partial x}\right) = 0.$$

Show that this partial differential equation admits the Lie symmetry vector fields

$$\frac{\partial}{\partial t}, \quad \frac{\partial}{\partial x}, \quad t\frac{\partial}{\partial t} + x\frac{\partial}{\partial x}, \quad t\frac{\partial}{\partial t} - 2u\frac{\partial}{\partial u}.$$

Calculate the commutators of these vector fields and classify the Lie algebra.

Problem 116. Let $\partial := \partial/\partial x$. The two *pseudodifferential operators*

$$L := \partial + \sum_{j=1}^{\infty} u_{j+1}\partial^{-j}, \qquad W := 1 + \sum_{j=1}^{\infty} w_j \partial^{-j}$$

are called the *Lax operator* and *gauge operator*, respectively. With $\partial^{-1}\partial = \partial\partial^{-1} = 1$ the *generalized Leibniz rule* is satisfied

$$\partial^k f \cdot = \sum_{j=0}^{\infty} \binom{k}{j}(\partial^j f)\partial^{k-j}.$$

for any $k \in \mathbb{Z}$ and $L = W\partial W^{-1}$. The *KP hierarchy* is given by the *Lax equations*

$$\partial_n L = [B_n, L]$$

where $\partial_n := \partial/\partial t_n$ and $B_n := L_+^n$ is the differential part of $L^n = L_+^n + L_-^n$ with

$$L_+^n = \sum_{j=0}^{\infty} c_j^n \partial^j, \qquad L_-^n = \sum_{j=-\infty}^{-1} c_j^n \partial^j.$$

Derive the first three equations of the KP hierarchy.

Problem 117. Let $\mathbf{v}(x,t) = (v_1(x,t), v_2(x,t))^T$. Consider the linear equations

$$\frac{\partial \mathbf{v}}{\partial x} = \begin{pmatrix} i|u|^2 - 2i\zeta^2 & i\partial u/\partial t + 2\zeta u \\ i\partial u^*/\partial t - 2\zeta^* u & -i|u|^2 + 2i\zeta^2 \end{pmatrix} \mathbf{v}$$

$$\frac{\partial \mathbf{v}}{\partial t} = \begin{pmatrix} -i\zeta & u \\ -u^* & i\zeta \end{pmatrix} \mathbf{v}.$$

Show that the compatibility condition of these two equations with $\partial\zeta/\partial t = 0$ provides the unstable nonlinear Schrödinger equation

$$i\frac{\partial u}{\partial x} + \frac{\partial^2 u}{\partial t^2} + 2|u|^2 u = 0.$$

Problem 118. Consider the two-dimensional sphere

$$S_1^2 + S_2^2 + S_3^2 = S^2$$

where $S > 0$ is the radius of the sphere. Consider the symplectic structure on this sphere with the symplectic differential two form

$$\omega := -\frac{1}{2S^2} \sum_{j,k,\ell=1}^{3} \epsilon_{jk\ell} S_j dS_k \wedge dS_\ell$$

($\epsilon_{123} = 1$) and the Hamilton vector fields

$$V_{S_j} := \sum_{k,\ell=1}^{3} \epsilon_{jk\ell} S_k \frac{\partial}{\partial S_\ell} \, .$$

The *Poisson bracket* is defined by

$$[S_j, S_k]_{\mathrm{PB}} := -V_{S_j} S_k \, .$$

(i) Calculate $[S_j, S_k]_{\mathrm{PB}}$.
(ii) Calculate $V_{S_j} \rfloor \omega$.
(iii) Calculate $V_{S_j} \rfloor V_{S_k} \rfloor \omega$.
(iv) Calculate the Lie derivative $L_{V_{S_j}} \omega$.

Problem 119. The special unitary group of degree n, denoted $SU(n)$, is the group of $n \times n$ unitary matrices with determinant 1. The group operation is matrix multiplication. The infinite-dimensional special unitary group $SU(\infty)$ is the limit of unitary groups $SU(n)$. One can consider elements of $SU(\infty)$ as transformations of the infinite dimensional separable *Hilbert space* $\ell_2(\mathbb{N})$. The sequence space $\ell_2(\mathbb{N})$ consists of all infinite sequences $\mathbf{z} = (z_1, z_2, ...)$ of complex numbers such that

$$\sum_{j=1}^{\infty} z_j \bar{z}_n = \sum_{j=1}^{\infty} |z_j|^2 < \infty \, .$$

The inner product (scalar product) in the Hilbert space $\ell_2(\mathbb{N})$ is defined by

$$\langle \mathbf{z} | \mathbf{w} \rangle = \langle \bar{\mathbf{w}} | \bar{\mathbf{z}} \rangle = \sum_{j=1}^{\infty} z_j \bar{w}_j .$$

The convergence of this last series follows from the Cauchy-Schwarz inequality. The space $\ell_2(\mathbb{N})$ is complete. This means that each Cauchy series of elements from $\ell_2(\mathbb{N})$ converges to an element of $\ell_2(\mathbb{N})$. Consider orthonormal sequences of $\mathbf{z}_j \in \ell_2(\mathbb{N})$. This means that if $\mathbf{z}_j$ and $\mathbf{z}_k$ are two arbitrary distinct elements of orthonormal sequence, then $|\mathbf{z}_j| = |\mathbf{z}_k| = 1$ and $\langle \mathbf{z}_j | \mathbf{z}_k \rangle = 0$. Consider linear operators A in the Hilbert space $\ell_2(\mathbb{N})$. The linear operator A is a function from the linear subset $D(A)$ of Hilbert space (the domain of the operator) to linear subset $R(A)$ (the range of the operator) that preserves the operations of vector addition and multiplication by scalar.

(i) Show that the set of unitary operators in the Hilbert space $\ell_2(\mathbb{N})$ form the group $U(\infty)$ under composition.

(ii) Let U be the unitary operator in $\ell_2(\mathbb{N})$ that maps

$$\mathbf{u} = \begin{pmatrix} u_1 & u_2 & u_3 & u_4 & u_5 & \cdots & u_{2n} & u_{2n+1} & \cdots \end{pmatrix}^T$$

onto

$$A\mathbf{u} = \begin{pmatrix} u_2 & u_4 & u_1 & u_6 & u_3 & \cdots & u_{2n+2} & u_{2n-1} & \cdots \end{pmatrix}^T.$$

Find the spectrum.

Problem 120.   (i) Consider the vector space of operators

$$\{ c_{j_1}^\dagger c_{j_2} c_{k_1}^\dagger c_{k_2} \}$$

where $c_j^\dagger$, c_j are Fermi creation and annihilation operators and $j_1, j_2, k_1, k_2 = 1, 2, \ldots, n$. Note that depending on j_1, j_2, k_1, k_2 the operator $c_{j_1}^\dagger c_{j_2} c_{k_1}^\dagger c_{k_2}$ could be the zero operator. Calculate the commutator

$$[c_{j_1}^\dagger c_{j_2} c_{k_1}^\dagger c_{k_2}, c_{m_1}^\dagger c_{m_2} c_{n_1}^\dagger c_{n_2}].$$

Thus show that we have a Lie algebra for this set of operators. Of course the Jacobi identity must also be checked.

(ii) Consider the *number operator* defined by

$$\hat{N} := \sum_{j=1}^n c_j^\dagger c_j.$$

Calculate the commutator $[\hat{N}, c_{j_1}^\dagger c_{j_2} c_{k_1}^\dagger c_{k_2}]$. Discuss.

Bibliography

Books

Aldous J. M. and Wilson R. J.
Graphs and Applications: An Introductory Approach, Springer (2000)

Anderson R. L. and Ibragimov N. H.
Lie-Bäcklund Transformations in Applications, SIAM (1979)

Armstrong M. A.
Groups and Symmetry, Springer (1988)

Baumslag B. and Chandler B.
Group Theory, Schaum's Outline Series, McGraw-Hill (1968)

Bäuerle G. G. A. and de Kerf E. A.
Lie Algebras, North-Holland (1990)

Bishop D. M.
Group Theory and Chemistry, Dover Publications (1993)

Bluman G. W. and Cole J. D.
Similarity Methods for Differential Equations, Springer (1974)

Bourbaki N.
Elements of Mathematics: Lie Groups and Lie Algebras, Addison-Wesley (1975)

Bredon G. E.
Introduction to Compact Transformation Groups, Elsevier (1972)

Bluman G. W. and Kumei S.
Symmetries and Differential Equations, Springer (1989)

329

Bronson R.
Matrix Operations, Schaum's Outlines, McGraw-Hill (1989)

Bump D.
Lie Groups, Springer (2000)

Carter R. W.
Simple Groups of Lie Type, John Wiley (1972)

Chern S. S., Chen W. H. and Lam K. S.
Lectures on Differential Geometry, World Scientific (1999)

Crampin M. and Pirani F. A. E.
Applicable Differential Geometry, Cambridge University Press (1984)

de Souza P. N. and Silva J.-N.
Berkeley Problems in Mathematics, Springer (1998)

Di Franceso P., Mathieu P. and Sénéchal D.
Conformal Field Theory, 2nd edition, Springer (1997)

Dixmier J.
Enveloping Algebras, North-Holland (1974)

Englefield M. J.
Group Theory and the Coulomb Problem, Wiley-Interscience (1972)

Erdmann K. and Wildon M.
Introduction to Lie Algebras, Springer (2006)

Flanders H.
Differential Forms, Academic Press (1963)

Frankel T.
The Geometry of Physics, 2nd edition, Cambridge University Press (2004)

Fuhrmann P. A.
A Polynomial Approach to Linear Algebra, Springer (1996)

Fulton W. and Harris J.
Representation Theory, Springer (1991)

Gallian J. A.
Contemporary Abstract Algebra, sixth edition, Houghton Mifflin (2006)

Gilmore R.
Lie Groups, Lie Algebras and Some of Their Applications, Wiley (1974)

Golub G. H. and Van Loan C. F.
Matrix Computations, third edition, Johns Hopkins University Press (1996)

Göckeler M. and Schücker T.
Differential geometry, gauge theories, and gravity, Cambridge University Press (1987)

Grossman S. I.
Elementary Linear Algebra, third edition, Wadsworth Publishing (1987)

Hall B. C.
Lie Groups, Lie Algebras, and Representations: an elementary introduction, Springer (2003)

Hamermesh M.
Group Theory and its Application to Physical Problems, Addison-Wesley (1962)

Helgason S.
Groups and Geometric Analysis, Integral Geometry, Invariant Differential Operators and Spherical Functions, Academic Press (1984)

Helgason S.
Differential Geometry, Lie Groups and Symmetric Spaces, Academic Press (1978)

Horn R. A. and Johnson C. R.
Topics in Matrix Analysis, Cambridge University Press (1999)

Humphreys J. E.
Introduction to Lie Algebras and Representation Theory, Springer (1972)

Ibragimov N. H.
Transformation Groups in Mathematical Physics, Nauka (1983)
(in Russian)

Inui T., Tanabe Y. and Onodera Y.
Group Theory and its Applications in Physics, Springer (1990)

Isham C. J.
Modern Differential Geometry, World Scientific (1989)

Jacobson N.
Lie Algebras, Interscience Publishers (1962)

James G. and Liebeck M.
Representations and Characters of Groups, 2nd edition, Cambridge University Press (2001)

Johnson D. L.
Presentation of Groups, Cambridge University Press (1976)

Jones H. F.
Groups, Representations and Physics, Adam Hilger (1990)

Kac V. G.
Infinite Dimensional Lie Algebras, Cambridge University Press (1985)

Kedlaya K. S., Poonen B. and Vakil R.
The William Lowell Putnam Mathematical Competition 1985–2000, The Mathematical Association of America (2002)

Kirillov A. A.
Elements of the Theory of Representations, Springer, 1976

Knapp A. W.
Lie Groups and Beyond: An Introduction, 2nd edition, Birkhäuser (2002)

Lang S.
Linear Algebra, Addison-Wesley, Reading (1968)

Lang S.
Introduction to Differentiable Manifolds, Wiley-Interscience (1962)

Lee Dong Hoon
The structure of complex Lie groups, Chapman and Hall/CRC (2002)

Lie M. S.
Untersuchungen über unendliche continuirliche Gruppen, Hirzel (1893)

Magnus W., Karrass A and Solitar D.
Combinatorial Group Theory, 2nd edition, Dover (1976)

McCrimmon K.
A Taste of Jordan Algebras, Springer (2004)

Miller W.
Lie Theory and Special Functions, Academic Press (1968)

Miller W.
Symmetry Groups and Their Applications, Academic Press (1972)

Ohtsuki T.
Quantum Invariants, World Scientific Publishing (2002)

Olver P. J.
Applications of Lie Groups to Differential Equations, Graduate Text in Mathematics, vol. 107, Springer (1986)

Ovsyannikov L. V.
Group Analysis of Differential Equations, Academic Press (1982)

Pontryagin L. S.
Topological Groups, Princeton University Press (1958)

Robinson D. J. S.
A course in the theory of groups, Springer (1996)

Schneider H. and Barker G. P.
Matrices and Linear Algebra, Dover Publications (1989)

Scott W. R.
Group Theory, Dover Publications (1987)

Sniatychi J.
Geometric Quantization and Quantum Mechanics, Springer (1980)

Spivak M.
Calculus on Manifolds, Benjamin (1966)

Spivak M.
Differential Geometry, Publish or Perish (1970)

Steeb W.-H. and Hardy Y.
Matrix Calculus and Kronecker Product, 2nd edition, World Scientific Publishing (2011)

Steeb W.-H.
Continuous Symmetries, Lie Algebras, Differential Equations and Computer Algebra, 2nd edition, World Scientific Publishing (2007)

Steeb W.-H.
Problems and Solutions in Theoretical and Mathematical Physics, third edition, Volume I: Introductory Level, World Scientific Publishing (2009)

Steeb W.-H.
Problems and Solutions in Theoretical and Mathematical Physics, third edition, Volume II: Advanced Level, World Scientific Publishing (2009)

Sternberg S.
Lie Algebras (2004)
http://www.math.harvard.edu/~shlomo/docs/lie_algebras.pdf

Tung W. K.
Group Theory Physics: Problems and Solutions by Michael Aivazis, World Scientific (1991)

Varadarajan V. S.
Lie Groups, Lie Algebras and Their Representations, Springer (2004)

von Westenholz C.
Differential Forms in Mathematical Physics, revised edition, North-Holland (1981)

Warner F. W.
Foundations of Differentiable Manifolds and Lie Groups, Scott Foresman (1971)

Wawrzynczyk A.
Group Representations and Special Functions, D. Reidel (1984)

Weyl H.
Gruppentheorie und Quantenmechanik, Hirzel (1928)

Wybourne B. G.
Classical Groups for Physicists, John Wiley (1974)

Publications

Agricola I., *Old and New on the Exceptional Group G_2*, Notices of the AMS, **55**, 922 (2008)

Boyer C. P. and Kalnins E. G., *Symmetries of the Hamilton-Jacobi equation*, J. Math. Phys. **18**, 1032 (1977)

Campoamor-Stursberg R., *Casimir operators induced by the Maurer-Cartan equations*, J. Phys. A : Math. Theor. **41**, 365207 (2008)

Campoamor-Stursberg R., *The structure of the invariants of perfect Lie algebras*, J. Phys. A: Math. Gen. 6709–6723 (2003)

Chinea F. J. and Guil F., *Local and non-local conserved currents for an equation related to the nonlinear σ-model*, J. Phys. A : Math. Gen. **15**, 2349 (1982)

DasGupta Ananda, *Disentanglement formulas: An alternative derivation and some applications to squeezed coherent states*, American J. Phys. **64** 1422 (1996)

Delduc F. and Fehér L., *Regular conjugacy classes in the Weyl group and integrable hierarchies*, J. Phys. A : Math. Gen. **28**, 5843 (1995)

Fairlie D. B., Nuyts J. and Zachos C. K., *A Presentation for the Virasoro and Super-Virasoro Algebras*, Comm. Math. Phys. **17**, 595 (1988)

Fernández F. M., *Time Evolution and Lie Algebras*, Phys. Rev. A **40**, 41 (1989)

Flaschka H. *A commutator representation of Painlevé equations*, J. Math. Phys. **21**, 1016-1018 (1980)

Fokas A. S., *Group Theoretical Aspects of Constants of Motion and Separable Solutions in Classical Mechanics*, Journal of Mathematical Analysis and Applications **68**, 347 (1979)

Franco Jos
Lie Algebras Exercises
http://www.doschivos.com/mate.htm

Garsia A., Musiker G., Wallach and Xin G., *Invariants, Kronecker Products and Combinatorics of some Remarkable Diophantine Systems*, arXiv:0810.0060 (2008)

Harrison B. K., *On Methods of Finding Bäcklund Transformations in Systems with more than Two Independent Variables*, Nonlinear Mathematical Physics, **2**, 201 (1995)

Harrison B. K., *The Differential Form Method for Finding Symmetries*, Symmetry, Integrability and Geometry: Methods and Applications **1**, 1 (2005)

Hsu L. and Kamran N., *Symmetries of Second-Order Ordinary Differential Equations and Elie Cartan's Method of Equivalence*, Letters in Mathematical Physics **15**, 91 (1988)

Kalnins E. G. and Miller W., *Lie theory and separation of variables. 11. The EPD equation*, J. Math. Phys. **17**, 369-377 (1976)

Koca Mehmet, E_8 *lattice with icosians and* Z_5 *symmetry*, J. Phys. A: Math. Gen. **22**, 4125-4134 (1989)

Kyriakopoulos E., *Lie algebra of the hypergeometric functions*, J. Math. Phys. **15**, 172-176 (1974)

Kyriakopoulos E., *Generating functions of the hypergeometric functions*, J. Math. Phys. **15**, 753 (1974)

Legaré M., *Symmetry Reductions of the Lax Pair of the Four-Dimensional Euclidean Self-Dual Yang-Mills Equations*, Nonlinear Mathematical Physics, **3**, 266 (1996)

Luan Pi-Gang, Lee H. C. and Zhang R. B., *Colored solutions of the Yang-Baxter equation from representations of* $U_q g\ell(2)$, J. Math. Phys. **41**, 6529 (2000)

Mahomed F. M. and Leach P. G. L., *Symmetry Lie algebras of nth order ordinary differential equations*, J. Math. Anal. Appl. **151**, 80 (1990)

McGehee R., *Triple Collision in the Collinear Three-Body Problem*, Inventiones Mathematica **27**, 191 (1974)

Morozov O. I., *Structure of Symmetry Groups via Cartan's Method: Survey of Four Approaches*, Symmetry, Integrability and Geometry: Methods and Applications **1**, 1 (2005)

Nattermann and Doebner H.-D., *Gauge Classification, Lie Symmetries and Integrability of a Family of Nonlinear Schrödinger Equations*, Nonlinear Mathematical Physics **3**, 302 (1996)

Olafsson S., *New non-linear evolution equations related to some superloop algebras*, J. Phys. A : Math. Gen. **22**, 157 (1989)

Patera J., Sharp R. T., Winternitz R. and Zassenhaus H., *Invariants of real low dimension Lie algebras*, **17**, 986 (1976)

Patera J., Sharp R. T., Winternitz R. and Zassenhaus H., *Subgroups of the Poincaré group and their invariants*, **17**, 977 (1976)

Perelomov A. M., *Instanton-like solutions in chiral models*, Physica 4D, 1 (1981)

Planat M. and Jorrand P., *Group theory for quantum gates and quantum coherence*, J. Phys. A: Math. Theor. **41**, 182001 (2008)

Saenkarun S. and Loutsiouk A., *Representations of G_2*, Thai Journal of Mathematics **5**, 309 (2007)

Sakovick S. Yu., *On conservation laws and zero-curvature representations of the Liouville equation*, J. Phys. A : Math. Gen. **27**, L125 (1994)

Sam Steven V., *Solutions to "Introduction to Lie Algebras and Representation Theory" by James E. Humphreys* (2010)

Shadwick W. F., *The KdV prolongation algebra*, J. Math. Phys. **21**, 454 (1980)

Suzuki Masuo, *Decomposition Formulas of Exponential Operators and Lie Exponentials with Some Applications to Quantum Mechanics and Statistical Physics*, J. Math. Phys. **26**, 601 (1985)

Swift J. W., *Hopf bifurcation with the symmetry of the square*, Nonlinearity **1**, 333 (1988)

Wahlquist H. D. and Estabrook F. B., *Prolongation structures of nonlinear evolution equations*, J. Math. Phys. **16**, 1 (1975)

Index